Is America Safe?

Is America Safe?

Terrorism, Homeland Security, and Emergency Preparedness

Second Edition

Don Philpott

Bernan Press

Lanham • Boulder • New York • London

Published by Bernan Press
An imprint of The Rowman & Littlefield Publishing Group, Inc.
4501 Forbes Boulevard, Suite 200, Lanham, Maryland 20706
www.rowman.com
800-865-3457; info@bernan.com

6 Tinworth Street, London SE11 5AL, United Kingdom

Library of Congress Cataloging-in-Publication Data

Names: Philpott, Don, author. | Jordan, Robert T. Is America safe?
Title: Is America safe? : terrorism, homeland security, and emergency preparedness / Don Philpott.
Description: Second edition. | Lanham : Bernan Press, [2019] | Includes bibliographical references and index.
Identifiers: LCCN 2018046172 (print) | LCCN 2018047959 (ebook) | ISBN 9781641433082 (electronic) | ISBN 9781641433075 (pbk. : alk. paper)
Subjects: LCSH: Terrorism—Government policy—United States. | Terrorism—United States—Prevention. | National security—United States. | Emergency management—United States. | United States—Politics and government.
Classification: LCC HV6432 (ebook) | LCC HV6432 .J67 2019 (print) | DDC 363.325/15610973—dc23
LC record available at https://lccn.loc.gov/2018046172

∞™ The paper used in this publication meets the minimum requirements of American National Standard for Information Sciences—Permanence of Paper for Printed Library Materials, ANSI/NISO Z39.48-1992.

Printed in the United States of America

Contents

Foreword

Is America Safe? is a much-needed, one-stop-shop "educational handbook" on what the average citizen should know and can do in both preparing for and responding to a terrorist attack on U.S. soil—an attack that most government officials who deal in this field feel is inevitable.

Where and when the next attack will take place cannot be predicted; however, the fact that the chances of a future terrorist attack are high, and the fact that a prime goal of most of these attacks is high civilian mortality obligates all of us to learn more about what actions we can take to help prevent, protect, or mitigate any such attempts.

Americans are, by nature, fighters when it comes to endangering our civil liberties. Americans are also aggressive, and sitting back and doing nothing while terrorist organizations plot and attempt different ways to harm us is not in our psyche.

Predictions of a hundred thousand or more innocent, civilian casualties, particularly when they might include our own families and loved ones, are a silent call to arms and reason enough to do all we can to fight back. *Is America Safe?* mentally equips us with those skills. The best way to prepare and to fight back is to first understand who the enemy is.

This book provides the reader with well-researched and well-documented background and detail on terrorist organizations. It starts with defining terrorism and its origins and then goes on to describe present-day organizations, where they operate from, what their philosophies entail, and what their motivation and objectives are. It then addresses different terrorist tactics and techniques that are being contemplated and their desired effects, including individual descriptions of their destructive capabilities. A sampling of the book's extensive list includes terrorist weaponization of chemical, biological, and nuclear agents. Knowing how to prepare for and survive during any of these attacks is a big step toward marginalizing the terrorists' intended high casualty, psychological impacts, and propaganda values. The added value is that these same lifesaving skills may also be used in naturally occurring catastrophic events such as hurricanes, tornados, floods, or earthquakes.

Is America Safe? empowers the average citizen with the knowledge and skills to understand terrorist strategies, which, in turn, allows each of us to contribute in disrupting the terrorists' intended goals through observing and reporting suspicious activity, reducing target vulnerabilities, and minimizing casualties through education and preparation. The better the chances we have to reduce the effects of an attack, the better we will be able to minimize the impact of the event, thereby denying the enemy the propaganda value he seeks from surprise, shock, and fear on an unprepared and panicky public. Prevention and preparedness are also in step with and a part of our national strategy.

Although not a terrorist attack, Katrina was a prime example of lack of preparation and response and an unrealistic expectation and reliance on outside authorities during the first 24–48 hours of a disaster. If any lesson is to be learned from Katrina, it is that the average citizen and local governments have to become more self-reliant. *You and I have_to make plans that will allow us to help ourselves.*

Is America Safe? is intended to help us help ourselves. It is intended to reduce some of our fear by educating us and providing us with the confidence that comes with not feeling powerless. In essence, it gives us all a chance to do our part to not only protect ourselves and our families; it gives us an opportunity to fight back by denying the enemy any chance of even partial success.

The Department of Homeland Security is spending considerable amounts of money and time trying to educate the public on basic survival skills. A primary concern is that certain public communications systems may be temporarily out of order and therefore public messaging, particularly on educating the public as to what type of method was used and how it acts, as well as what personal measures should be initiated to reduce casualties, may not get through to all of those in the impact area(s). Simple precautions and a little knowledge may be the difference between survival and death to large numbers of unsuspecting individuals, particularly in scenarios involving weapons of mass destruction—the terrorists' preferred method of attack.

Understanding the enemy, reporting suspicious activity, reducing vulnerabilities, knowing how to survive, and initiating preparatory actions is one way for each individual American to do his or her part in helping to win the war on terrorism.

Brigadier General Matthew E. Broderick, U.S. Marine Corps (Retired)
Vice President for DHS and DoD, Abraxas Corporation
Former Director of Operations for Homeland Security

Introduction

Governments are instituted among Men, deriving their just Powers from the Consent of the Governed, that whenever any Form of Government becomes destructive to these Ends, it is the Right of the People to alter or to abolish it, and to institute new Government, laying its foundation on such Principles and organizing its Powers in such Form, as to them shall seem most likely to affect **their Safety and Happiness** . . . *their just Powers from the Consent of the Governed, that whenever any Form of Government becomes destructive to these Ends, it is the Right of the People to alter or to abolish it, and to institute new Government, laying its foundation on such Principles and organizing its Powers in such Form, as to them shall seem most likely to effect* **their Safety and Happiness**. . . .

—Thomas Jefferson, U.S. Constitution

Terror may strike anywhere . . . anytime . . . to anyone.

When terror strikes it is indifferent to whom its victims may be.

Terrorists come in all ages, all religious persuasions, both sexes, all races, and all nationalities.

Terrorist goals include intimidation, subjugation, and elimination of anyone who opposes their particular view of the world.

Terrorist strategies include murder and mayhem, attacking our core values, and destroying our economy.

Do you feel safe?

How secure do you feel your family is today as we face attacks from international terrorists, school shootings, carjackings, and home invasions in our cities, as well as natural disasters of historical proportions around the globe?

Do you feel in control? . . . Or do you feel so helpless that you feel unable to protect yourself or your loved ones?

Are you among those who feel that it is the federal government's primary responsibility to ensure your safety and security? Or do you feel that there are times when you would forego safety and security for the sake of freedom and liberty?

What were our founding fathers' intentions when they included the term "safety and happiness" in our Constitution? Old Ben Franklin said, "The Constitution only guarantees the American people the right to pursue happiness. You have to catch it yourself."

It was also Franklin who insisted, "They that can give up essential liberty to obtain a little temporary safety deserve neither liberty nor safety." And yet we hear a loud vocal minority asking us to do just that in every generation without regard to the ultimate cost.

To many of us, *safety* and *security* are interchangeable terms. And that's true to some degree. However, for the purposes in this book, think of safety in more personal terms and security in terms of community and national interests.

Think about it. You may control much of what goes on in your own life: whether to smoke or not, whether to overindulge with food or alcohol, whether to swim in unknown waters or to only take your family to a guarded and protected beach. All of these things have to do with your own and your family's personal safety and health—and, for the most part are within your personal choices and within your own control. But, what about natural disasters like Hurricane Katrina, Irma, or Maria; terrorist attacks such as 9/11 and the London bombings; or just seemingly random acts of violence like the Parkland school shooting? What can you do about these if they threaten your neighborhood?

Regardless of what you think, *you have the power* to enhance your security by discouraging potential terrorist attacks and by mitigating the terrible effects inflicted on our communities by catastrophic natural or man-made disasters. Think about it—can you ever be truly "safe" if your nation or your community is at risk?

This book is dedicated to protecting Americans and securing our homeland—but that needs your help.

IRRATIONAL FEAR . . . OR INFORMED DECISIONS?

Ignorance, prejudice, unfound fear, and apathy compromise both our safety and our security. Each of us needs to be fully informed as to what threat we face, become engaged in educating and helping others, and actively becoming involved in viable solutions rather than being part of the problem.

For example, David Meyers points out in an article for the American Psychological Society *Observer* that many people refused to fly in commercial aircraft following the September 11th terrorist attacks in New York City, Pennsylvania, and Washington, DC. In fact, so many people cancelled flights that airlines, travel agencies, and hotels all suffered economic losses for months to come. You and I became unsuspecting terrorist victims because of other people's fears and the negative economic fallout.

"Indeed, the terrorists may still be killing us, in ways unnoticed," Meyers explained. "If we now fly 20 percent less and instead drive half those unflown miles, we will spend 2 percent more time in motor vehicles. This translates into 800 more people dying as passengers and pedestrians. So, in just the next year the terrorists may indirectly kill three times more people on our highways than died on those four fated planes."

Thus, irrational fear and ignorance sometimes kill more people than do bullets and bombs.

> *Christopher Robin says to Pooh, "Promise me that you'll always remember: You're braver than you believe . . . and stronger than you seem . . . and smarter than you think."*
>
> —A. A. Milne

Terrorists count on horror to intimidate their victims to the point that they become frozen by fear, immobilized like trapped animals incapable of response. As philosopher Edmund Burke said, all that is necessary for evil to succeed is for good men to do nothing. Fortunately that is not our history. We are a blend of the sons and daughters of the world of nations—stepbrothers and stepsisters, if you will, many of whom fled tyranny and who share a common bond of loving freedom and liberty.

That doesn't mean, however, that all Americans understand the threats to our way of life, nor does it mean that all our citizens are motivated by "brotherly love" to assist one another in times of crisis. Attacks at home and abroad accent the sad fact that home-grown "domestic terrorists" may be of more concern that those who attack us from abroad.

WHO SHOULD READ THIS BOOK?

Every tax-paying citizen should read this book.

In 2006, the combined budget for the Department of Defense, Homeland Security, and Veteran's Affairs was $601 billion of which over $200 billion was used to support the War on Terror. The combined budget for the three departments in fiscal year 2018–2019 is $886 billion and this figure is expected to rise again in 2019–2020.

The $161 billion cost for the post-Katrina cleanup and rebuilding bill for Louisiana, Mississippi, and Alabama by far exceeded the $70 billion spent to recover from the terrorist attacks of September 11, 2001. When you take into account the $125 billion cost of Hurricane Harvey which slammed into Texas in August 2017 and the tens of billions of dollars damage caused in recent years by devastating droughts, wildfires, and hurricanes you can understand why there is an ongoing "beans or bullets" discussion about where priorities should lie.

Which will you choose: (1) continue supporting the War on Terror, (2) demand that more resources be devoted to domestic issues and infrastructure, or (3) insist that "we cut the political pork," tighten our belts, and try to do both?

Scholars are invited to use this book to spark debate, enhance understanding of the choices that Americans face, and as background and a resource for serious students of the history and impact of terror upon international dynamics.

Similarly, professional and volunteer "first responders" should read this book. It's important to note that experience shows that 95 percent of all first responders are either bystanders or the victims themselves. Unfortunately, many of these individuals have minimal to no training and even fewer understand the nature of the danger that they face. *Is America Safe?* is a source not only of valuable information, but an asset also in determining what priorities you need to set within your scope of influence or service.

Once you've read it, put it in your library for future reference . . . then send copies to others whom you feel would benefit from what you've learned. Sign up for an American Red Cross first aid class and consider volunteering in your local Neighborhood Watch. Help elevate your neighborhood's awareness to a point that no terrorist may operate there without raising suspicion. Join your fellow Americans to win the War on Terror.

I'm not afraid of storms, for I'm learning how to sail my ship.

—Louisa May Alcott

APPLYING LESSONS LEARNED

The sad story about Hurricane Katrina is that much of the tragedy was avoidable. Lessons learned from previous killer hurricanes and from a computer-generated training exercise, called Pam, offered New Orleans years of lead time to prepare for the worse.

In the aftermath, it is apparent that the majority of the responsibility for failing to implement those lessons lay at the feet of New Orleans city and Louisiana state officials. But their lapses in planning and judgment also highlighted weaknesses in homeland security and identified a need for greater federal involvement much earlier in the planning and preparation phase whenever level 3 through level 5 hurricanes threaten heavily populated areas such as the Gulf Coast. But the biggest lesson of all is that you must never expect your government to provide for your every need, nor to keep you safe and secure in every circumstance.

Unfortunately, one wonders if almost two decades later, we are applying lessons learned in the wake of 9/11 any more effectively than we have applied the lessons drawn from these recent natural disasters. Of course, you probably don't work in a 110-story building, like the World Trade Center. But what would you do differently today if you did?

Do you work in a "lucrative target" for foreign or domestic terrorists?

If not, is your workplace or home in close proximity to a prospective target?

Do you understand what motivates terrorists to attack innocent civilians?

What would you do if tomorrow you found yourself a hostage aboard a commercial aircraft?

How would you react if terrorists stormed into the local school where your children attend and threatened to blow it up if their terms weren't met?

If you heard that a terrorist detonated a "dirty bomb" in an area of town where your family lives, would you know what to do?

Do you have a plan for your family to survive destructive weather events?

How should you go about assessing the natural and man-made dangers about you?

What can you do to protect yourself, your family, or your business from terrorist attacks or natural disasters?

What are the most likely types of weapons of mass destruction (WMD) that a terrorist might use?

INFORMATION IS POWER

The goal of this book is to help you answer these and other questions that will equip you to make intelligent decisions about how best to keep your loved ones, yourself, and your neighbors safe and secure.

Hopefully, having read *Is America Safe?* you will:

- Appreciate that terror in America is not a new phenomenon. Our forefathers endured it and, in some cases, perpetuated it. But terror's history is one of being ineffective except in the short term.
- Various forms of domestic terrorism continue to plague us even today . . . and no doubt will continue to frustrate us for years to come.
- International terrorism, once practiced by weak states against stronger states, is now "asymmetric"—transnational by design. No nation is immune.
- Our homeland is under constant threat. The fact that we have not been struck since 9/11 is a combination of preemptive strikes overseas, extra measures of diligence at home . . . and a whole lot of good luck.
- As potential targets, such as military facilities, symbols of democracy, government buildings, and infrastructure, are "hardened," terrorists will shift to softer targets such as churches, schools, malls, mass entertainment centers, high-rise apartments, transportation centers, and energy facilities. Their goal will be to disrupt or destroy our economy, impose fear and uncertainty, break our national will, and deflect our attention and support from the Middle East.
- Safety and security are nebulous concepts requiring eternal vigilance, intuitive thinking, intelligent planning, and personal involvement at every level.

It is not the intent of this book to exaggerate the threats to your safety and security. It is rather to equip you with information that empowers you not to overreact. The hope is that you'll use this book as a resource to intelligently put things in proper perspective . . . so you and your loved ones may fully enjoy your birthright as Americans to pursue your right to seek the "security and happiness" that is promised in our Constitution.

It is worth remembering that every year, an estimated 270 million occupational accidents and 160 million work-related illnesses are reported throughout the world. In many developing nations, occupational safety still remains a largely reactive approach to injuries and illnesses sustained in the workplace.

In 2017, almost 5,000 people died in workplace accidents in the United States, and over 40,100 people died in road accidents, according to the National Safety Council. According to the nonprofit watchdog Health Watch USA, doctors' errors in hospital claim about 250,000 lives a year, the third-largest cause of death in the United States, after heart disease (614,000) and cancer (591,000).

Put that into perspective with the U.S. State Department's (DOS) report that there were about 22,400 acts of international terrorism in 2017, which killed 18,473 civilians, a 33 percent drop over 2016. Syria accounted for almost one-third of these attacks and the Islamic State is the most violent terrorist group being responsible for the deaths of 4,500 people in more than 4,500 attacks.

Most attacks were perpetrated by terrorists applying conventional fighting methods such as bombs and weapons including small arms. However, technology continues to empower terrorists and effective methods of attack are offsetting countermeasures against terrorism. Terrorists continued their practice of coordinated attacks including secondary attacks on first responders at attack sites with uniquely configured weapons and other materials to create improvised explosive devices (IED), including chemical IEDs.

Global and regional cooperation must be stepped up to combat the increasing use by terrorists of the Internet and satellite communications for recruitment, training, intelligence, and resource sharing.

There has also been an increase in attacks by Islamic supporters using improvised weapons such as motor vehicles and trucks, as witnessed in the last year in London, Canada, Brussels, and elsewhere. Many of these people had no known ties with terrorist organizations but were sympathetic to the cause. Security experts warn that this trend probably means a larger number of smaller attacks, less meticulously planned, and local rather than transnational in scope.

It is a fact, however, that anywhere you live may be subject to life-threatening natural disasters, such as wildfires, earthquakes, and mudslides on the West Coast; tornadoes and flooding almost anywhere; and hurricanes along our coastal areas.

However these are exceptional times. In 2002, there were only 205 reported acts of terrorism worldwide which claimed 725 lives. In 2017, there were 22,400 terrorist attacks which killed 18,473 people. This increase reflected the numerous indiscriminate attacks on "soft targets," such as places of worship, hotels, and commercial districts, intended to produce mass casualties.

Though any death is regrettable, it's obvious that most of us have more to fear from our workplaces, our doctors, our homes, and natural disasters than we do from terrorists who are upset with our lifestyle. However, we must remain always mindful of the nearly almost 3,000 souls killed and 6,000 injured on September 11, 2001—and that in the span of only a few hours.

Never forget that terrorists are patient. They don't operate on a predictable timeline and they don't give warnings.

Knowledge will help you assess the level of risk that you may face. Awareness will minimize the terrorists' ability to act. Involvement will ensure that we all maximize our potential to protect those we love and to preserve the freedom that we enjoy. Smart choices will mitigate the potential damage that a terrorist attack may generate.

Part I

TERRORISM TODAY

Chapter 1

Twenty-First-Century Threats to the Homeland

The number one threat to American national security during this long war is neither anthrax nor truck bombs . . . it is uncontrolled spending. We cannot afford to put guards on every bridge and at every critical node of our infrastructure. The outcome of this war will determine the type of nation our grandchild will know. I do not want that to be a nation that is bankrupt.

—Randall Larsen, Director, ANSER Institute for Homeland Security

The history of the United States is full of examples of our citizens overcoming enormous challenges to form our democracy from the crucibles of conquest, exploitation, internal and external strife, pandemic disease, and global war.

Though we have emerged as the most powerful democracy the world has known, it was not without sacrifice. America has never been "safe"—it is not presently safe—and the odds are that it will never be safe. But that doesn't mean that we need live in constant fear. It means that we do need to be aware of what threats we face and make informed decisions as to how we as a nation, as a community, and as individuals respond the those threats.

What are your worst fears about your future?
Is it another 9/11 attack?
Is it being caught in an underwater tunnel when a terrorist explodes a truck bomb within?
Is it being taken hostage at a large entertainment event?

What about your family and your children?
Do you fear that terrorists will attack your children's school?

Are you afraid that your home will be invaded and your loved ones threatened or killed?

Is it possible that you or your family will suffer from some natural or man-made disaster? Or—have you ever considered what you would do if the U.S. economy should be seriously affected by man-made or natural disasters?

You should be concerned about these and a myriad of other security and safety concerns. But you would do well to put them into a rational perspective. Some things you can control, others you can avoid, minimize, or mitigate—but there are some things over which you have little or no control. You also need to determine how likely certain things are to be a threat to you and which things should only be of little concern.

Hopefully, *Is America Safe?* will help you to sort through the threats to the homeland that you and your family may face and then assist you in developing a plan to eliminate or minimize your exposure.

The federal government recognizes that the Constitution requires it to "provide for the common defense" and to "repel invasions." Many would interpret those requirements to justify homeland security and related counterterrorism activities as an inherently governmental obligation. However, they point out that the vast majority of the targets that require protection are those owned by the private sector—critical infrastructure such as water and power sources and information systems, port facilities, railroads, air and ground transportation systems. This amplifies the need for a coalition to be formed between industry, the private sector, and government to address overarching homeland security issues.

How do you see your role in defending against terrorist threats or preparing for natural disasters?

The Department of Homeland Security (DHS) incorporates several initiatives in its strategic planning that we, as individuals, should also use:

1. *Awareness*—Identify and understand threats, assess vulnerabilities, determine potential impacts, and disseminate timely information to our community leaders, law enforcement, neighbors, and families.
2. *Prevention*—Detect, deter, and mitigate threats to our nation, our community, our family, and ourselves.
3. *Protection*—Safeguard our fellow citizens and their freedoms, and defend our critical infrastructure, public and private property, and the economy of our nation from acts of terrorism, natural disasters, or other threats.
4. *Response*—Work with our community leaders, our first responders, law enforcement, volunteer agencies, our churches, our schools, and each other to learn how to properly prepare for and to respond to acts of terrorism, natural disasters, or other emergencies.

5. *Recovery*—Support national, state, local, and private sector efforts to restore services and rebuild our communities after acts of terrorism, natural disasters, or other emergencies.

VULNERABILITY AUDIT

The first step to prepare your defense against attack, injury, or disease is to enhance your *awareness* of potential threats. That's what this book is all about. It's impossible to defend yourself against anything until you know:

1. The nature of the threat,
2. The probability that the threat may be directed toward you, your family, your community, or your nation, and
3. The means at your disposal to address the threat to either avoid it, mitigate it, or defend against it.

President George W. Bush succinctly defined the individual's role in homeland security when he said: "Defending our nation against its enemies is the first and fundamental commitment of the federal government. Today, that task has changed dramatically. Enemies in the past needed great armies and great industrial capabilities to endanger America. *Now, shadowy networks of individuals can bring great chaos and suffering to our shores for less than it costs to purchase a single tank.*"

It will also be the collective awareness of and response to these threats by individual citizens that will ultimately defeat the asymmetrical threat from global terrorism. Each and every one of us has a role to play in combating terror, in responding to human needs in the aftermath of disaster, and to recover from adversity, destruction, and disease.

HOW DO OTHERS RESPOND?

In a lifetime of global travel, one human characteristic stands out: *resiliency.* Whether innocent victims of war in Vietnam, terror in Beirut or Iraq and Afghanistan, townships like Soweto or tsunamis in Southeast Asia, the human instinct to survive is amazing.

It manifests itself in the United States following a disaster. People who have lost their homes to hurricane, fire, or flooding, try to salvage what they can and immediately start the task of rebuilding their lives. People rally round to help those less able to fend for themselves and communities come together as never before and emerge stronger than ever.

Around the world similar scenes are repeated daily. In Iraq and Afghanistan, pockets of insurgency terrorize local residents, but a majority of citizens are concentrating on rebuilding and establishing infrastructures of water, electricity, sewage treatment, schools, and hospitals that promise a brighter future.

It is the same in our own history. Some of our citizens may be briefly intimidated by terror or deterred and depressed by nature's wrath—but not for long. Our founders and each generation since demonstrate a will not only to survive but also to overcome adversity and to thrive.

WHAT ARE THE THREATS?

First and foremost, we must be concerned with *economic threats*. Whether we are threatened by a terrorist or by a series of natural disasters, we must both as a nation and as individuals, guard against economic chaos. If we spend so much of our national budget defending the homeland that we have to cut health care services and social security, are we winners or losers?

Cyber terrorism is part of this economic threat and a growing concern. From thousands of miles away, a small terrorist cell could hack into our critical infrastructure and close down all power lines. They could infiltrate all the major banks and destroy financial records. Every day, our government and critical infrastructure providers face thousands of attacks by hackers, many of them working on behalf of their own governments.

Second, in terms of national concern, we must be aware of the threat of *Weapons of Mass Destruction*, or *WMD*. Armed with advanced technology, fanatics could effectively threaten us physically, economically, and psychologically with nuclear, biological, or chemical WMDs. The end of the Cold War and disintegration of the Soviet Empire left stockpiles of radioactive material vulnerable. Corrupt officials and porous borders would make it relatively easy for terrorist groups with enough money to acquire this material and using the latest technology, they could pack a radioactive "dirty" bomb into a large suitcase or backpack. These devices are not nuclear weapons but they do release radioactive materials which in high enough doses can contaminate large areas and would cause widespread panic with hundreds or thousands of people needing medical attention.

The attack on 9/11 demonstrated that the homeland is vulnerable—and that thousands may be slaughtered—and many more severely injured—by the willful acts of just a few zealots.

Third is the threat of *natural disasters* that range from horrific earthquakes, to monster storms and tornadoes, to the possibility of a pandemic outbreak of disease. As we witnessed recently during Hurricanes Harvey, Irma, and Maria, when destruction is so widespread and inclusive, it may become impossible to draw upon government, our friends and neighbors—or anyone else besides ourselves for sustenance, shelter, or anything else for days, weeks, and possibly months.

Then, finally, there are those more *personal threats* such as anarchy, murder, hostage taking, assault, carjacking, home invasion, robbery, accidents, and disease.

ECONOMIC THREATS

The then DHS Secretary Michael Chertoff, testifying before congress in early March of 2006, stated that he wanted his department to move from a population-based funding formula to a risk-based one, which depends on criteria such as the nature of critical infrastructure and proximity to borders. Chertoff pointed out that "The issue is not where state lines are drawn. The issue is where are the vulnerabilities, where are the consequences."

As Randall Larsen, director of the ANSER Institute for Homeland Security, said, "We cannot afford to put guards on every bridge and at every critical node of our infrastructure. Nor can we afford a sophisticated chemical and biodetector in every government building.

"America cannot afford a risk-free society in a world of global terrorism," said Larsen. "*The enemy's strategy is to destroy our economy.* We must not facilitate their efforts. America will need to spend considerable sums of money to ensure our security—but we must do it wisely—there will be no money to waste on irrational fear and unconscionable pork. We must develop a strategic plan to guide our efforts. This must include federal, state and local governments, plus the private sector."

Here we are seventeen years later and yet we see continuing reports of pork-barrel politics and inappropriate spending of the resources dedicated to protecting the homeland.

"How does that affect me?" you might ask.

Wasteful spending affects every American citizen. It raises our taxes, thus depriving us of personal choices as to how we might spend our income. It depresses the economy in critical areas where those tax dollars might better be applied. More importantly—it dilutes the security for you and me for which it was intended.

Taken to the extreme, the pork-barrel politicians and the greedy state, regional, and local representative who misspend homeland security funds give aid and comfort to the fanatical enemy that is dedicated to destroying our infrastructure and bankrupt our society.

On a moral and humanitarian level, misuse of these funds restrict our capability and capacity to respond to both man-made and natural disasters, such as terrorist sabotage or attacks and hurricanes, tornadoes, flooding, and earthquakes.

WEAPONS OF MASS DESTRUCTION

I'm not too sure that all our citizens know what WMDs really are. From the rhetoric being spouted in the media, it's apparent that most people only think of nuclear bombs or intercontinental missiles as weapons of mass destruction. Nuclear devices are high on the list of WMDs, but they are only part of the mix. Civil Defense categorizes WMDs as Chemical, Biological, Radiological, Nuclear, and Explosive (CBRNE), defined as:

1. *Any explosive, incendiary, poison gas, bomb, grenade, or rocket having a propellant charge of more than four ounces [113 g], missile having an explosive or incendiary charge of more than one-quarter ounce [7 g], or mine or device similar to the above.*
2. *Poison gas.*
3. *Any weapon involving a disease organism.*
4. *Any weapon that is designed to release radiation at a level dangerous to human life.* (18 U.S.C. Section 2332a)

Among the most insidious of these are the biological weapons targeted toward inflicting hideous diseases upon thousands of innocent civilians who are unaware and unprotected. How would you protect yourself and your family from being a victim of a biological attack?

The media continues to present near nightly reminders of how deadly and innovative our terrorist enemies can be in launching indiscriminate suicide attacks on hotels, hospitals, police stations, restaurants, open air markets, and any other target where people congregate in large numbers. Are you apt to run over an improvised explosive device here in the United States? Well, it's relatively unlikely at present. However, it is quite likely that such a device might be employed in a theater, shopping center, school, apartment building, or sports arena. Remember the Boston Marathon bombings. The perpetrators had spent hours studying websites to see how they could make improvised bombs and had even discussed acquiring radioactive materials to make a dirty bomb.

High on the list of threats that the DHS is guarding against is the "dirty bomb." How would you respond if you were working on one side of the town and a dirty bomb was exploded over another area of the town where your family was threatened? What do you need to know to minimize danger to you and your loved ones from the radiated dust that such an attack would spread by air currents passing over your community?

You probably recall the anthrax scare that the United States experienced just a few years ago as the deadly chemical arrived in personal mail of various politicians and average citizens. A few deaths caused by the agent created waves of fear across the country, causing disruption of mail service and costing huge sums to detect and decontaminate.

Do you recall the symptoms? Do you know what to look for? Do you know who to call if you receive a suspicious piece of mail?

NATURAL DISASTERS

Hurricanes

Obviously, hurricanes form at sea, gain strength in warm tropical waters, and then continue to travel along a certain course until either weather conditions and cooler water dissipates their power, or they blow themselves out over land. June through November is hurricane season on the East Coast. These storms usually form in the Gulf and then move north, east, or west. But any point along the Eastern seaboard may experience their fury.

Despite the massive cost of three major Hurricanes—Harvey, Irma, and Maria—the 2017 hurricane season was only the 17th deadliest since 1900. The 1900 Galveston hurricane claimed 8,000 lives and is still the most deadly hurricane in U.S. history.

The second most deadly hurricane caused 1,836 deaths during September 6–20, 1928, as it made landfall in southeastern Florida at West Palm Beach, then plowed through the area of Lake Okeechobee. The barometer dropped to 27.57, which is believed to be the lowest reading ever recorded in the United States. Thousands were injured, 15,000 were homeless, and property losses of $50 million to $75 million were estimated.

The 2017 season was, however, the costliest on record with damage totaling more than $200 billion. One consolation is that despite the enormous damage these monster storms cause and the resulting price tag, the number of people killed by them is far fewer than would have been the case twenty or thirty years ago.

It wasn't until 1953 that tropical storms received formal names. But, as early as September of 1775, we have reports about a huge hurricane that spread destruction from North Carolina to Nova Scotia, called the "Hurricane of Independence," that killed an estimated 4,170 in the United States and Canada.

Approximately 90 percent of deaths attributed to deaths by hurricanes are caused by storm surge—the dome of water created by the low pressure center of a hurricane. This storm surge quickly floods low-lying coastal areas with anywhere from 3 feet for a Category 1 storm to over 19 feet for a Category 5 storm. As ocean front properties attracted more dense populations, fatalities and physical damage increased.

By August 23, 1992, when a Category 5 storm named Andrew slammed ashore near South Miami, early warning systems that rely on weather satellites and computer models could detect hurricanes in advance. That, combined with proactive evacuation plans, minimized human exposure so that only twenty-three deaths were directly attributed to the storm. But more than 135,000 single family and mobile homes were destroyed or damaged, 160,000 people were left homeless, and 86,000 lost their jobs. Total damage caused by Hurricane Andrew was estimated approximately $25 billion.

So what can you expect for the future?

Unfortunately, experts say that nearly a third of all Floridians and nearly half of all Americans have no hurricane preparedness plan, according to a recent survey. Forecasters hope that residents pay attention to news reports when storms brew, and that they follow advice to evacuate when necessary.

The experts attribute the high number of tropical storms and hurricanes to a cyclical pattern that is repeated over time. *That cycle is not expected to end for several years to come.*

Tornadoes

Somewhat related to hurricanes, tornadoes are also cyclonic in nature. And though most of the roughly 1,200 tornadoes reported each year occur in early spring and fall, tornadoes may appear almost anywhere and at any time throughout the year. They tend to form primarily in the southern and central regions of the United States. Although tornadoes cause billions in damage each year, only an average of about 60 result in death. Annually about 80 people are killed by tornadoes and 1,500 injured.

Earthquakes

According to the U.S. Geological Survey, there are several million earthquakes worldwide each year as the earth's tectonic plates and fault line shift and grind, releasing pent-up energy and then settling quietly for a period of time. About 500,000 of these are detectable. Southern California has about 10,000 earthquakes although most of them are so small they cannot be felt. Overall, there are about 3,000 detectable earthquakes annually in the United States. The majority of these are rated between magnitudes of 2.0 and

4.9—which are very minor in terms of potential damage. Still, four or five earthquakes occur each year in the United States that measure 6.0 or above on the seismograph—and these should be of some concern if they activate in any population center. The largest recorded earthquake in the United States was of a magnitude of 9.2 that struck Prince William Sound, Alaska, on Good Friday, March 28, 1964. The worst earthquake recorded in the United States was the 1906 San Francisco earthquake in which 3,000 people died as a result of the earthquake and from extensive fires that erupted in its aftermath. On average, less than ten people are killed each year by earthquakes with the toll periodically rising to sixty or seventy.

Tsunamis

The Japanese name for "harbor waves," tsunamis are relatively rare, but present the potential for massive destruction and death. The waves may travel in the open sea as fast as 450 miles per hour. As the big waves approach shallow waters along the coast they grow to a great height and smash into the shore. They can be as high as 100 feet. They can cause a lot of destruction on the shore. They are sometimes mistakenly called "tidal waves," but tsunamis have nothing to do with the tides. Most tsunamis—about 80 percent—happen within the Pacific Ocean's "Ring of Fire," a geologically active area where tectonic shifts make earthquakes and volcanic eruptions common.

Hawaii is the state at greatest risk for a tsunami. It gets about one a year, with a damaging tsunami happening about every seven years. Alaska is also at high risk. California, Oregon, and Washington experience a damaging tsunami about every eighteen years. An Alaskan earthquake generated a tsunami with waves in 1964 that reached between 10 and 20 feet high along parts of the California, Oregon, and Washington coasts. In 1946, a tsunami with waves of 20 to 32 feet crashed into Hilo, Hawaii, flooding the downtown area.

More than 230,000 people were killed and 125,000 injured by the tsunami that struck following a massive seaquake of around 9 on the Richter scale, with an epicenter off the Indonesian island of Sumatra in South Asia on December 26, 2004. It displaced 1.5 million more people in coastal areas and caused enormous loss of property, livelihoods, and infrastructure. Over 47,000 victims in Indonesia's Aceh province share a mass grave and many victims remain missing.

Experts say that it's not a matter of "if" but "when" and "where" a cataclysmic tsunami will hit the United States. Warning systems of delicate sensors lie on the ocean floor. These could offer Hawaii hours of advance warning time in which to evacuate. Unfortunately, the U.S. West Coast may only have the advantage of minutes instead of hours to react—but every minute can make a huge difference if you are prepared.

Floods

The National Safety Council (NSC) calculates that your chance of dying in a flood is about one in 32 million for any given year and about one in 414,000 throughout your lifetime. To put that into perspective, NSC sets your odds of drowning in your bathtub at one out of 818,000 each year and one out of 10,582 during your lifetime. So perhaps you should worry more about taking a bath than avoiding a flood? Not really. But the choices that you make and the lifestyle that you choose obviously impact for good or bad upon your chances of being killed or injured.

Pandemic Diseases

There is a historical record of influenza pandemics of varying severity at 20–40-year intervals. The "Spanish flu" began in August 1918 in three disparate locations: Brest; Boston, Massachusetts; and Freetown. An unusually severe and deadly strain of influenza spread worldwide. The disease spread across the world, killing 25 million in the course of six months—some estimates put the total of those killed worldwide at over twice that number. An estimated 17 million died in India, 500,000 in the United States, and 200,000 in the UK. It vanished within 18 months and was caused by the H1N1 type of flu virus, which has since been reconstructed at the CDC.

The "Asian Flu," a (H2N2) virus, caused about 70,000 deaths in the United States. First identified in China in late February 1957, the Asian flu rapidly spread throughout the United States by June 1957.

The "Hong Kong Flu" (H3N2) caused about 34,000 deaths in the United States. This virus was first detected in Hong Kong in early 1968 and spread to the United States later that year. Influenza (H3N2) viruses still circulate today.

In 2003, avian influenza virus (H5NI) was detected in Vietnam, increasing fears of new variant strains emerging. After a lull, several new cases were reported in Egypt in April 2009—raising fears that the virus had mutated into a form that could more easily be transmitted from people to people. The first and only human infection with an Asian H5NI virus in the Americas was reported in January 2014 in Canada in a traveler returning from China.

Health officials fear, however, that if the avian influenza virus combines with a human influenza virus, the new subtype created could be both highly contagious and highly lethal in humans. Such a subtype could cause a global influenza pandemic, similar to the Spanish Flu, or the lower mortality pandemics such as the Asian Flu and the Hong Kong Flu.

What about the Future?

Diseases that can attain pandemic proportions include Lassa fever, Rift Valley fever, Marburg virus, Ebola virus, and hemorrhagic fever. However, emergence of these diseases into the human population in recent years has shown their virulence is so high they dissipated in geographically confined areas. Thus their effect on humans is currently considered limited.

HIV—the virus that causes AIDS—may be considered a global pandemic, but it is currently most extensive in southern and eastern Africa. It is restricted to a small proportion of the population in other countries, and is only spreading slowly in those countries. If there were to be a true destruction-of-life pandemic it would be likely to be similar to HIV, i.e., a constantly evolving disease.

There is a fear that antibiotic-resistant "super bugs" may also revive diseases previously regarded as "conquered." This treat has come about in part by an insistence by patients that their doctors prescribe antibiotics for every sniff and sniffle when they are both inappropriate and ineffective for most viruses or colds.

Personal Threats

How safe are you in your own home? It may depend on where you live, what lifestyle you pursue, what circles you move within, what sex you are, and where you are on an economic scale.

Anarchy is indiscriminate. Unleashed by ineffective or absence of authority, individuals who would otherwise be law abiding and civil may deteriorate into a mob mentality, destroying public and private property, inflicting harm on others and giving into immoral and unlawful conduct of every type.

The Federal Bureau of Investigation (FBI) reported that there were 1,248,185 violent crimes reported nationwide in 2016—the latest published figures. That was 4 percent higher than in 2015 and equates to 386.3 violent crimes per 100,000 inhabitants.

There were a reported 17,250 murders and this figure represents a 16.1 percent increase over 2015. The data showed that most murder victims were adults and most were males. Almost a third of female murder victims were killed by a husband or boyfriend.

Aggravated assault and rape (legacy definition) offenses increased 5.1 percent and 4.9 percent, respectively, and robbery increased 1.2 percent. Nationwide, there were an estimated 7,919,035 property crimes and burglaries dropped 4.6 percent. Collectively, victims of property crimes (excluding arson) suffered losses estimated at $15.6 billion in 2016.

Carjacking is defined as completed or attempted robbery of a motor vehicle by a stranger to the victim. It differs from other motor vehicle theft because the victim is present and the offender uses or threatens to use force. Carjacking victimization rates are highest in urban areas, followed by suburban and rural areas. Ninety-three percent of carjackings occurred in cities or suburbs. A weapon was used in 74 percent of carjacking victimizations. Firearms were used in 45 percent of carjackings, knives in 11 percent, and other weapons in 18 percent of the crimes.

Home invasion is the residential equivalent of an automobile carjacking and it is on the rise. Like the crime of carjacking, most police agencies don't track home invasions as a separate crime. Most police agencies and the FBI will statistically record the crime as a residential burglary or a robbery. Without the ability to track the specific crime of home invasion, little can be done to alert the public as to the frequency of occurrence in their community or devise a law enforcement plan of action to control it.

Home invasion robbers, in contrast to burglars, work more often at nights and on weekends when residences are more likely to be occupied. The home invaders will sometimes target the resident as well as the residence. The selection process may include a woman living alone, a senior citizen or a known drug dealer, for example. It is not unheard of for a robber to follow you home based on the value of the car you are driving or the jewelry you are wearing.

Many home robbers may have been in your home before as a delivery person, installer, or repair vendor. Home robbers rarely work alone and rely on an overwhelming physical confrontation to gain control and instill fear in their victims. The greatest violence usually occurs during the initial confrontation and home invaders often come prepared with handcuffs, rope, duct tape, and firearms. Some in-home robbers appear to enjoy the intimidation, domination, and violence and some even claim it's a "rush."

The FBI's Crime Clock estimates that:

Every 26.3 seconds a violent crime is committed.
Every 33.5 minutes someone is murdered.
Every 4.2 minutes someone commits a forcible rape.
Every 1.6 minutes someone is robbed.
Every 41.3 seconds someone is the victim of an aggravated assault.
Every 20 seconds someone commits a burglary.
Every 5.5 seconds someone is the victim of larceny-theft.
Every 44.6 seconds someone's automobile is stolen.

Chapter 2

Counterterrorism Tactics

STRATEGY FOR VICTORY

President George W. Bush developed a multiple-approach strategy to defeating terrorists both at home and overseas and this plan of action was continued by President Obama. It has been further refined by President Trump who also advocated a more isolationist approach by restricting immigration and sealing borders.

Within the United States, law enforcement and intelligence agencies are tasked with protecting the homeland although every citizen has his or her role to play. Overseas, the U.S. military and secret service are working with foreign governments, their military, and their intelligence communities as part of the global war on terrorism.

"In this new century freedom is once again under assault," President Bush said in 2005. "In the four years since September 11, the United States and the world have been committed to winning the war on terror. The evil that came to our shores on September 11 has reappeared on other days and in other places. In cities across the world, we have seen images of destruction and suffering that can seem like random acts of madness but are part of a larger terrorist threat. To combat this evil, we must remember the calling of September 11—we will confront this mortal danger to all humanity and not tire or rest until the war on terror is won."

Bush insisted that Muslims are now playing a vital role in the war on terror. "Muslim leaders are denouncing terrorism, and the time has come for all responsible Islamic leaders to denounce an ideology that exploits Islam for political ends and defiles a noble faith. Many Muslims have joined the fight against extremism. The United States is proud to stand by those who stand up for the liberty, justice, and humanity of their own tradition," he said.

> We are here to do the work that ensures no other family members have to lose a loved one to a terrorist who turns a plane into a missile, a terrorist who straps a bomb around her waist and climbs aboard a bus, a terrorist who figures out how to set off a dirty bomb in one of our cities. This is why we are here: to make our country safer and make sure the nearly 3,000 who were taken from us did not die in vain; that their legacy will be a more safe and secure Nation. (Barack Obama, speech in the U.S. Senate, March 6, 2007)

THE BUSH GOVERNMENT'S RATIONALE

1. The Militant Threat and the Importance of Winning the War on Terror

The ideology known as Islamic radicalism, militant Jihadism, or Islamo-fascism—different from the religion of Islam—exploits Islam to serve a violent political vision that calls for the murder of all those who do not share it. The followers of Islamic radicalism are bound together by their shared ideology, not by any centralized command structure. Although they fight on scattered battlefields, these terrorists share a similar ideology and vision for the world openly stated in videos, audiotapes, letters, declarations, and websites.

First, these extremists are determined to end American and Western influence in the Middle East. Because the United States is spreading democracy and peace, we stand in the way of their radical ambitions.

Al-Qaeda's leader, Osama Bin Laden, has called on Muslims to dedicate their "resources, sons and money to driving the infidels out of their lands."

"To meet this goal, they hit us and expect us to run in the hope that we will repeat the sad history of Beirut in 1983 and Mogadishu in 1993—only this time on a larger scale and with greater consequences," President Bush explained.

Second, the militant network wants to use the vacuum created by retreat to expand its power. As they once did in Afghanistan and as they are now trying to do in Iraq, they will seek to gain control of an entire country in order to establish a base from which to launch attacks and conduct their war against non-radical Muslim governments.

Third, from their new base, these militants will seek to establish a radical Islamic Empire—a *Caliphate*. The militants believe that controlling one country will rally the Muslim masses, enabling them to overthrow moderate governments in the Middle East and establish a radical Islamic empire that spreads from Spain to Indonesia.

To accomplish their extremist agenda, the militants thrive on the suffering and frustrations of others. The radicals exploit local conflicts to build a culture of victimization, exploit resentful and disillusioned young men and women, and use modern technology to multiply their destructive power. They are sheltered and supported by authoritarian regimes—allies of convenience like Syria and Iran—that share the goal of hurting America and moderate Muslim governments, and that use terrorist propaganda to blame their own failures on the West, America, and Jews.

The terrorists are strengthened by front operations—such as corrupted charities—and those who aggressively fund the spread of radical, intolerant versions of Islam. The militants are aided by elements of the Arab news media that incite hatred and anti-Semitism, feed conspiracy theories, and speak of a so-called American war on Islam—with seldom a word about American action to protect Muslims in Afghanistan, Bosnia, Somalia, Kosovo, Kuwait, and Iraq.

This extremism may not be given concessions, bribed, or appeased. The Free World must never accept anything less than victory.

To those who argue that extremism has been strengthened by the actions of our Coalition in Iraq, we must remember that we were not in Iraq on September 11, 2001. The radicals' hatred existed before Iraq was an issue, and it will exist after Iraq is no longer an excuse. The government of Russia did not support Operation Iraqi Freedom, and yet the Islamic militants killed more than 180 Russian schoolchildren in Beslan.

Islamic radicalism is another in a line of bankrupt ideologies. The murderous ideology of the Islamic radicals is the great challenge of our century. Yet, in many ways, this fight resembles the struggle against communism in the last century. Like communism, Islamic radicalism is led by an elitist self-appointed vanguard that presumes to speak for the Muslim masses. Bin Laden says his role is to tell Muslims "what is good for them and what is not." What this man who grew up in wealth and privilege considers good for poor Muslims is that they become killers and suicide bombers. He assures them that this is the road to paradise—though he never offers to go along for the ride.

This explains their cold-blooded contempt for human life seen in the murders of Daniel Pearl, Nicholas Berg, Margaret Hassan, and many others. These militants have shown themselves to be enemies of not only America and Iraq, but enemies of Islam and humanity as well. The world has witnessed this kind of cruelty before—the heartless zealotry that led to the gulags, the Cultural Revolution, and the killing fields.

Islamic radical leaders have endless ambitions of imperial domination, and they wish to make everyone powerless except themselves. While promising a future of justice and holiness, the terrorists are preparing a future of oppression and misery—banning dissent and books, brutalizing women, and controlling every aspect of life.

Zarqawi has said that Americans are "the most cowardly of God's creatures." However, it is cowardice that kills children and the elderly with car bombs, cuts the throat of a bound captive, or targets worshipers leaving a mosque. It is courage that liberated 50 million people from tyranny, keeps an untiring vigil against the enemies of a rising democracy, and will once again destroy the enemies of freedom.

Like communism, Islamic radicalism contains inherent contradictions that doom it to failure. By fearing freedom, distrusting human creativity, punishing change, and limiting the contributions of half the population, this ideology declares war on the idea of progress itself. The only thing modern about the militants' vision is the weapons they seek to use. The outcome of this war is not in doubt—those who despise freedom and progress have condemned themselves to isolation, decline, and collapse. Because free people believe in the future, they will own the future.

THE U.S. STRATEGY TO WIN

Defeating a broad and adaptive network requires patience, constant pressure, and strong partners. Working with these partners, the United States is disrupting militant conspiracies, destroying their ability to make war, and giving millions a hopeful alternative to resentment and violence.

Prevent Attacks

Our leaders are reorganizing government for a broad and coordinated homeland defense, reforming intelligence agencies for the difficult task of tracking enemy activity, and acting, along with governments from many countries, to destroy the terrorist networks and incapacitate their leaders. Our defenders of the homeland have prevented terrorist plots and Al-Qaeda efforts to case targets in the United States. Because of strikes against terrorist leaders and disruption of their plots, the enemy is wounded but still capable of deadly operations around the globe. Our commitment must be clear: we must not relent until the organized international terror networks are exposed and broken and their leaders held to account for their acts of murder.

Weapons of Mass Destruction Denied to Terrorist Allies

Working with Great Britain, Pakistan, and other nations, the United States shut down A.Q. Khan's black-market operation in nuclear technology. Libya abandoned its chemical and nuclear weapons programs, as well as its long-range ballistic missiles. In the last year, America and our partners in the Proliferation Security Initiative stopped more than a dozen shipments of suspect weapons technology—including equipment for Iran's ballistic missile programs. This progress significantly reduced the danger, but did not remove it entirely. Evil men still work to gain these weapons, and the United States and its allies are working urgently to keep them out of terrorist hands.

Radical Groups Denied the Support and Sanctuary of Outlaw Regimes

State sponsors like Syria and Iran have a long history of collaboration with terrorists—and they deserve no patience from the victims of terror. The United States makes no distinction between those who commit acts of terror and those who support and harbor terrorists, because they are equally guilty of murder.

Deny Militant Control of Any Nation

The United States is fighting beside our Afghan partners against the remnants of the Taliban and its Al-Qaeda allies. We are working with President Musharraf to isolate the militants in Pakistan. We are fighting the terrorists and regime remnants in Iraq who seek to overthrow a democracy, claim a strategic country as a haven of terror, destabilize the Middle East, and strike America and other free nations with ever increasing violence. Our goal is to defeat the terrorists and their allies at the heart of their power—and we are achieving this goal.

Coalition and Iraqi Forces Clear Terrorists from Iraqi Cities

Coalition forces are improving the lives of Iraqi citizens by conducting offensive operations to clear out enemy forces and leave Iraqi units behind to prevent the enemy from returning. With U.S. help, the Iraqi military is gaining new capabilities and new confidence with every passing month. At the time of the Fallujah operations of eleven months ago, there were only a few Iraqi battalions in combat—today there are more than eighty. The progress is not easy, but it is steady.

Political Progress

The extremists in Iraq are not patriots nor resistance fighters but murderers at war with the Iraqi people. In contrast, the elected leaders of Iraq are proving to be strong and steadfast. In the space of two and a half years, Iraq has made incredible progress on the path to becoming a democracy. It is true that the seeds of freedom have only recently been planted, but we are confident that if our Coalition and the Iraqi people each does their part, Iraqi democracy will succeed.

Cut-and-Run Not an Option—Nor a Strategy

Having removed a dictator who hated free people, the United States must not stand by as a new set of killers seize control of Iraq. There is always a temptation, in the middle of a long struggle, to seek the quiet life, escape the duties and problems of the world, and hope the enemy grows weary. However, this enemy is never tired and considers every retreat an invitation to greater violence. In Iraq, there is no peace without victory. We must keep our nerve and win that victory.

Advancing Democracy and Hope

If the region is left to grow in bitterness and misery, while radicals stir the resentments of millions, then that part of the world will be a source of endless conflict and mounting danger. However, the extremists will be marginalized and their violence ended if the people of that region are permitted to choose their own destiny and advance by their own energy and participation as free men and women. America is making this stand in practical ways.

The United States encourages our friends in the Middle East, including Egypt and Saudi Arabia, to take the path of reform to strengthen their own societies in the fight against terror by respecting the rights of people. The United States supports dissidents and exiles who stand against oppressive regimes, because we know that the dissidents of today will be the democratic leaders of tomorrow. Our nation makes its case through public diplomacy—stating clearly and confidently our belief in self-determination, religious freedom, and equal rights for women. By standing for the hope and freedom of others, we will make our own freedom more secure.

U.S. Counterterrorism Policy

1. Make no concessions to terrorists and strike no deals;
2. Bring terrorists to justice for their crimes;
3. Isolate and apply pressure on states that sponsor terrorism to force them to change their behavior; and
4. Bolster the counterterrorism capabilities of those countries that work with the United States and require assistance.

Congress authorized the **Antiterrorism Assistance** (ATA) program in 1983 as part of a major initiative against international terrorism. Since that time, ATA has provided training for more than 36,000 students from 142 countries. The ATA program provides training and related assistance to law enforcement and security services of selected friendly foreign governments. Assistance to the qualified countries focuses on the following objectives:

- Enhancing the antiterrorism skills of friendly countries by providing training and equipment to deter and counter the threats of terrorism.
- Strengthening the bilateral ties of the United States with friendly, foreign governments by offering concrete assistance in areas of mutual concern.
- Increasing respect for human rights by sharing with civilian authorities modern, humane, and effective antiterrorism techniques.

ATA courses are developed and customized in response to terrorism trends and patterns. The training may be categorized into four functional areas:

- Crisis Prevention
- Crisis Management
- Crisis Resolution, and
- Investigation.

Countries needing assistance are identified on the basis of the threat or actual level of terrorist activity they face.

Antiterrorism assistance and training may be conducted either in-country or within the United States. This arrangement provides flexibility to maximize the effectiveness of the program for countries of strategic importance in the global war on terrorism.

ATA programs may take the form of advisory assistance, such as police administration and management of police departments, how to train police instructors or develop a police academy, and modern interview and investigative techniques. This approach enables the program to provide a narrow focus to solutions for country-specific problems that are not resolved in the classroom-training environment. Equipment or explosive-detection trained dogs may also be included in the assistance package.

The ability of the United States to assist friendly governments to master the detection and prevention of terrorist activities will clearly enhance the mutual security of all the participating nations. Detecting and eliminating terrorist cells at the root before their violence can cross borders and oceans will ensure a safer world for all nations.

ATA continues its efforts to familiarize ambassadors, regional security officers, and other U.S. officials with the program offerings. The success of these efforts is evidenced by the fact that every frontline nation has requested antiterrorist assistance in some form. U.S. diplomats report that the ability of the United States to offer immediate, specific, and intensive training, along with technical tools and equipment, has succeeded in breaking down barriers and building trust.

ATA is responding to the growing demand for training and services not only by expanding course selection but also by pursuing development of the Center for Antiterrorism and Security Training. ATA is already offering training at a variety of venues, to include other government agencies and state and local facilities—both within and outside the Washington, DC, metropolitan area.

The **Rewards for Justice Program** is one of the most valuable U.S. government assets in the fight against international terrorism. Established by the 1984 Act to Combat International Terrorism—Public Law 98-533—the Program is administered by the U.S. Department of State's Bureau of Diplomatic Security.

Under the Program, the Secretary of State may offer rewards of up to $5 million for information that prevents or favorably resolves acts of international terrorism against U.S. persons or property worldwide. Rewards may also be paid for information leading to the arrest or conviction of terrorists attempting, committing, and conspiring to commit—or aiding and abetting in the commission of—such acts.

The U.S. Patriot Act of 2001 authorizes the Secretary to offer or pay rewards of greater than $5 million if he determines that a greater amount is necessary to combat terrorism or to defend the United States against terrorist acts. Secretary Powell has authorized a reward of up to $25 million for the information leading to the capture of Bin Laden and other key Al-Qaeda leaders.

In November 2002, the State and Treasury Departments announced a $5 million rewards program that will pay for information leading to the disruption of any terrorism financing operation.

Diplomatic Security has fully supported the efforts of the private business sector and citizens to establish a Rewards for Justice fund, a nongovernmental, nonprofit 501(c)(3) charitable organization administered by a group of private U.S. citizens. One hundred percent of all donated funds will be used to supplement reward payments only.

Diplomatic Security has forged a strong relationship with the private business and U.S. citizen representatives of the Rewards for Justice Fund.

Diplomatic Security has embarked on a much closer relationship with the U.S. public and private businesses in the U.S. government's continuing efforts to bring those individuals responsible for the planning of the September 11 attacks to justice and to prevent future international terrorist attacks against the United States at home or abroad.

Since its inception, the Rewards for Justice Program has been very effective. In the past seven years, the Secretary of State has authorized payments for more than $52 million to 33 people who provided credible information that put terrorists behind bars or prevented acts of international terrorism worldwide. The program played a significant role in the arrest of international terrorist Ramzi Yousef, who was convicted in the 1993 bombing of the World Trade Center and most recently in the efforts to locate Uday and Qusay Hussein.

The U.S. government will make no concessions to individuals or groups holding official or private U.S. citizens hostage. The United States will use every appropriate resource to gain the safe return of U.S. citizens who are held hostage. At the same time, it is U.S. government policy to deny hostage takers the benefits of ransom, prisoner releases, policy changes, or other acts of concession.

Basic Premises

It is internationally accepted that governments are responsible for the safety and welfare of persons within the borders of their nations. Aware of both the hostage threat and public security shortcomings in many parts of the world, the United States has developed enhanced physical and personal security programs for U.S. personnel and established cooperative arrangements with the U.S. private sector. It has also established bilateral assistance programs and close intelligence and law enforcement relationships with many nations to prevent hostage-taking incidents or resolve them in a manner that will deny the perpetrators benefits from their actions. The United States also seeks effective judicial prosecution and punishment for hostage takers victimizing the U.S. government or its citizens and will use all legal methods to these ends, including extradition. U.S. policy and goals are clear, and the U.S. government actively pursues them alone and in cooperation with other governments.

U.S. GOVERNMENT RESPONSIBILITIES WHEN PRIVATE U.S. CITIZENS ARE TAKEN HOSTAGE

On the basis of past experience, the U.S. government concluded that making concessions that benefit hostage takers in exchange for the release of hostages increased the danger that others will be taken hostage. U.S. government policy is, therefore, to deny hostage takers the benefits of ransom, prisoner releases, policy changes, or other acts of concession. At the same time, the U.S. government will make every effort—including contact with representatives of the captors—to obtain the release of hostages without making concessions to the hostage takers.

Consequently, the United States strongly urges U.S. companies and private citizens not to accede to hostage-taker demands. It believes that good security practices, relatively modest security expenditures, and continual close cooperation with Embassy and local authorities can lower the risk to U.S. citizens living in high-threat environments.

The U.S. government is concerned for the welfare of its citizens but cannot support requests that host governments violate their own laws or abdicate their normal enforcement responsibilities.

If the employing organization or company works closely with local authorities and follows U.S. policy, U.S. Foreign Service posts can be involved actively in efforts to bring the incident to a safe conclusion. This includes providing reasonable administrative services and, if desired by local authorities and the U.S. entity, full participation in strategy sessions. Requests for U.S. government technical assistance or expertise will be considered on a case-by-case basis. The full extent of U.S. government participation must await an analysis of each specific set of circumstances.

The host government and the U.S. private organizations or citizen must understand that if they wish to follow a hostage-resolution path different from that of U.S. government policy, they do so without U.S. government approval. In the event a hostage-taking incident is resolved through concessions, U.S. policy remains steadfastly to pursue investigation leading to the apprehension and prosecution of hostage takers who victimize U.S. citizens.

LEGAL CAUTION

Under current U.S. law, 18 USC 1203 (Act for the Prevention and Punishment of the Crime of Hostage-Taking, enacted October 1984 in implementation of the UN convention on hostage taking), seizure of a U.S. citizen as a hostage anywhere in the world is a crime, as is any hostage-taking action in which the U.S. government is a target or the hostage taker is a U.S. national. Such acts are, therefore, subject to investigation by the Federal Bureau of Investigation and to prosecution by U.S. authorities. Actions by private persons or entities that have the effect of aiding or abetting the hostage taking, concealing knowledge of it from the authorities, or obstructing its investigation may themselves be in violation of U.S. law.

The U.S. government established four primary counterterrorism lists to serve as tools in the fight against terrorism: the State Sponsors of Terrorism, Foreign Terrorist Organizations (FTO), Executive Order 13224, and the Terrorist Exclusion List (TEL). Each list has its individual mechanisms, but they all serve to prevent terrorism, punish terrorists and their supporters, and pressure changes in the behavior of designated states and groups.

Because these lists are a means to fight terrorism rather than an end unto themselves, they are not designed or intended to be immutable. The U.S. government encourages states and organizations to take the necessary actions to get out of the terrorism business. The bar for a state or group being removed from a terrorism list is and must be high—it must end all involvement in any facet of terrorism, including passive support, and satisfy all U.S. government counterterrorism concerns.

STATE SPONSORS OF TERRORISM

The Secretary of State is authorized to designate a government as a state sponsor of terrorism if that government "has repeatedly provided support for acts of international terrorism." U.S. law requires the imposition of various sanctions on a state so designated. A number of U.S. laws and sanctions affect countries whose governments have been designated as state sponsors of terrorism.

The four main categories of sanctions include a ban on arms-related exports and sales; restrictions on exports of dual use items; prohibitions on official U.S. government economic assistance (except humanitarian assistance), including a requirement that the U.S. government oppose multilateral bank assistance; and imposition of miscellaneous trade and other restrictions, including a prohibition on imports and liability in U.S. courts for officials of the country that engages in terrorist activity. The original list of State Sponsors of Terrorism—announced in December 1979—named Iraq, Libya, Pakistan, South Yemen, and Syria.

Inclusion on the State Sponsors of Terrorism list also targets a country for other sanctions laws that penalize persons and countries engaging in certain trade with state sponsors. Currently, there are four countries on the list: Cuba, Iran, North Korea, and Sudan (Iraq was removed from the list in October 2004 and Libya in May 2006).

FOREIGN TERRORIST ORGANIZATIONS

The Secretary of State is authorized to designate as foreign terrorist organizations (FTOs) groups that conduct terrorism and threaten the interests of the United States. Designation allows the U.S. government to block designees' assets in U.S. financial institutions, criminalize witting provision of material support to designated groups, and block visas for members of FTOs without having to show that the individual was involved in specific terrorist activities. FTO designation also sends a strong signal that any group that engages in terrorism—regardless of its purported goals—will be condemned and penalized for its actions.

TERRORIST FINANCING

President Bush signed Executive Order 13224 on September 23, 2001, giving the U.S. government a strong tool for eliminating the financial supporters and networks of terrorism. EO 13224 enables the United States to block designees' assets in any financial institution in the United States or held by any U.S. citizen. It also expands government authority to permit the designation of individuals and organizations that provide support or financial or other services to, or associate with, designated terrorists. EO 13224 designations have allowed the USG, as well as Coalition partners acting in concert, to block tens of millions of dollars intended to bankroll the murderous activities of Al-Qaeda and other terrorist groups.

On October 26, 2001, President Bush signed into law a comprehensive counterterrorism bill (Public Law 107-56, also known as USA PATRIOT). The new law strengthened enforcement tools and made improvements to the last major terrorism bill, the Antiterrorism and Effective Death Penalty Act of 1996. The USA Patriot Act also created a Terrorist Exclusion List (TEL) with immigration consequences for groups named therein. Designation on the TEL allows the United Sates to exclude or deport aliens who provide material assistance to, or solicit it for, designated organizations, giving the Department of State and U.S. law enforcement agencies a critical tool for bolstering homeland security. In May 2011, President Obama signed a four-year extension of three key provisions of the Act—roving wiretaps, searches of business records, and conducting surveillance of "lone wolves." Many parts of the Patriot Act expired on June 1, 2015, after failing to get Congressional approval but the USA Freedom Act was approved on June 2, 2015, which reinstated the expired and other sections and runs through 2019.

TERRORISM AND IMMIGRATION IN EUROPE

Recent developments have seen a divergence in social and political responses to terrorism between the United States and Western Europe. The September 11, 2001, attacks were carried out by foreigners who entered the United States to deliberately carry out terrorist attacks, on behalf of a foreign organization, operating from bases in a remote country. Western European countries, on the other hand, now confront domestic terrorism based within a domestic religious minority, some recent immigrants, but many native-born citizens as the London suicide bombings illustrated.

Although Muslims are a relatively small minority in the United States, in some European cities they are approaching a majority. In many cases they are concentrated in areas of poor housing and low employment which has led to increasing frustration and bitterness.

Outbreaks of rioting in France were concentrated in communities dominated by North African and Middle Eastern immigrants and with a high Muslim population. These areas have long been ignored by politicians and many have almost become no-go areas for the police with gangs and drug dealers controlling the streets.

Following 9/11, and Madrid and London bombings, there was a backlash against the Muslim population and many Muslims were assaulted and mosques attacked. Defusing potential backlash is now a standard item of European counterterrorism policy.

The direction of European responses to terrorism was indicated by new policies, proposed by Tony Blair in August 2005 following the deadly London mass transit bombings, including:

- Deportation and exclusion on grounds of fostering hatred, advocating violence to further a person's beliefs or justifying or validating such violence a criminal offence of condoning or glorifying terrorism;
- Refusal of asylum to anyone with a connection to terrorism;
- New pre-trial procedures and extending detention pre-charge of terrorist suspects;
- Extended use of control orders for those who are British nationals and who cannot be deported, with imprisonment for any breach of the order; and
- New power to order closure of a place of worship which is used as a "center for fomenting extremism."

Target-Hardening

Common targets of terrorists are areas of high population concentration, such as mass transit vehicles (metro, bus, and trains), aircraft, office buildings, and crowded restaurants. Whatever the target of terrorists, there are multiple ways of hardening the targets so as to prevent the terrorists from hitting their mark. Perhaps the single most effective of these is bag-searching for explosives, which is only effective if it is conducted before the search subjects enter an area of high population concentration.

Another method is to place concrete barriers a sufficient distance outside buildings to prevent truck bombing. Aircraft cockpits are kept locked during flights, and have reinforced doors, which only the pilots in the cabin are capable of opening.

Telework Dispersal

In 1999, the federal government authorized telework—working in your home, at a hotel, or other offsite location rather than at your office—as a means of reducing fuel consumption, improving employee quality of life by reducing commuting time and reducing the need for childcare. Telework has now become an integral component of continuity of operations (COOP) planning. By equipping employees to work from home or other remote locations, the government reduces the number of people concentrated in one office facility, thus making that structure less attractive as a prospective terrorist target. It also enhances the government's capacity to maintain operations in the wake of man-made or natural disasters.

Preemptive Neutralization

Some countries see preemptive attacks as a legitimate strategy. This includes capturing, killing, or disabling suspected terrorists before they can mount an attack. Israel, the United States, and Russia have taken this approach, while Western European states are generally more cautious.

In July 2005, Brazilian Jean Charles de Menezes was shot dead by police at Stockwell Tube station in London, because he was misidentified as a suspected suicide bomber, and police feared he had a bomb ready for detonation. The shooting led to public concern and diplomatic protest.

Another major method of preemptive neutralization is interrogation of known or suspected terrorists to obtain information about specific plots, targets, the identity of other terrorists, and whether the interrogation subject himself is guilty of terrorist involvement. Sometimes methods are used to induce fear such as water boarding or increase suggestibility, such as sleep deprivation, or drugs. Human rights and legal objections apart, such methods may lead captives to offer false information in an attempt to stop the treatment, or because of confusion brought on by it.

Domestic Intelligence and Surveillance

Most counter-terrorism strategies involve an increase in standard police and domestic intelligence. The central activities are traditional: interception of communications, and the tracing of persons. New technology has, however, expanded the range of such operations. Domestic intelligence is often directed at specific groups, defined on the basis of origin or religion, which is a source of political controversy. Mass surveillance of an entire population raises objections on civil liberties grounds.

Military Intervention

Countries often use terrorism to justify military intervention in countries where terrorists are said to be based. That was the main stated justification for the U.S. invasion of Afghanistan and one reason for the 2003 invasion of Iraq. It was also a stated justification for the second Russian invasion of Chechnya.

> *How will we fight and win this war? We will direct every resource at our command—every means of diplomacy, every tool of intelligence, every instrument of law enforcement, every financial influence, and every necessary weapon of war—to the disruption and to the defeat of the global terror network.*
>
> —George W. Bush

The president's call to arms outlines a global campaign of unprecedented scale and complexity along multiple fronts:

Diplomatic

Diplomatic action for the campaign began within minutes of the attack. State Department officials immediately began working with foreign officials around the world to forge a coalition to support our response. The fruits of that labor have been assessed by President Bush as having resulted in the "greatest worldwide coalition in history."

Since September 11, the president has met with leaders from more than fifty nations, and Secretary Powell has met with even more foreign ministers and other representatives of our Coalition partners. Senior members of the Departments of State and Defense and the Central Intelligence Agency have also met with numerous foreign officials in Washington. He and his staff traveled to every continent to help fashion the diplomatic framework needed to wage the campaign to combat terrorism with a global reach.

Diplomacy abroad is also the leading edge of every nation's homeland security, and the global Coalition against terrorism has required—and will continue to require—intensive and innovative effort in that arena. Since September 11, for instance, the Department of State has begun formal dialogues with China and Pakistan on terrorism, and Department officials have brought their expertise to numerous conferences around the world, such as the one hosted by Polish president Kwasniewski, designed to strengthen the capabilities of our global partners in defeating terrorism.

In addition, numerous multilateral organizations, such as the EU, OAS, NATO, G-7, G-8, and others, have taken substantive steps to enhance information sharing, tighten border security, and combat terrorist financing. The

UN Security Council adopted Resolution 1373 which requires all states to prevent and suppress the financing of terrorist acts, including freezing funds and other financial assets. The resolution also obliges all states to improve border security, clamp down on the recruitment of terrorists, intensify information sharing and law enforcement cooperation in the international campaign against terrorism, and deny terrorists and their supporters any support or safe haven. This resolution augments the positive trend of Security Council resolutions 1267 and 1333 (passed in 1999 and 2000, respectively) which imposed targeted or "smart" sanctions against the Taliban in Afghanistan.

The existing twelve UN Conventions against terrorism represent a solid international foundation for nations to support this global struggle. In December, the United States ratified the two newest, the UN Convention for the Suppression of the Financing of Terrorism and the UN Convention for the Suppression of Terrorist Bombings. All nations should become parties to all twelve of these conventions so they may more readily apprehend and prosecute terrorists wherever they are located.

Public diplomacy has been an important aspect of our efforts as well. The Department of State has aggressively sought to counter distorted views of the United States overseas, to emphasize that the war on terrorism is not a war against Islam, and to underscore that terrorists are not martyrs but cowards and criminals. Senior U.S. officials conveyed these messages in hundreds of media interviews, and ambassadors have organized thousands of outreach activities around the world to build and maintain an international Coalition. The United States maintains Coalition Information Centers in London and Islamabad. The Department of State has an active speakers' program to explain U.S. policies to foreign and domestic audiences. The Office of International Information Programs maintains an informative and frequently visited website featuring publications such as "The Network of Terrorism" and "Islam in the United States."

Intelligence

Cooperation among intelligence agencies around the world has expanded to unprecedented levels. Sharing of intelligence about terrorists, their movements, and their planned attacks is an absolute prerequisite for successful interdiction. Governments in every region of the world have been able to use this information to expose the criminal netherworld in which terrorists operate. Undoubtedly, planned attacks have been prevented, and lives have been saved.

Our military campaign in Afghanistan as well as law enforcement and intelligence operations by Coalition members have yielded a wealth of intelligence that will require further exploitation for action. Such information will be extremely valuable in identifying and interdicting other terrorist cells around the world. Effective intelligence exchange allows countries to act preemptively to counter terrorists before they act. It closes an important seam that terrorists exploit to their advantage. There is room for continued improvement, but the initial results have been very encouraging.

Law Enforcement

The world's law enforcement professionals have launched a global dragnet to identify, arrest, and bring terrorists to justice.

In the United States, the FBI leads the law enforcement engagement, working with all federal, state, and local law-enforcement agencies. More than 7,000 FBI agents and support personnel have worked diligently with their U.S. and foreign law enforcement partners to unravel the planning leading to the execution of the 9/11 operation, as well as to interdict other Al-Qaeda cells and operatives in the United States and around the world. Their enhanced law enforcement efforts, and cooperative work by officials around the world, have resulted in the apprehension of more than 1,000 suspected terrorists and the breakup of Al-Qaeda and other terrorist cells.

During 2001 and through March 2002, Secretary Powell designated or redesignated thirty-three groups as foreign terrorist organizations under the Immigration and Nationality Act, as amended by the Antiterrorism and Effective Death Penalty Act of 1996. The designations make it a criminal offense to provide funds or other material support to such organizations, require U.S. financial institutions to block the funds of the groups, and make members of the groups ineligible for U.S. visas and, if they are aliens, deportable.

On October 26, the U.S. Congress enacted the USA PATRIOT Act, which significantly expands the ability of U.S. law enforcement to investigate and prosecute persons who engage in terrorist acts. On December 5, in accordance with the USA PATRIOT Act, Secretary Powell designated thirty-nine groups as "terrorist exclusion list" organizations. The legal consequences of the designations relate to immigration, and they strengthen the U.S. ability to exclude supporters of terrorism from entering the country or to deport them when found within our borders.

The United States brought to conclusion the prosecution of four Al-Qaeda members for the bombing of the U.S. embassies in Kenya and Tanzania. In May, in a courtroom within sight of the World Trade Center, the bombing suspects received guilty verdicts on all 302 counts in the trial and all were sentenced to life in prison.

Economic

Money is like oxygen to terrorists, and it must be choked off. When President Bush signed Executive Order 13224 on September 23, he imposed dramatic penalties on those who provide financial support to terrorist organizations.

The Order blocks the assets of designated organizations and individuals linked to global terrorism. It prohibits transactions with terrorist groups, leaders, and corporate and charitable fronts listed in the Order. It also establishes America's ability to block the U.S. assets of, and deny access to U.S. markets to, those foreign banks that refuse to freeze terrorist assets. As of March 2002, the Order contained the names of 189 groups, entities, and individuals. Accordingly, approximately 150 countries and independent law-enforcement jurisdictions (for example, Hong Kong, Taiwan) issued orders freezing the assets of suspected terrorists and organizations.

The U.S. Department of Treasury has been leading the war against terrorist financing and has worked with all relevant agencies and departments to identify terrorist financing networks and to find ways to disrupt their operations.

Many nations and independent law-enforcement jurisdictions have made changes in their laws, regulations, and practices in order to suppress terrorism financing more effectively. UNSCR 1373 mandates worldwide improvements, and we are working with the UN Counter Terrorism Committee (CTC) and others to help improve the capability of countries to meet their obligations under the resolution to combat terrorist financing.

The first step has been to deny terrorists access to the world's organized financial structures; simultaneously, we have been moving to prevent the abuse of informal money-transfer systems and charities. Both lines of attack have produced results.

President Bush launched the first offensive in the war on terrorism by signing Executive Order 13224, freezing the U.S.-based assets of those individuals and organizations involved with terrorism. All but a handful of the countries in the world have expressed their support for the financial war on terror.

Approximately 150 countries and jurisdictions have issued orders freezing terrorist assets, and the international community was helping others improve their legal and regulatory systems so they can move effectively to block terrorist funds.

International organizations are key partners in the war on financial terrorism. Over 100 nations have submitted reports to the United Nations on the actions they have taken to block terrorist finances, as required under United Nations Security Council resolution 1373, which calls on all nations to keep their financial systems free of terrorist funds.

The Financial Action Task Force—a twenty-nine-nation group promoting policies to combat money laundering—adopted strict new standards to deny terrorist access to the world financial system.

The G-20 and IMF member countries have agreed to make public the list of terrorists whose assets are subject to freezing, and the amount of assets frozen.

Military

The terrorist attacks of September 11 were acts of war against the United States and a grievous affront to all humanity. The international community has responded accordingly:

On September 12, 2001, the UN Security Council condemned the attacks and reiterated the inherent right of collective self-defense in accordance with the UN Charter. On September 21, Foreign Ministers of the OAS invoked the collective self-defense clause of the Inter-American Treaty of Reciprocal Assistance ("Rio Treaty").

In Brussels on October 5, NATO invoked Article 5 of the Washington Treaty, which states that an armed attack on one or more of the allies in Europe or North America shall be considered an attack against them all.

- 136 countries offered a range of military assistance.
- 89 countries granted over flight authority for U.S. military aircraft.
- 76 countries granted landing rights for U.S. military aircraft.
- 23 countries agreed to host U.S. and Coalition forces involved in military operations in Afghanistan.

THE OBAMA STRATEGY

Barack Obama and Joe Biden's strategy for securing the homeland against twenty-first-century threats is focused on preventing terrorist attacks on our homeland, preparing and planning for emergencies, and investing in strong response and recovery capabilities. Obama and Biden will strengthen our homeland against all hazards—including natural or accidental disasters and terrorist threats—and ensure that the federal government works with states, localities, and the private sector as a true partner in prevention, mitigation, and response.

Defeat Terrorism Worldwide

- **Find, Disrupt, and Destroy Al-Qaeda:** Responsibly end the war in Iraq and focus on the right battlefield in Afghanistan. Work with other nations to strengthen their capacity to eliminate shared enemies.

- **New Capabilities to Aggressively Defeat Terrorists:** Improve the American intelligence apparatus by investing in its capacity to collect and analyze information, share information with other agencies, and carry out operations to disrupt terrorist networks.
- **Prepare the Military to Meet Twenty-First-Century Threats:** Ensure that our military becomes more stealthy, agile, and lethal in its ability to capture or kill terrorists. Bolster our military's ability to speak different languages, navigate different cultures, and coordinate complex missions with our civilian agencies.
- **Win the Battle of Ideas:** Defeat Al-Qaeda in the battle of ideas by returning to an American foreign policy consistent with America's traditional values, and work with moderates within the Islamic world to counter Al-Qaeda propaganda. Establish a $2 billion Global Education Fund to work to eliminate the global education deficit and offer an alternative to extremist schools.
- **Restore American Influence and Restore Our Values:** Stop shuttering consulates and start opening them in the tough and hopeless corners of the world. Expand our foreign service, and develop the capacity of our civilian aid workers to work alongside the military.

Prevent Nuclear Terrorism

Barack Obama and Joe Biden have a comprehensive strategy for nuclear security that will reduce the danger of nuclear terrorism, prevent the spread of nuclear weapons capabilities, and strengthen the nuclear nonproliferation regime. They will:

- **Secure Nuclear Weapons Materials in Four Years and End Nuclear Smuggling:** Lead a global effort to secure all nuclear weapons materials at vulnerable sites within four years—the most effective way to prevent terrorists from acquiring a nuclear bomb. Fully implement the Lugar-Obama legislation to help our allies detect and stop the smuggling of weapons of mass destruction.
- **Strengthen Policing and Interdiction Efforts:** Institutionalize the Proliferation Security Initiative (PSI), a global initiative aimed at stopping shipments of weapons of mass destruction, their delivery systems, and related materials worldwide.
- **Convene a Summit on Preventing Nuclear Terrorism:** Convene a summit in 2009 (and regularly thereafter) of leaders of Permanent Members of the UN Security Council and other key countries to agree on preventing nuclear terrorism.

- **Eliminate Iran's and North Korea's Nuclear Weapons Programs through Tough, Direct Diplomacy:** Use tough diplomacy—backed by real incentives and real pressures—to prevent Iran from acquiring nuclear weapons and to eliminate fully and verifiably North Korea's nuclear weapons program.
- **Strengthen the International Atomic Energy Agency (IAEA):** Seek to ensure that the Agency gets the authority, information, people, and technology it needs to do its job.
- **Control Fissile Materials:** Lead a global effort to negotiate a verifiable treaty ending the production of fissile materials for weapons purposes.
- **Prevent Nuclear Fuel from Becoming Nuclear Bombs:** Work with other interested governments to establish a new international nuclear energy architecture—including an international nuclear fuel bank, international nuclear fuel cycle centers, and reliable fuel supply assurances—to meet growing demands for nuclear power without contributing to proliferation.
- **Set the Goal of a Nuclear-Free World:** Show the world that America believes in its existing commitment under the Nuclear Nonproliferation Treaty to work to ultimately eliminate all nuclear weapons. America will not disarm unilaterally.
- **Seek Real, Verifiable Reductions in Nuclear Stockpiles:** Seek deep, verifiable reductions in all U.S. and Russian nuclear weapons and work with other nuclear powers to reduce global stockpiles dramatically.
- **Work with Russia to Increase Warning and Decision Time:** Work with Russia to end dangerous Cold War policies like keeping nuclear weapons ready to launch on a moment's notice, in a mutual and verifiable manner.
- **Appoint White House Coordinator for Nuclear Security:** Appoint a deputy national security advisor to be in charge of coordinating all U.S. programs aimed at reducing the risk of nuclear terrorism and weapons proliferation.
- **Strengthen Nuclear Risk Reduction Work at Defense, State, and Energy Departments:** Expand our foreign service, and develop the capacity of our civilian aid workers to work alongside the military. Thwarting terrorist networks requires international partnerships in military, intelligence, law enforcement, financial transactions, border controls, and transportation security.

Strengthen American Biosecurity

Biological weapons pose a serious and increasing national security risk. Barack Obama and Joe Biden will work to prevent bioterror attacks and mitigate consequences. They will:

- **Prevent Bioterror Attacks:** Strengthen U.S. intelligence collection overseas to identify and interdict would-be bioterrorists before they strike.
- **Build Capacity to Mitigate the Consequences of Bioterror Attacks:** Ensure that decision-makers have the information and communication tools they need to manage disease outbreaks by linking health care providers, hospitals, and public health agencies. A well-planned, well-rehearsed, and rapidly executed epidemic response can dramatically diminish the consequences of biological attacks.
- **Accelerate the Development of New Medicines, Vaccines, and Production Capabilities:** Build on America's unparalleled talent to create new drugs, vaccines, and diagnostic tests and to manufacture them more quickly and efficiently.
- **Lead an International Effort to Diminish Impact of Major Infectious Disease Epidemics:** Promote international efforts to develop new diagnostics, vaccines, and medicines that will be available and affordable in all parts of the world.

Protect Our Information Networks

Barack Obama and Joe Biden—working with private industry, the research community, and our citizens—will lead an effort to build a trustworthy and accountable cyber infrastructure that is resilient, protects America's competitive advantage, and advances our national and homeland security. They will:

- **Strengthen Federal Leadership on Cyber Security:** Declare the cyber infrastructure a strategic asset and establish the position of national cyber advisor who will report directly to the president and will be responsible for coordinating federal agency efforts and development of national cyber policy.
- **Initiate a Safe Computing R&D Effort and Harden our Nation's Cyber Infrastructure:** Support an initiative to develop next-generation secure computers and networking for national security applications. Work with industry and academia to develop and deploy a new generation of secure hardware and software for our critical cyber infrastructure.
- **Protect the IT Infrastructure That Keeps America's Economy Safe:** Work with the private sector to establish tough new standards for cyber security and physical resilience.
- **Prevent Corporate Cyber-Espionage:** Work with industry to develop the systems necessary to protect our nation's trade secrets and our research and development. Innovations in software, engineering, pharmaceuticals and other fields are being stolen online from U.S. businesses at an alarming rate.

- **Develop a Cyber Crime Strategy to Minimize the Opportunities for Criminal Profit:** Shut down the mechanisms used to transmit criminal profits by shutting down untraceable Internet payment schemes. Initiate a grant and training program to provide federal, state, and local law enforcement agencies the tools they need to detect and prosecute cybercrime.
- **Mandate Standards for Securing Personal Data and Require Companies to Disclose Personal Information Data Breaches:** Partner with industry and our citizens to secure personal data stored on government and private systems. Institute a common standard for securing such data across industries and protect the rights of individuals in the information age.

Improve Intelligence Capacity and Protect Civil Liberties

- **Improve Information Sharing and Analysis:** Improve our intelligence system by creating a senior position to coordinate domestic intelligence gathering, establishing a grant program to support thousands more state and local level intelligence analysts, and increasing our capacity to share intelligence across all levels of government.
- **Give Real Authority to the Privacy and Civil Liberties Board:** Support efforts to strengthen the Privacy and Civil Liberties Board with subpoena powers and reporting responsibilities. Give the Board a robust mandate designed to protect American civil liberties and demand transparency from the Board to ensure accountability.
- **Strengthen Institutions to Fight Terrorism:** Establish a Shared Security Partnership Program overseas to invest $5 billion over three years to improve cooperation between U.S. and foreign intelligence and law enforcement agencies.

Protect Americans from Terrorist Attacks and Natural Disasters

- **Allocate Funds Based on Risk:** Allocate our precious homeland security dollars according to risk, not as pork-barrel spending or a form of general revenue sharing. Eliminate waste, fraud and abuse that cost the nation billions of Department of Homeland Security dollars.
- **Prepare Effective Emergency Response Plans:** Further improve coordination between all levels of government, create better evacuation plan guidelines, ensure prompt federal assistance to emergency zones, and increase medical surge capacity.
- **Support First Responders:** Increase federal resources and logistic support to local emergency planning efforts.

- **Improve Interoperable Communications Systems:** Support efforts to provide greater technical assistance to local and state first responders and dramatically increase funding for reliable, interoperable communications systems. Appoint a National Chief Technology Officer to ensure that the current non-interoperable plans at the federal, state, and local levels are combined, funded, implemented, and effective.
- **Working with State and Local Governments and the Private Sector:** Make the federal government a better partner to states and localities, one that listens to local concerns and considers local priorities. Reach out to the private sector to leverage its expertise and assets to protect our homeland security.

Protect Critical Infrastructure

- **Create a National Infrastructure Protection Plan:** Develop an effective critical infrastructure protection and resiliency plan for the nation and work with the private sector to ensure that targets are protected against all hazards.
- **Secure our Chemical Plants:** Work with all stakeholders to enact permanent federal chemical plant security regulations.
- **Improve Airline Security:** Redouble our efforts to adequately address the threats our nation continues to face from airplane-based terrorism.
- **Monitor our Ports:** Redouble our efforts to develop technology that can detect radiation and work with the maritime transportation industry to deploy this technology to maximize security without causing economic disruption.
- **Safeguard Public Transportation:** Work to protect the public transportation systems Americans use to get to work, school, and beyond every day.
- **Improve Border Security:** Support the virtual and physical infrastructure and manpower necessary to secure our borders and keep our nation safe.

Modernize America's Aging Infrastructure

- **Build-in Security:** Ensure that security is considered and built into the design of new infrastructure, so that our critical assets are protected from the start and more resilient to naturally occurring and deliberate threats throughout their life-cycle.
- **Create a National Infrastructure Reinvestment Bank:** Address the infrastructure challenge by creating a National Infrastructure Reinvestment Bank of $60 billion over 10 years, to expand and enhance, not supplant, existing federal transportation investments. This independent entity will be directed to invest in our nation's most challenging transportation infrastructure needs, without the influence of special interests.

- **Invest in Critical Infrastructure Projects:** Invest in our nation's most pressing short and long-term infrastructure needs, including modernizing our electrical grid and upgrading our highway, rail, ports, water, and aviation infrastructure. Establish a Grid Modernization Commission to facilitate adoption of Smart Grid practices to improve efficiency and security of our electricity grid.

THE TRUMP STRATEGY

President Trump's 2018 national security strategy is based on putting America First and in doing so, he is chartering a new course.

- The 2018 National Security Strategy provides the strategic vision necessary to protect the American people, preserving our way of life, promoting our prosperity, maintaining peace through strength, and advancing American influence in the world.
- The administration recognizes that economic security is national security which is why policies are being implemented that jump-start job creation and strengthens our country's economic base.
- America is leading again, proving that America First does not mean America alone by revitalizing relationships with countries that share our goals and interests while combating radicalism and extremism.
- The strategy pursues "peace through strength" by modernizing and rebuilding America's armed forces. In the words of our nation's first president, George Washington, "To be prepared for war is one of the most effectual means of preserving peace."
- The National Security Strategy will modernize our nuclear forces and deploy state-of-the-art ballistic missile defenses.

Combatting Terrorism

Under President Trump's leadership, the United States is defeating the terrorists who threaten our homeland and citizens.

- By working with local forces and giving our commanders on the ground increased flexibility, we have almost completely defeated ISIS.
- Since President Trump took office, ISIS has lost nearly 100 percent of the territory it once controlled in Iraq and Syria.
- Along with our partners, we will continue to pursue the thousands of ISIS fighters who remain at large, and we will not rest until we have achieved the enduring defeat of ISIS and decimated global jihadists and their networks.

- President Trump will relentlessly pursue threats to America's security, prioritizing efforts to confront and defeat radical Islamic terror and the ideology that sustains it.
- President Trump's conditions-based South Asia Strategy provides commanders with the authority and resources needed to deny terrorists the safe haven they seek in Afghanistan and Pakistan.
- The president continues to reassert American leadership among our Middle Eastern allies and partners.
- This includes bringing America's Middle Eastern allies and partners into the struggle, empowering credible voices to counter terrorist propaganda, and promoting our open society as the great alternative to repression and extremism.
- President Trump is making clear to our allies that they cannot be America's friend while supporting or condoning terror.
- The president has suspended security assistance to Pakistan, sending a long overdue message to aid recipients that we expect them to fully join us in combatting terrorism.

Securing the Homeland

President Trump is securing our borders and eliminating violence from our neighborhoods.

- The administration has instituted extreme vetting and enhanced screening to make it harder for terrorists, transnational criminals, and other dangerous actors to reach our country.
- Routes into the United States are now more secure, regardless of whether people travel here as tourists, business visitors, immigrants, or refugees.
- The president will continue to work with international partners to ensure the safety of our citizens.
- All foreign countries are now required to share critical information with the United States about their nationals, so we know who is entering our country, or face consequences. To enhance our security, the administration has raised the baseline of security cooperation, and most countries have complied.
- After the president highlighted alarming threats against Jewish schools last year, the Trump administration worked with Israeli counterparts to find and end the source of those threats.
- In addition to the domestic arrests of almost 800 MS-13 gang members, the Trump administration has worked with Central American partners to arrest and charge roughly 4,000 members of MS-13.

Reassuring Allies and Deterring Enemies

The president is restoring critical alliances and working with allies and partners to confront our enemies.

- President Trump is redefining America's relationships in the Indo-Pacific by securing business and trade that benefit American workers and our economy.
- President Trump will continue to rally the world to apply maximum pressure to convince North Korea to de-nuclearize.
- President Trump is restoring American leadership in the Middle East, rebuilding relations with Israel and taking strong action to confront aggressors in the region.
- In December, the president officially recognized Jerusalem as the capital of Israel and announced the relocation of the U.S. embassy in Israel to that city.
- The Trump administration is confronting Iran's destabilizing behavior and the threat of its nuclear program, working to fix the serious flaws in the Iran nuclear deal, and sanctioning dozens of Iranian officials and entities.
- The president is revitalizing the NATO alliance by successfully encouraging our partners to meet their defense spending commitments and share the burden of collective defense.
- President Trump is pressing for reform at the United Nations, and will consider withholding funds to United Nations entities that perpetuate the dependency of millions.

Citizen Support

What can you do?

Devastating acts, such as the terrorist attacks on the World Trade Center and the Pentagon, and horrific incidents overseas, have left many concerned about the possibility of future incidents in the United States and their potential impact. They raise uncertainty about what might happen next, which increases stress levels. Nevertheless, you can do many things to prepare for the unexpected that will reduce the stress that you may feel now—and later should any emergency arise. Taking preparatory action can reassure you and your children so you may exert a measure of control even in the face of such terrible events.

For detailed information on emergency preparedness and what you can do to protect yourself and your loved ones, after you read the following chapter, go to Part III.

Chapter 3

Are We Winning the War?

The war on terror has been a hugely costly one—in terms of lives lost and money spent. But has it made America and Americans safer? Before we look at what is happening at home, it is worth examining what is happening beyond our borders.

OVERSEAS

Overseas, democracies are being restored to some extent in Afghanistan and Iraq although progress is slow. While there are hundreds of "good news" stories occurring every day in both these countries, the overwhelming image in the minds of most Americans is of countries still split by rivalries and struggling to secure a permanent peace.

AFGHANISTAN

In Afghanistan—officially the Islamic Republic of Afghanistan—the ousted Taliban is now a militia openly operating in many areas of the country outside the control of the government in Kabul, especially in the south and east. Remnants of the Taliban continue to launch attacks on government and U.S. troops, police, and overseas workers and are increasingly focusing on soft targets.

In January 2006, following two suicide bomb attacks which killed twenty-six people and wounded fifty others, then Afghan president Hamid Karzai, said, "We are in a joint struggle against terrorism, for us and for the international community."

"If you don't defend yourself here, you will have to defend yourself back home, in European capitals and Americans' capitals. It will take many, many more years before we can defend ourselves with our own means, before we can feed ourselves or work for our development with our own means," Karzai added.

While some democratic structures have been installed, Afghanistan is still desperate for foreign aid and overseas investment to help rebuild the country where war lords still control vast tracts of land and the poppy fields. Today, opium still accounts for over 10 percent of the nation's economic output and more than 3.3 million Afghans are involved in producing opium and its derivatives—morphine and heroine.

According to the United Nations Office on Drugs and Crime, opium production has grown every year since the Taliban were ousted in 2001. Afghanistan now produces over 90 percent of the world's opium, most of which is smuggled to Europe. One major problem is that many of the drug lords are also now members of the central government.

The United States started to pull its troops out of Afghanistan in 2011 and in June 2013 the official handover of security from NATO to Afghan forces was completed. However, some 100,000 NATO troops, including U.S. military, remained on in support roles, mainly training army Afghan recruits. However, due to the deteriorating military situation, more U.S. troops were sent to Afghanistan in 2014 to carry out missions against the Taliban and by 2015 there were more than 10,800 stationed there. The United States still maintains four garrisons outside Kabul and a strong counterterrorism capability with drones and Special Operations forces and under President Trump, the U.S. military has been given the authority to increase the current troop level cap of 8,400, although in reality it was already about 3,000 over that limit. In August 2017, it was announced that another 4,000 troops were being deployed.

IRAQ

Much has been achieved in Iraq since the ousting of Saddam Hussein. There have been democratic elections, a constitution was approved in 2005 and parliament meets regularly. There was also a steady transfer of security from the international coalition to the Iraqi army and police. There has been considerable reconstruction in some parts of the country and many provinces enjoyed relative calm. However, attacks on military and civilian personnel and targets continued and the situation deteriorated throughout 2006.

In 2007, President Bush ordered an additional 21,500 troops to Iraq. This "surge" in security activity led to significant reductions in overall violence, although on August 14, 2007, the deadliest single attack of the war took place when almost 800 civilians were killed in a coordinated suicide bomb attacks in non-Muslim villages in the north of the country.

Over the next year, the number of attacks reduced and U.S. troop deaths led to their lowest level since 2003.

In February 2009, President Obama announced an 18-month withdrawal window for "combat forces," leaving behind 30,000 to 50,000 troops "to advise and train Iraqi security forces and to provide intelligence and surveillance." The final contingent of U.S. troops exited the country in December 2011.

Their departure led to political instability and an increase in the insurgency. Not only did Shiite and Sunni militias clash in Iraq in escalating violence, they also crossed into Syria to fight in the civil war there.

In 2014 Sunni insurgents fighting under the Islamic State of Iraq and the Levant (ISIL) banner, seized large areas of the country including the major cities of Mosul, Tikrit, and Fallujah. They displaced hundreds of thousands of residents and committed numerous atrocities, including the barbaric execution of 1,700 Iraqi soldiers who had surrendered. In June 2014, ISIL announced a new caliphate under the leadership of Abu Bakr al-Baghdadi and changed its name to the Islamic State (IS).

In August 2014, President Obama ordered airstrikes against ISIL positions and air drops of aid to thousands of refugees stranded in nearby mountains. The United States and British sent in small contingents but continued the air strikes leaving it to Iraqi and Kurdish soldiers to do most of the ground fighting. ISIL continued to take ground for the rest of 2014 but were slowly forced back in a series of major offensives in 2015 and 2016.

On November 17, 2017, Iraqi forces captured ISIL's last two strongholds and victory was declared by Prime Minister Nouri al-Maliki.

Many ISIL fighters and supporters still remain in Iraq especially around their traditional hideouts in the hilly Wadi Hauran area and in January 2018 after a period of calm, twenty-seven people were killed by suicide bombers in Baghdad. IS claimed responsibility. Mopping up operations continue to hunt down the remaining ISIL fighters.

The Iraqi people themselves hold the key to their future and on several occasions, they have been able to pull back from the brink and resume talking to each other. If peace and common sense prevail, Iraq has a future; if not, it could slip into sectarian warfare that could create divisions that will last for years.

THE REST OF THE WORLD

The U.S. invasion of Afghanistan and Iraq was not responsible for creating the various terrorist groups. Most were active in the 1990s but certainly the invasion of Iraq provided them—and especially the radical Islamic groups, with a tangible cause and a battleground on which to fight it. Terrorists flooded into the Middle East and then into Iraq and to a lesser extent, Afghanistan. Ironically, as the security forces have started to gain the upper hand in Iraq, many of the insurgents have moved to the porous Pakistan-Afghanistan border.

Terror groups though have become increasingly active around the world. The number of attacks has increased in the last few years and the attacks have become more deadly as the terrorists become more sophisticated with their bomb making and attack strategies.

Some of the most deadly terrorist attacks in the last 25 years include:

- October 23, 1983, suicide bombing of the Multinational Peacekeeping Force in Beirut, Lebanon, housing U.S. Marines and French military—241 Marines, 58 French peacekeepers, and 6 civilians were killed. Islamic Jihad claimed responsibility.
- December 21, 1988, bombing of Pan Am flight 103 over Lockerbie, Scotland, killing all 243 passengers and 16 crew. The Libyan government accepted responsibility for the bombing in 2003 and paid compensation to the families of the victims.
- March 12, 1993, Mumbai bombings—270 dead and 70+ injured.
- April 19, 1995, Oklahoma City bombing—168 people killed and 680 injured.
- September 22–23, 1997, Bentalha massacres, Algeria—up to 400 killed. Armed Islamic groups claimed responsibility.
- August 7, 1998, bombings of U.S. Embassies in Nairobi and Dar-es-Salaam—224 killed and 4,000 wounded.
- September 11, 2001, bombing of World Trade Center, New York—2,996 dead and 6,000+ wounded.
- October 12, 2002, Bali bombings—202 dead.
- October 23, 2002, Moscow Theatre hostage situation—170 killed and 700 wounded.
- February 27, 2004, Superferry 14 bombing, Manila, 116 killed.

Table 3.1 Terror Attacks Worldwide, 2014–2016

	2014	*2015*	*2016*
Terror attacks worldwide	13,463	11,774	11,072
People killed	32,727	28,328	25,621
People injured	34,791	35,320	33,814
People kidnapped	9,428	12,189	15,543

(2016 is the last year for which complete date is available.)

- March 11, 2004, Madrid train bombing—almost 200 killed.
- September 1–3, 2004, Beslan, Russia, school hostage siege—385 killed and 783 wounded.
- July 7, 2005, London transport bombings—52 killed and 700 injured.
- March 6, 2007, bombing of Shia pilgrims at Al Hillah, Iraq—almost 200 killed.
- July 11, 2006, Mumbai train bombing—209 dead and 700+ injured.
- November 23, 2006, Sadr City bombing and mortar attack, Baghdad—215 killed.
- April 18, 2007, Baghdad, five car bombs—198 dead.
- August 14, 2007, Yazidi communities bombing, near Mosel, Iraq—796 dead and 1,562 injured.
- October 18, 2007, Karachi bombings—139 dead.
- February 14, 2014, Borno Massacre, Nigeria, by Boko Haram Islamists—200 killed.
- August 14, 2014, Deir ez-Zor, Syria, massacre by ISIL—700+ men killed.
- January 8, 2015, Boko Haram attack on town of Baga, Nigeria—200+ killed, 2,000 unaccounted for.
- June 26–30, 2015, Boko Haram killing spree in northern Nigeria—200 killed.
- October 31, 2015, bomb on Russian jet over Syria—224 killed.
- July 3, 2016, car bombings, Baghdad—300+ killed and 220 injured.
- January 27, 2018, Kabul bombing by Taliban outside Interior Ministry—103 killed and 235 injured.

If the number of attacks doesn't present an accurate measure of success in the war on terrorism, which numbers do? The following facts are also important in the equation:

- Number of *Taliban-style* states created since 9/11: 0
- Number of countries that have recognized Al-Qaeda: 0
- Number of nations that have adopted "state-sponsored" terrorism as an official policy: 0
- Number of states that have voluntarily given up weapons of mass destruction programs since 9/11: 1
- Number of transnational nuclear smuggling networks broken up since 9/11: 1
- Number of Middle Eastern states that have moved closer to democracy: 5

Terrorist attacks, however, continue to take place worldwide. In the Far East, radical Islamic terrorist groups are active in many countries. There is an increasing threat from radical Islamic groups in India and Pakistan and despite efforts by the Pakistan government, leaders of Al-Qaeda and other terrorist groups are being harbored by sympathetic tribesmen in the mountainous border region between Pakistan and Afghanistan.

Russia's southern flanks are threatened by almost a dozen countries—all former Soviet territories—that now have large and militant Muslim populations. Ongoing attacks in Russia that were attributed to Chechen rebels are now believed to be the work of home-grown Muslim terrorists.

Terrorism is rife throughout the Middle East with deadly attacks in Jordan, Egypt, Kuwait, Saudi Arabia, and elsewhere. Almost all these attacks have been carried out by Islamic extremists who want to topple the existing regimes and replace them with an Islamic Caliphate that would encompass all of Middle East.

Europe is embroiled with terrorism from separatist groups to Muslim terror cells in Britain. France suffered weeks of rioting, arson, and destruction when tens of thousands of young immigrants, almost all of them Muslim, vented their anger and frustration at being treated, as they saw it, as second class citizens. That rioting spread to surrounding counties, especially Belgium, although it was quickly contained.

Deadly bombings in Madrid brought down the Spanish government and suicide bombings in London killed and maimed hundreds. Both were carried out by Islamic extremists. The London bombings sent a shock wave through the Western world because the terrorists, all Muslim, were almost all British born and seemingly leading normal, respectable lives in their communities. The subsequent Euro-wide police hunt found that sleeper cells existed through Europe and North Africa. Intelligence agencies are convinced that similar sleeper cells exist in North America.

The enemy in the war on terror is not limited to ISIS, Al-Qaeda, and its associated movements, but includes a global network of extremist groups, according to a U.S. Central Command general.

"If we declared victory and walked away from Iraq and Afghanistan tomorrow, we would be fighting this fight for years and years," said Army Brig. Gen. Mark T. Kimmitt, CENTCOM's deputy director of plans and policy, speaking in March 2006. "We are fighting an insurgency, a terrorist movement that is represented by al Qaeda, but it is far more than al-Qaeda."

The terrorist network the United States is facing in this war includes extremist groups around the world, and it is a network the nation has been fighting since before September 11, 2001, Kimmitt said. The extremist groups are made up of primarily Sunni Muslims whose goal is to reclaim what they see as the holy lands in the Middle East and to remove Western influence, he added. Their ultimate goal is to establish a Caliphate in the region, where Sharia, or Islamic law, rules, and the people are oppressed, he said.

To defeat this network, CENTCOM is pursuing an aggressive campaign to defeat terrorists where they are active and to prevent the spread of their ideology, Kimmitt said.

One of the important aspects of the CENTCOM strategy is that it relies on its own network to defeat the terrorist network. The global terror network uses people, the Internet, smuggling, nongovernmental organizations that are sympathetic to their cause, legitimate governments, front companies, and safe havens to achieve its goals.

CENTCOM's own network uses all elements of national power to defeat terrorists. It relies on federal and state law enforcement agencies, the State Department, and federal intelligence agencies.

Another important aspect of CENTCOM's strategy is building the capacity of other nations to fight terrorism themselves. Islamic terrorists have been very clear in stating that their goal is to rid the region of Western influence and then go after surrounding countries, which they call apostate or secular governments. These clear declarations have made countries such as Saudi Arabia, Qatar, Jordan, and Kuwait more willing to work with the United States and more prepared to fight terrorism on their own. These countries get it—they understand that it's not simply a problem the Islamic terrorists have with the West; it is simply a problem that these organizations have with anybody that does not believe in their extremist ideology.

Most of the high-value target operations in these neighboring countries have been conducted by the countries' own national forces and the United States continues to work with them to help them prevent terrorist activity inside their borders.

Also important to CENTCOM's strategy is denying safe havens or sanctuaries to terrorists. The United States has to ensure that as the Islamic terrorists and their related organizations are pushed out of Iraq, Syria, and Afghanistan; they don't just resettle somewhere else.

Combined Joint Task Force Horn of Africa, with headquarters in Djibouti, is the model for this preventive strategy. American and coalition forces there focus primarily on civil affairs and humanitarian missions to establish relationships with the people and foster cooperation, should terrorists try to move into the area.

To make sure a victory in the war on terror lasts, the United States has tailored its post-conflict strategy for the area.

Military experts accept that it would be a mistake for the United States to garrison the Middle East like it did to Western Europe after World War II. The correct strategy for the Middle East is to maintain a sufficient capability in the region to do necessary tasks, such as training, maintaining lines of communication, and maintaining access. CENTCOM recognizes that the war on terror will be a long fight, perhaps lasting a generation, but it is one that must be fought, and the support of the American public is crucial.

INCREASING SOPHISTICATION

Another alarming trend is the increasing number of terrorist groups that are seeking weapons of mass destruction, said the U.S. State Department in a report on global terrorism.

"Although ISIS remains the primary concern regarding possible WMD threats, the number of groups expressing interest in such material is increasing, and WMD technology and know-how is proliferating in the *jihadist* community," the report said.

Terrorist groups use the Internet to communicate with each other. They post bomb-making information on websites and they use all the latest multimedia skills to get their message across.

At Home

There have been no serious terrorist attacks on U.S. soil since 9/11 and a number of potential terror attacks have been foiled (far more than have been made public). Law enforcement and intelligence agencies have not released the full details of some of these attacks but the perpetrators were in most cases of Middle Eastern or North African origin and most were Muslim. Some were born in this country and many were recruited while in prison. It is known that there are strong radical Muslim gangs in many of the nation's prisons and these are fertile recruiting grounds. Prisons not only make great places to recruit new members, they provide an environment without distractions when recruits can be taught their terrorism skills. And what they don't acquire inside, they can get from activists and the web outside.

What the intelligence agencies do not know, however, is how many sleeper cells may exist in the United States.

Islam is the fastest growing religion in the United States. In Illinois, a growth of about 25 percent in the Muslim population was due to conversion, but immigration, mainly from southwest Asia, accounts for most of the growth. While the exact number of Muslims living in the United States is not known, estimates place the number at between 7 and 9 million, of which almost 80 percent are aged between 16 and 65.

"Typically, young males prone to be recruited by Islamist activists generally fall in the 16–45 age bracket. The younger ones—those in their teens and early twenties— are typically the most vulnerable," said Claude Salhani,

UPI International Editor in Washington, in a white paper he wrote for www.isamericasafe.com.

"Just as Muslims make up the majority of the incarcerated population of Europe's jails, so too, do Muslims comprise the majority of inmates crowding American prisons. Much of the conversion takes place inside jails and usually by Imams espousing radical thinking. Often, moderate Muslims who enter jail leave incarceration far more radicalized," he said.

A survey of British Muslims found that 6 percent—about 100,000 people—believed the London bombings were justified and 1 percent felt that "the decadent and immoral" Western society should be brought to an end by violence if necessary.

"Using the figure of 7 million Muslims living in the U.S., one percent would amount to 70,000. That is a staggering 70,000 potential people supporting violence against their adopted homeland. Break that down even further to one percent of one percent—and you have 700 probably extremists," he warned.

"The danger posed by sleeper cells, by al-Qaeda sympathizers and wannabe jihadists is not to be underestimated. The threat is real," he said.

While President Trump has softened his tone on Muslims—but not Islamic extremists—since his time on the campaign travel, his travel ban does largely affect people coming from predominantly Muslim countries that have a history of terrorism or supporting terrorism.

Post-9/11

There is no doubt that as a result of 9/11 we are, as a nation, better able to defend ourselves and slowly but surely an infrastructure has been put in place that is better able to identify threats and respond to them.

However, the Patriot Act and its successor, and other anti-terrorism measures, such as the domestic eaves-dropping on the telephone calls of U.S. citizens, have eroded some of the very freedoms we are fighting so hard to protect. Fighting wars in Afghanistan and Iraq put enormous strains on the military and their families, especially members of the National Guard, many of whom never expected to be deployed overseas.

The war on terror, which is still being waged at home and overseas, has also placed a colossal economic burden on the shoulders of the nation's taxpayers. While all the government spending has not gone on homeland security and the military, the United States has gone from a budget surplus to a massive deficit. The current outstanding public debt of the United States as of February 19 was $20,710,203,610,249.74. Every man, woman, and child owes $68,165 for their share!

Apart from the cost of fighting tyranny and terrorism, the war on terror has impacted on us in many ways:

Homeland Defense

The creation of the Department of the Homeland Defense was the largest reorganization in federal government history. While not unexpected, when you try to merge 180,000 people from 22 departments into four directorates of one entity you are going to have problems.

It was not simply a matter of merging the manpower and responsibilities of the different departments. Each department had its own computer systems, its own software, and its own communications network. Billions have been and continue to be spent trying to make all the systems compatible.

Travel

Most Americans have patiently endured the longer waits that have resulted from more stringent security checks at airports. Before 9/11 only 5 percent of passengers' bags were checked, now 4.9 million carry-on bags are checked every day. Sky marshals travel on an undisclosed number of flights and a greater percentage of check-in luggage and air freight are screened. Cockpit doors have been strengthened and groups are not allowed to congregate in the aisles or near the toilets during flights.

Ports

Despite the furor over foreign companies taking over U.S. ports there is concern that they still represent a weak link in the nation's security. About 90 percent of the world's trade is carried in cargo containers and almost half of incoming U.S. trade arrives by containers aboard ships. Every year, 16 million containers move through America's 327 ports. Only 2 percent are physically inspected—although other forms of screening may be carried out using x-ray machines and detection technology. Pre-screening at the port of departure is another tool used at the larger ports worldwide. Coast Guard and U.S. Customs and Border Protection (CBP) personnel are based at many of these larger ports to do pre-screening and assist the local authorities with seaport security.

Two important initiatives have been introduced:

> **Container Security Initiative (CSI)**, to increase security for container cargo shipped to the United States. One purpose is to "extend the zone of security outward so that American borders are the last line of defense, not the first."

The core elements of CSI are:

- Using intelligence and automated information to identify and target containers that pose a risk for terrorism.

- Pre-screening those containers that pose a risk at the port of departure before they arrive at U.S. ports.
- Using detection technology to quickly pre-screen containers that pose a risk.
- Using smarter, tamper-evident containers.

Customs-Trade Partnership Against Terrorism (C-TPAT)

This is a supply-chain security program led by CBP. Companies can get certification if they have a documented process for determining and alleviating risk throughout their international supply chain. This allows the companies to be considered low risk, resulting in expedited processing of their cargo and fewer customs examinations.

The Department of Homeland Security was mandated by Congress to have 100 percent screening of all containers by 2012 but this goal has proved elusive because of the cost of installing scanners and radiation detection equipment at all foreign ports. However, scanning is only one aspect of seaport security. Effective security also requires improvements in intelligence, cooperation with foreign ports, and tracking of containers while in transit.

Essential Infrastructure (Water, Nuclear, Food Chain)

Without doubt, much of our essential infrastructure remains vulnerable. However, our strategy of layered defenses aims to restrict access of potential terrorist threats by dealing with them at a distance. More cargo is now checked at ports of origin and manifests are screened for suspicious points of origin or dubious shippers. Ship crews are screened and cargoes are tracked by satellite. Naval and Coast Guard ships stand ready to board ships far at sea to examine suspicious crew members and cargoes.

Our water supplies are very accessible. However, it would take huge volumes of pollutants or poison to be a threat to users after the water flows through testing and treatment procedures. Hackers are known to have penetrated one water company's servers and increase the amounts of fluoride added to the water. Luckily other safety mechanisms built into the system were able to detect this and prevent any affected water leaving the facility.

Our nuclear power and chemical plants are in the process of hardening their sites. But many are still open to aerial attacks or a dedicated ground attack. Since most of these facilities are in remote areas, they probably rank lower on the terrorist's target list than one might think.

The same is true of our food supply. Small, isolated attacks are very feasible, but targeting mass casualties through our food supply chain is less likely that other more spectacular and more effective terrorist alternatives.

WMT or Biochemical Attack

Could we cope? Most experts agree that weapons of mass destruction (WMD) should remain our prime concern from either foreign or domestic terrorist threats. Such attacks may be launched with minimal resources and by very few individuals being involved. WMD attacks present all the attributes that terrorists seek to achieve: fear, confusion, mass casualties, and notoriety. But most WMD weapons are difficult to reliably employ. Delivery systems, wind directions, weather conditions, all must be just right to target huge groups of people in a contained situation, such as a meeting place or sporting event. The same is true for chemical and biological weapons. While they could have devastating consequences if released, there are many factors that could mitigate against their potential use.

Mass Evacuation

Some years ago, the UK government came up with a scheme to evacuate London should it be necessary. Unfortunately the plan was only able to get 1 million people out of the capital within 24 hours and that assumed the railways would still be operating. Apparently, the rest of the capital's population—about 7 million people—would have to fend for themselves. Could we do any better in evacuating a major U.S. city? Hurricane Katrina demonstrated how difficult it was to evacuate cities like New Orleans and Houston.

Tolerance

Are we more tolerant as a nation than we were before 9/11? Are we more sympathetic or more suspicious of our Muslim neighbors?

SUMMARY

The war on terror is high on most Americans' list of concerns. It promises to be with us for at least another generation to come. We are relatively safe from terror within our borders—now. But experts agree that it's not if, but when we may be attacked in the future. Much more needs to be done in many areas, such as protecting our borders, screening foreign visitors, and getting the public more involved in neighborhood watch programs on land and at sea.

For most of us, however, medical malpractice, smoking, disease, automobile accidents, and deaths and injuries in the home should be of greater concern—and within greater control of individual actions.

Man-made and natural disasters are also a greater threat to most of us than terrorism—and cost U.S. taxpayers and victims dearly. Still, there is much that one may do to minimize exposure to such threats and to mitigate the damage done to their loved ones and their property if they are caught in one of these disasters.

Of primary concern should be the billions that we are spending to keep America safe. We must realize that no amount of money will keep us perfectly safe. The nature of our free society makes us vulnerable—and a target for those who hate freedom. But we must charge our leadership to be good stewards of the money that we send them. There must be a balance between what we spend our tax dollars on and what we get in return in terms of safety, security, and quality of life.

Is America safe? What do you think?

Chapter 4

Foreign Terrorist Organizations

Foreign Terrorist Organization (FTO) designations are an important element of the U.S. counterterrorism efforts. Designations of foreign terrorist groups expose and isolate these organizations, deny them access to the U.S. financial system, and create significant criminal and immigration consequences for their members and supporters. Moreover, designations can assist or complement the law enforcement actions of other U.S. agencies and other governments.

Legal Criteria for Designation under Section 219 of the INA as amended:

- It must be a *foreign organization.*
- The organization must *engage in terrorist activity*, as defined in section 212 (a)(3)(B) of the INA (8 U.S.C. § 1182(a)(3)(B)), or *terrorism*, as defined in section 140(d)(2) of the Foreign Relations Authorization Act, Fiscal Years 1988 and 1989 (22 U.S.C. § 2656f(d)(2)), *or retain the capability and intent to engage in terrorist activity or terrorism.*
- The organization's terrorist activity or terrorism must threaten the security of U.S. nationals or the national security (national defense, foreign relations, or the economic interests) of the United States.

U.S. GOVERNMENT DESIGNATED FOREIGN TERRORIST ORGANIZATIONS

Abdallah Azzam Brigades (AAB)
Abu Nidal Organization (ANO)
Abu Sayyaf Group (ASG)

Al-Aqsa Martyrs Brigade (AAMB)
Ansar al-Dine (AAD)
Ansar al-Islam (AAI)
Ansar al-Shari'a in Benghazi (AAS-B)
Ansar al-Shari'a in Darnah (AAS-D)
Ansar al-Shari'a in Tunisia (AAS-T)
Army of Islam (AOI)
Asbat al-Ansar (AAA)
Aum Shinrikyo (AUM)
Basque Fatherland and Liberty (ETA)
Boko Haram (BH)
Communist Party of Philippines/New People's Army (CPP/NPA)
Continuity Irish Republican Army (CIRA)
Gama'a al-Islamiyya (IG)
Hamas
Haqqani Network (HQN)
Harakat ul-Jihad-i-Islami (HUJI)
Harakat ul-Jihad-i-Islami/Bangladesh (HUJI-B)
Harakat ul-Mujahideen (HUM)
Hizballah
Indian Mujahedeen (IM)
Islamic Jihad Union (IJU)
Islamic Movement of Uzbekistan (IMU)
Islamic State of Iraq and Syria (ISIS)
Islamic State's Khorasan Province (ISIS-K)
ISIL-Libya
ISIL Sinai Province (ISIL-SP)
Jama'atu Ansarul Muslimina Fi Biladis-Sudan (Ansaru)
Jaish-e-Mohammed (JeM)
Jaysh Rijal Al-Tariq Al-Naqshabandi (JRTN)
Jemaah Ansharut Tauhid (JAT)
Jemaah Islamiya (JI)
Jundallah
Kahane Chai
Kata'ib Hizballah (KH)
Kurdistan Workers' Party (PKK)
Lashkar e-Tayyiba (LeT)
Lashkar i Jhangvi (LJ)
Liberation Tigers of Tamil Eelam (LTTE)
Mujahidin Shura Council in the Environs of Jerusalem (MSC)
Al-Mulathamun Battalion (AMB)
National Liberation Army (ELN)

Al-Nusrah Front (ANF)
Palestine Islamic Jihad (PIJ)
Palestine Liberation Front—Abu Abbas Faction (PLF)
Popular Front for the Liberation of Palestine (PFLP)
Popular Front for the Liberation of Palestine-General Command (PFLP-GC)
Al-Qa'ida (AQ)
Al-Qa'ida in the Arabian Peninsula (AQAP)
Al-Qa'ida in the Indian Subcontinent (AQIS)
Al-Qa'ida in the Islamic Maghreb (AQIM)
Real IRA (RIRA)
Revolutionary Armed Forces of Colombia (FARC)
Revolutionary People's Liberation Party/Front (DHKP/C)
Revolutionary Struggle (RS)
Al-Shabaab (AS)
Shining Path (SL)
Tehrik-e Taliban Pakistan (TTP)

Abdallah Azzam Brigades

Also known as Abdullah Azzam Brigades; Ziyad al-Jarrah Battalions of the Abdallah Azzam Brigades; Yusuf al'Uyayri Battalions of the Abdallah Azzam Brigades

Description: Designated as a Foreign Terrorist Organization on May 30, 2012, the Abdallah Azzam Brigades (AAB) formally announced its establishment in a July 2009 video statement claiming responsibility for a February 2009 rocket attack against Israel. The group is divided into two branches: the Arabian Peninsula–based Yusuf al-'Uyayri Battalions of the Abdullah Azzam Brigades, named after the deceased founder of al-Qa'ida in Saudi Arabia, Yusef al-'Uyayri; and the Lebanon-based Ziyad al-Jarrah Battalions of the Abdallah Azzam Brigades, named after Lebanese citizen Ziad al Jarrah, one of the planners of the September 11 attacks on the United States.

Activities: After its initial formation, AAB relied primarily on rocket attacks against Israeli civilians. It is responsible for numerous rocket attacks fired into Israeli territory from Lebanon targeting—among other things—population centers, including Nahariya and Ashkelon.

In November 2013, AAB began targeting Hizballah. AAB claimed responsibility for a suicide bombing outside the Iranian Embassy in Beirut, Lebanon, which killed 23 people and wounded more than 140. The group warned it would carry out more attacks if Hizballah did not stop sending fighters to Syria in support of Syrian government forces.

In February 2014, AAB again attacked Hizballah for its involvement in the Syrian conflict, claiming a twin suicide bomb attack against the Iranian cultural center in Beirut, which killed four people. AAB was also believed to be responsible for a series of bombings in Hizballah–controlled areas around Beirut. In June 2014, AAB was believed to have targeted Lebanese General Security head Major General Abbas Ibrahim in a suicide bombing at a police checkpoint on the Beirut-Damascus highway; Ibrahim narrowly escaped. AAB also was blamed for a suicide bombing in June 2014 in the Beirut neighborhood of Tayyouneh, which killed a security officer and wounded 25 people.

In February 2015, Lebanon's military court charged four AAB suspects for their involvement in the June 2014 suicide bombings. During 2015 and 2016, AAB continued its involvement in the Syrian conflict. In June 2015, the group released photos of a training camp for its "Marwan Hadid Brigade" camp in Syria, likely located in Homs province.

Strength: Unknown

Location/Area of Operation: AAB is based in Lebanon but operates in both Lebanon and Syria.

Funding and External Aid: Unknown

Abu Nidal Organization

Also known as ANO; Arab Revolutionary Brigades; Arab Revolutionary Council; Black September; Fatah Revolutionary Council; Revolutionary Organization of Socialist Muslims

Description: The Abu Nidal Organization (ANO) was designated as a Foreign Terrorist Organization on October 8, 1997. The group was formed by Sabri al-Banna (a.k.a. Abu Nidal) in 1974 after splitting from the Palestine Liberation Organization. In August 2002, Nidal died in Baghdad. Leadership of the organization after Nidal's death remains unclear. The ANO has advocated for the elimination of Israel and sought to derail diplomatic efforts in support of the Middle East peace process.

Activities: The ANO has carried out terrorist attacks in 20 countries, killing or injuring almost 900 people. It has not staged a major attack against Western targets since the late 1980s and was expelled from its safe haven in Libya in 1999. Major attacks include those on the Rome and Vienna airports in 1985, the 1986 attack on the Neve Shalom Synagogue in Istanbul, the hijacking of Pan Am Flight 73 in Karachi in 1986, and the 1988 attack on the City of Poros day–excursion ship in Greece. The ANO was suspected of assassinating PLO Deputy Chief Abu Iyad and PLO Security Chief Abu Hul in Tunis in 1991, and a senior Jordanian diplomat in Beirut in 1994. In 2008, a Jordanian official apprehended an ANO member planning to carry out attacks in Jordan. The ANO has been dormant since 2014.

Strength: Unknown

Location/Area of Operation: ANO associates may still exist in Lebanon, although they are likely inactive.

Funding and External Aid: Unknown

Abu Sayyaf Group

Also known as al Harakat al Islamiyya (the Islamic Movement)

Description: The Abu Sayyaf Group (ASG) was designated as a Foreign Terrorist Organization on October 8, 1997. The ASG split from the Moro Islamic Liberation Front in the early 1990s and has since become the most violent terrorist group in the Philippines. The group claims to promote an independent Islamic state in western Mindanao and the Sulu Archipelago.

Activities: The ASG has committed kidnappings-for-ransom, bombings, ambushes of security personnel, public beheadings, assassinations, and extortion. In April 2000, an ASG faction kidnapped 21 people, including 10 Western tourists, from a resort in Malaysia. In May 2001, the group kidnapped three U.S. citizens and 17 Philippine citizens from a tourist resort in Palawan, Philippines; several hostages were murdered, including U.S. citizen Guillermo Sobero. A hostage rescue operation in June 2002 freed U.S. hostage Gracia Burnham; her husband, U.S. national Martin Burnham was killed during the operation. In October 2002, Philippine and U.S. authorities blamed the ASG for a bombing near a military base in Zamboanga that killed a U.S. soldier, and in February 2004, the ASG bombed SuperFerry 14 in Manila Bay, killing 116 people. On July 28, 2014, ASG militants armed with assault rifles opened fire on civilians celebrating the end of Ramadan, killing at least 21 people—including six children and at least four members of a Talipao security force—and wounding 11 others.

In a July 2014 video, senior ASG leader and Federal Bureau of Investigation most-wanted terrorist Isnilon Hapilon swore allegiance to ISIS and ISIS leader Abu Bakr al-Baghdadi. Since then, several ASG and other violent extremist factions have also pledged allegiance to ISIS. Hapilon is now thought to be the leader of all ISIS-affiliated groups in the Philippines.

Throughout 2015, the ASG was responsible for multiple attacks, including the February abduction of a 73-year-old Korean businessman (freed after 10 days in captivity) in Lanao del Sur and the May kidnappings of two civilians, two Philippine Coast Guard personnel, and a Philippine city official during various incidents; some of these hostages were later beheaded. In September, the ASG was blamed for the armed abduction of two Canadians, a Norwegian, and a Philippine woman from the Holiday Oceanview Samal Resort on Samal Island. The ASG set ransom at US $60 million. In 2016, the two Canadian citizens were beheaded when ransom deadlines were not met.

The group continues its kidnapping-for-ransom operations.

Strength: The ASG is estimated to have 400 members.

Location/Area of Operation: Primarily in the Philippine provinces of the Sulu Archipelago—namely Basilan, Sulu, and Tawi-Tawi—and on the Zamboanga Peninsula. The group has also conducted cross-border operations into eastern Malaysia.

Funding and External Aid: The ASG is funded through kidnapping-for-ransom operations and extortion. It may receive funding from external sources, including remittances from supportive overseas Philippine workers and Middle East-based sympathizers. In the past, the ASG has also received assistance from regional terrorist groups such as Jemaah Islamiya, whose operatives have provided training to ASG members and have helped facilitate several ASG terrorist attacks.

Al-Aqsa Martyrs Brigade

Also known as al-Aqsa Martyrs Battalion

Description: Designated as a Foreign Terrorist Organization on March 27, 2002, the al-Aqsa Martyrs Brigade (AAMB) is composed of small cells of Fatah-affiliated activists that emerged at the outset of the al-Aqsa Intifada in September 2000. AAMB strives to drive the Israeli military and West Bank settlers from the West Bank to establish a Palestinian state loyal to Fatah.

Activities: During the 2000 Intifada, AAMB carried out primarily small-arms attacks against Israeli military personnel and settlers. By 2002, the group was striking Israeli civilians inside Israel. In January 2002, AAMB claimed responsibility for the first female suicide bombing in Israel. In 2010 and 2011, the group launched numerous rocket attacks on communities in Israel, including the city of Sederot and areas of the Negev desert. In November 2012, two men recruited by AAMB were arrested in connection with the stabbing of a student in Beersheba, Israel. That same year, AAMB claimed it had fired more than 500 rockets and missiles into Israel during Operation Pillar of Defense, the week-long Israel Defense Forces operation in Gaza.

In February 2015, AAMB declared an open war against Israel. In a television broadcast in June 2015, AAMB asked Iran for funds to help the group in its fight against Israel. In the same broadcast, an AAMB fighter displayed a new two-mile tunnel crossing the border beneath Gaza and Israel, which the leader claimed would be used in the next rounds of battle. In October 2015, the group announced it had developed a new rocket, the K60, with a 40-mile firing range. Throughout 2015, AAMB continued attacking Israeli soldiers and civilians; the group was responsible for the October 2015 fatal shooting of husband and wife, Rabbi Eitam and Na'ama Henkin.

In March 2016, armed confrontation in Nablus between Palestinian youths and Palestinian security officials broke out following the arrest of an AAMB associate on charges of murder; seven youths and six Palestinian security officials were injured in the unrest.

Strength: Estimated at a few hundred members.

Location/Area of Operation: Most of AAMB's operational activity is in Gaza but it has also planned and conducted attacks inside Israel and the West Bank. AAMB has members in Palestinian refugee camps in Lebanon.

Funding and External Aid: Iran has provided AAMB with funds and guidance, mostly although Hizballah facilitators.

Ansar Al-Dine

Also known as Ansar Dine; Ansar al-Din; Ancar Dine; Ansar ul-Din; Ansar Eddine; Defenders of the Faith

Description: The Mali-based group Ansar al-Dine (AAD) was designated as a Foreign Terrorist Organization on March 22, 2013. AAD was created in late 2011 after AAD's leader Iyad ag Ghali failed in his attempt to take over another secular Tuareg organization. Following the March 2012 coup that toppled the Malian government, AAD was among the organizations (including al-Qa'ida in the Islamic Maghreb (AQIM) and Movement for Unity and Jihad in West Africa) to take over northern Mali, destroy UNESCO World Heritage sites, and enforce a severe interpretation of Sharia law upon the civilian population living in the areas under its control.

Beginning in January 2013, French and allied African forces conducted operations in northern Mali to counter AAD and other violent extremist groups, eventually forcing AAD and its allies out of the population centers they had seized. Ghali, however, remained free, and has appeared in AAD videos in 2015 and 2016 threatening France and the UN Multidimensional Integrated Stabilization Mission in Mali (MINUSMA).

Activities: AAD received backing from AQIM in its fight against the Government of Mali, including for its capture of the Malian towns of Agulhok, Gao, Kidal, Tessalit, and Timbuktu, between January and April 2012. In March 2012, AAD members were reportedly among the Tuareg rebels responsible for killing 82 Malian soldiers and kidnapping 30 others in an attack against Agulhok. Before the French intervention in January 2013, Malian citizens in towns under AAD's control who refused to comply with AAD's laws allegedly faced harassment, torture, and death.

AAD was severely weakened by the 2013 French intervention, but increased its activities in 2015 and 2016. In 2016, AAD claimed attacks targeting the Malian army and MINUSMA. In February, AAD launched a suicide and rocket attack on a UN base in Kidal that killed six peacekeepers. In April, the group kidnapped three workers from the International Committee of the Red Cross. In July, AAD attacked an army base, leaving 17 soldiers dead and six missing. On August 16 alone, the group claimed three attacks: two improvised explosive device attacks on French forces and a rocket or mortar attack on a joint UN-French base near Tessalit. In October, AAD claimed credit for a series of attacks on UN forces near Agulhok, and in November, took responsibility for several attacks on French and UN forces. A November attack on a military convoy in the central region of Mopti killed a UN peacekeeper and two Malian civilians, and injured seven peacekeepers.

Strength: The group's exact membership numbers were unknown at the end of 2016.

Location/Area of Operation: AAD is active in Mali, and has also threatened to attack Mauritania and the Ivory Coast.

Funding and External Aid: AAD cooperates closely with and has received support from AQIM since its inception. AAD is also said to receive funds from foreign donors and through smuggling.

Ansar Al-Islam

Also known as Ansar al-Sunna; Ansar al-Sunna Army; Devotees of Islam; Followers of Islam in Kurdistan; Helpers of Islam; Jaish Ansar al-Sunna; Jund al-Islam; Kurdish Taliban; Kurdistan Supporters of Islam; Partisans of Islam; Soldiers of God; Soldiers of Islam; Supporters of Islam in Kurdistan

Description: Ansar al-Islam (AAI) was designated as a Foreign Terrorist Organization on March 22, 2004. AAI was established in 2001 in the Iraqi Kurdistan region with the merger of two Kurdish violent extremist factions that traced their roots to the Islamic Movement of Kurdistan. On May 4, 2010, AAI's leader Abu Abdullah al-Shafi'i was captured by U.S. forces in Baghdad; he remained in prison at the end of 2016. On December 15, 2011, AAI announced a new leader: Abu Hashim Muhammad bin Abdul Rahman al Ibrahim. AAI seeks to expel western interests from Iraq and establish an independent Iraqi state based on its interpretation of Sharia law.

In March 2012, a Norwegian court convicted Iraqi citizen and AAI founder Mullah Krekar (aka Najmuddin Faraj Ahmad) of issuing threats and inciting terrorism, and sentenced him to six years in prison. Living in Norway on a long-term resident permit, Krekar appealed, and in December 2012, an appeals court affirmed his convictions for issuing threats and intimidating witnesses, but reversed his conviction for inciting terrorism. The appeals court reduced his prison sentence to two years and 10 months. He was released from prison in late January 2015, but was arrested again shortly after for praising the January 2015 Charlie Hebdo attacks in Paris in a television interview.

Activities: AAI has conducted attacks against a wide range of targets including Iraqi government and security forces, and U.S. and Coalition Forces. AAI has carried out numerous kidnappings, murders, and assassinations of Iraqi citizens and politicians. The group has either claimed responsibility, or is believed to be responsible, for attacks in 2011 that resulted in 24 deaths and wounded 147. In 2012, the group claimed responsibility for the bombing of the Sons of Martyrs School in Damascus, which was occupied by Syrian security forces and pro-government militias; seven people were wounded in the attack. In 2014, AAI claimed responsibility for attacks near Kirkuk, Tikrit, and Mosul, Iraq, primarily directed against Iraqi police and security forces and, in one instance, an oil field.

During summer 2014, part of AAI issued a statement pledging allegiance to ISIS. In 2015, however, reports suggested that a faction of AAI in Syria continued to oppose ISIS. In June 2015, for example, AAI urged factions in Aleppo to confront ISIS "militarily and religiously." In 2016, AAI continued to operate in Syria.

Strength: Precise numbers are unknown.

Location/Area of Operation: AAI is active primarily in northern Iraq, but also maintains a presence in western and central Iraq. AAI expanded its operations into Syria in 2011.

Funding and External Aid: AAI receives assistance from a loose network of associates in Europe and the Middle East.

Ansar Al-Shari'a in Benghazi

Also known as Ansar al-Sharia in Libya; Ansar al-Shariah Brigade; Ansar al-Shari'a Brigade; Katibat Ansar al-Sharia in Benghazi; Ansar al-Shariah-Benghazi; Al-Raya Establishment for Media Production; Ansar al-Sharia; Soldiers of the Sharia; Ansar al-Shariah; Supporters of Islamic Law

Description: Designated as a Foreign Terrorist Organization on January 13, 2014, Ansar al-Shari'a in Benghazi (AAS-B) was created after the 2011 fall of the Qadhafi regime in Libya. It has been involved in terrorist attacks against civilian targets, and assassinations and attempted assassinations of security officials and political actors in eastern Libya.

Activities: Members of AAS-B were involved in the September 11, 2012, attacks against the U.S. Special Mission and Annex in Benghazi, Libya. Four U.S. citizens were killed in the attack: Glen Doherty, Sean Smith, Tyrone Woods, and the U.S. Ambassador to Libya, J. Christopher Stevens.

During 2016, AAS-B continued its fight against the "Libyan National Army" in Benghazi that resulted in the deaths of both Libyan security forces and civilians. The group is believed to have cooperated with Ansar al-Shari'a in Darnah in some of its attacks targeting Libyan security forces in Benghazi.

In addition to its attacks, AAS-B controlled several terrorist training camps in Libya, and has trained members of other terrorist organizations, some which operate in Syria, Iraq, and Mali.

Strength: Unknown

Location/Area of Operation: Benghazi, Libya

Funding and External Aid: In the past, AAS-B has obtained funds from al-Qa'ida in the Islamic Maghreb, charities, donations, and criminal activities.

Ansar Al-Shari'a in Darnah

Also known as Supporters of Islamic Law; Ansar al-Sharia in Derna; Ansar al-Sharia in Libya; Ansar al-Sharia; Ansar al-Sharia Brigade in Darnah

Description: Designated as a Foreign Terrorist Organization on January 13, 2014, Ansar al-Shari'a in Darnah (AAS-D) was created after the 2011 fall of the Qadhafi regime in Libya. It has been involved in terrorist attacks against civilian targets, and assassinations and attempted assassinations of security officials and political actors in eastern Libya.

Activities: Members of AAS-D were involved in the September 11, 2012, attacks against the U.S. Special Mission and Annex in Benghazi, Libya. Four U.S. citizens were killed in the attack: Glen Doherty, Sean Smith, Tyrone Woods, and the U.S. Ambassador to Libya, J. Christopher Stevens.

In 2013 and 2014, AAS-D is believed to have cooperated with Ansar al-Shari'a in Benghazi in multiple attacks and suicide bombings targeting Libyan security forces in Benghazi. In 2016, AAS-D continued its involvement in fighting in and around Darnah.

In addition to its attacks, AAS-D maintained several terrorist training camps in Darnah and Jebel Akhdar, Libya, and has trained members of other terrorist organizations operating in Syria and Iraq.

Strength: Unknown

Location/Area of Operation: Darnah, Libya

Funding and External Aid: Unknown

Ansar Al-Shari'a in Tunisia

Also known as Al-Qayrawan Media Foundation; Supporters of Islamic Law; Ansar al-Sharia in Tunisia; Ansar al-Shari'ah; Ansar al-Shari'ah in Tunisia; Ansar al-Sharia

Description: Designated as a Foreign Terrorist Organization on January 13, 2014, Ansar al-Shari'a in Tunisia (AAS-T) was founded by Seif Allah Ben Hassine in early 2011. AAS-T has been implicated in attacks against Tunisian security forces, assassinations of Tunisian political figures, and attempted suicide bombings of locations frequented by tourists. AAS-T has also recruited youth in Tunisia to fight in Syria.

Activities: AAS-T was involved in the September 14, 2012 attack against the U.S. Embassy and American school in Tunis, which threatened the safety of more than 100 U.S. Embassy employees. In February and July 2013, AAS-T members were implicated in the assassination of Tunisian politicians, Chokri Belaid and Mohamed Brahmi. In October 2013, a bomber associated with AAS-T blew himself up outside a hotel in Sousse, Tunisia, resulting in no other fatalities. On the same day, police prevented a suicide bombing in Monastir, Tunisia, when they arrested a would-be bomber at the Tomb of Habib Bourguiba.

Strength: Unknown

Location/Area of Operation: Tunisia and Libya

Funding and External Aid: Unknown

Army of Islam

Also known as Jaysh al-Islam; Jaish al-Islam

Description: Designated as a Foreign Terrorist Organization on May 19, 2011, the Army of Islam (AOI), founded in late 2015, is a Gaza-based terrorist organization responsible for numerous terrorist acts against the Israeli and Egyptian governments and British, New Zealand, and U.S. citizens. The group, led by Mumtaz Dughmush, subscribes to a violent extremist Salafist ideology.

Activities: AOI is responsible for a number of rocket attacks on Israel, the 2006 kidnapping of a U.S. and a New Zealander journalist in Gaza, and the 2007 kidnapping of a British citizen in Gaza. AOI also carried out the early 2009 attacks on Egyptian civilians in Cairo and Heliopolis, Egypt, and planned the January 1, 2011, attack on a Coptic Christian church in Alexandria that killed 25 and wounded 100. On July 28, 2012, AOI released a statement that its member Nidal al ‘Ashi was killed fighting in Syria, and in November 2012 announced that it had launched rocket attacks on Israel in a joint operation with the Mujahidin Shura Council in the Environs of Jerusalem. In August 2013, an Israeli official reported that AOI leader Dughmush was running training camps in Gaza. In August 2014, Israeli forces reportedly intercepted and killed several AOI fighters allegedly planning to attack Israel.

In September 2015, AOI reportedly released a statement pledging allegiance to ISIS. In the short post attributed to the group, AOI declared itself an inseparable part of ISIL-Sinai Province, the ISIS-branch in Egypt’s Sinai Peninsula.

In 2016, AOI continued to express support for ISIS. The group released a video in October in an effort to encourage ISIS fighters defending Mosul.

Strength: Membership is estimated in the low hundreds.

Location/Area of Operation: Egypt, Gaza, and Israel.

Funding and External Aid: AOI receives much of its funding from a variety of criminal activities in Gaza.

Asbat Al-Ansar

Also known as AAA; Band of Helpers; Band of Partisans; League of Partisans; League of the Followers; God's Partisans; Gathering of Supporters; Partisan's League; Esbat al-Ansar; Isbat al-Ansar; Osbat al-Ansar; Usbat al-Ansar; Usbat ul-Ansar

Description: Designated as a Foreign Terrorist Organization on March 27, 2002, Asbat al-Ansar (AAA) is a Lebanon-based violent Sunni terrorist group composed primarily of Palestinians. Linked to al-Qa'ida and other Sunni violent extremist groups, AAA aims to thwart perceived anti-Islamic and pro-Western influences in the country. AAA's base is largely confined to Lebanon's refugee camps.

Activities: AAA first emerged in the early 1990s. In the mid-1990s, the group assassinated Lebanese religious leaders and bombed nightclubs, theaters, and liquor stores. The group has also plotted against foreign diplomatic targets. In October 2004, AAA member Mahir al-Sa'di was sentenced in absentia to life imprisonment for his 2000 plot to assassinate then–U.S. Ambassador to Lebanon David Satterfield. Between 2005 and 2011, AAA members traveled to Iraq to fight Coalition Forces. AAA has been reluctant to involve itself in operations in Lebanon due in part to concerns of losing its safe haven in the Ain al-Hilwah refugee camp. AAA did not publicly claim any attacks in 2016.

Strength: The group has at least 650 members.

Location/Area of Operation: The group's primary base of operations is the Ain al-Hilwah Palestinian refugee camp near Sidon in southern Lebanon.

Funding and External Aid: It is likely the group receives money through international Sunni extremist networks.

Aum Shinrikyo

Also known as A.I.C. Comprehensive Research Institute; A.I.C. Sogo Kenkyusho; Aleph; Aum Supreme Truth

Description: Aum Shinrikyo (AUM) was designated as a Foreign Terrorist Organization on October 8, 1997. It was established in 1987 by now-jailed leader Shoko Asahara and gained legal status in Japan as a religious entity in 1989. The Japanese government revoked its recognition of AUM as a religious organization following AUM's deadly 1995 sarin gas attack in Tokyo. Despite claims it has renounced violence and Asahara's teachings, members of AUM continue to adhere to the violent and apocalyptic teachings of its founder.

Activities: In March 1995, AUM members simultaneously released the chemical nerve agent sarin on several Tokyo subway trains, killing 13 and causing up to 6,000 people to seek medical treatment. Subsequent investigations by the Japanese government revealed the group was responsible for other mysterious chemical incidents in Japan in 1994, including a sarin gas attack on a residential neighborhood in Matsumoto that killed seven and injured approximately 500. Japanese police arrested Asahara in May 1995; in February 2004, authorities sentenced him to death for his role in the 1995 attacks, but have yet to carry out the sentence. In 2010 and 2011, several death sentences for other AUM senior members were finalized or affirmed by Japanese courts. In 2012, the final three AUM fugitives were arrested after 17 years on the run. In April 2015, the Tokyo District Court sentenced former AUM sect leader Katsuya Takahashi to life in prison for his involvement in the 1995 Tokyo subway sarin attack and three other attacks carried out by the group in the 1990s.

Since 1997, the group has split into two factions, both of which have recruited new members, engaged in commercial enterprises, and acquired property. In July 2000, Russian authorities arrested a group of Russian AUM followers who planned to detonate bombs in Japan as part of an operation to free Asahara from jail and smuggle him to Russia. In August 2012, a Japan Airlines flight to the United States turned back after receiving a bomb threat demanding the release of Asahara.

Although AUM has not conducted a terrorist attack since 1995, concerns remain regarding its continued adherence to the violent teachings of Asahara. In March 2016, Montenegro expelled 58 people (43 from Russia, seven from Belarus, three from Ukraine, and one from Uzbekistan) associated with AUM found holding a conference at a hotel in Danilovgrad. One month later, Russian authorities carried out dozens of raids on 25 AUM properties and opened a criminal investigation into an AUM cell, stating the group's activities involved violence and posed a danger to civilian health; 10 people were detained in the raids.

Strength: In 2016, the group was estimated to have 1,500 followers. AUM continues to maintain facilities in Japan and Russia.

Location/Area of Operation: Japan and Russia

Funding and External Aid: Funding primarily comes from member contributions and grou–prun businesses.

Basque Fatherland and Liberty

Also known as ETA; Askatasuna; Batasuna; Ekin; Euskal Herritarrok; Euzkadi Ta Askatasuna; Herri Batasuna; Jarrai-Haika-Segi; K.A.S.; XAKI

Description: Designated as a Foreign Terrorist Organization on October 8, 1997, Basque Fatherland and Liberty (ETA) was founded in 1959 with the aim of establishing an independent homeland based on Marxist principles in the Spanish Basque provinces of Alava, Guipuzcoa, and Viscaya; the autonomous region of Navarre; and the southwestern French territories of Labourd, Lower-Navarre, and Soule.

Activities: ETA has primarily conducted bombings and assassinations. Targets typically include Spanish government officials, businessmen, politicians, judicial figures, and security and military forces; however, the group has also targeted journalists and major tourist areas. ETA is responsible for killing 829 civilians and members of the armed forces and police, and injuring thousands since it formally began its campaign of violence in 1968.

ETA has committed numerous attacks in the last four decades, including the February 2005 car bombing in Madrid that wounded more than 20 people at a convention center where Spanish King Juan Carlos and then-Mexican president Vicente Fox were scheduled to appear. In December 2006, ETA exploded a massive car bomb destroying much of the covered parking garage at Madrid's Barajas International Airport. ETA marked its fiftieth anniversary in 2009 with a series of high profile and deadly bombings, including the July 2009 attack on a Civil Guard Barracks that injured more than 60 men, women, and children.

ETA has not launched any attacks since it announced a "definitive cessation of armed activity" in October 2011.

At the end of 2016, the group had yet to formally disband or give up its weapons arsenal. Authorities seized ETA weapons in 2016, including a cache found in a forest north of Paris. In 2016, the presumed top ETA leader was also captured in a joint French and Spanish operation near the French border with Spain.

Strength: Since 2004, more than 900 ETA militants have been arrested both in Spain and abroad. It is unknown how many ETA members remain.

Location/Area of Operation: Primarily in the Basque autonomous regions of northern Spain and southwestern France, and Madrid.

Funding and External Aid: ETA is probably experiencing financial shortages given that the group announced publicly in September 2011 that it had ceased collecting "revolutionary taxes" from Basque businesses. Sources of ETA funding are unknown.

Footnote: On April 20, 2018, ETA publicly apologized to its victims and the people it has hurt since it began is violent insurgency in the early 1960s. The group which had already started to hand over its weapons and explosives under a French government brokered deal, said it planned to officially disband on the first weekend of May.

Boko Haram

Also known as Nigerian Taliban; Jama'atu Ahlus-Sunnah Lidda'Awati Wal Jihad; Jama'atu Ahlis Sunna Lidda'awati wal-Jihad; People Committed to the

Prophet's Teachings for Propagation and Jihad; Sunni Group for Preaching and Jihad

Description: Nigeria-based Boko Haram (BH) was designated as a Foreign Terrorist Organization on November 14, 2013. The group is responsible for numerous attacks in northern and northeastern Nigeria, and the Lake Chad Basin in Cameroon, Chad, and Niger that have killed thousands of people since its reemergence in 2009.

In March 2015, BH pledged allegiance to ISIS in an audiotape message; ISIS accepted the group's pledge and the group began calling itself ISIS-West Africa. In August 2016, ISIS announced that Abu Musab al-Barnawi was to replace Abubakar Shekau as the new leader of the group. Shekau continues to maintain a group of followers concentrated in the Sambisa Forest; in a recent video, Shekau referred to his faction as Jama'atu ahlus-sunnah lidda await wal jihad (Boko Haram). The Government of Nigeria routinely calls both Boko Haram.

Activities: BH was responsible for the August 26, 2011, bomb attack on the UN building in Abuja that killed at least 21 people and wounded dozens more. The group was also responsible for a series of bomb attacks in 2012 in Kano, Nigeria.

Boko Haram has increasingly crossed Nigerian borders to evade pressure and conduct operations. In February 2013, BH claimed responsibility for kidnapping seven French tourists in the Far North of Cameroon. Security forces from Chad and Niger also reportedly participated in skirmishes against suspected BH members along Nigeria's borders. In 2013, the group also kidnapped eight French citizens in northern Cameroon and obtained ransom payments for their release.

In 2014, BH killed approximately 5,000 Nigerian civilians in various attacks. The kidnapping of 276 female students from a secondary school in Chibok, Borno State, brought global attention to Boko Haram's activities and highlighted its deliberate targeting of noncombatants, including children. In 2015, the group continued to abduct women and girls in the northern region of Nigeria, some of whom it later subjected to domestic servitude, other forms of forced labor, and sexual servitude through forced marriages to its members.

Between January 3 and 7, 2015, BH carried out a massacre in Baga, Borno State; reported casualties ranged from 150 to more than 2,000 killed, injured, or disappeared. The January 2015 attacks and other BH operations in surrounding smaller villages in 2015 displaced an estimated 35,000 people and allowed BH to gain control of Borno State. In February, BH expanded into Cameroon with an attack on the northern town of Fotokol, where it murdered residents inside their homes and in a mosque. On April 6, BH militants disguised as Islamic preachers killed at least 24 people and wounded several others in an attack near a mosque in Borno State; the attackers gathered people in the village of Kwajafa, offering to preach Islam, then opened fire. In mid-October, BH conducted a coordinated attack on the Baga Sola market and a refugee camp in N'Djamena, Chad; 33 people were killed and 51 others wounded. On November 18, two young female suicide bombers detonated explosives just before afternoon prayers in Kano, killing 15.

BH continued its operational tempo through 2015 and into early 2016 under the name ISIS-West Africa, with January operations in Cameroon that killed 16 people. In February, the group resumed attacks in Nigeria, killing 30 people on February 13 in a spate of attacks in Borno State that included forcing worshipers into a mosque and killing them. On April 14, the group released a video of select teenage girls abducted during the Chibok incident as a "proof of life." BH released 21 Chibok schoolgirls to Nigerian authorities in exchange for the release of selected BH members in October; it was the first mass release of Chibok hostages since the 2014 abductions.

The attacks and kidnappings continue.

Strength: Membership is estimated to be several thousand fighters.

Location/Area of Operation: Northeastern Nigeria, northern Cameroon, southeast Niger, and areas of Chad along the border with Nigeria.

Funding and External Aid: BH largely self-finances through criminal activities such as looting, extortion, kidnapping-for-ransom, and bank robberies.

Communist Party of Philippines/New People's Army

Also known as CPP/NPA; Communist Party of the Philippines; the CPP; New People's Army; the NPA

Description: The Communist Party of the Philippines/New People's Army (CPP/NPA) was designated as a Foreign Terrorist Organization on August 9, 2002. The military wing of the Communist Party of the Philippines (CPP)—the New People's Army (NPA)—is a Maoist group formed in March 1969 with the aim of overthrowing the government through protracted guerrilla warfare. NPA's founder, Jose Maria Sison, reportedly directs CPP/NPA activity from the Netherlands, where he lives in self-imposed exile. Luis Jalandoni, a fellow Central Committee member and director of the CPP's overt political wing, the National Democratic Front (NDF), also lives in the Netherlands. Although primarily a rural-based guerrilla group, CPP/NPA has an active urban infrastructure to support its terrorist activities and, at times, uses city-based assassination squads.

Senior CPP/NPA leaders Wilma and Benito Tiamzon were arrested in the Philippines in 2014. NPA second-in-command Adelberto Silva was arrested in 2015.

Activities: CPP/NPA primarily targets Philippine security forces, government officials, local infrastructure, and businesses that refuse to pay extortion, or "revolutionary taxes." CPP/NPA also has a history of attacking U.S. interests in the Philippines. In 1987, for example, CPP/NPA conducted direct actions against U.S. personnel and facilities, killing three U.S. soldiers in four separate attacks in Angeles City. In 1989, the group issued a press statement claiming responsibility for the ambush and murder of Colonel James Nicholas Rowe, chief of the Ground Forces Division of the Joint U.S.-Military Advisory Group.

Over the past few years, CPP/NPA has continued to carry out killings, raids, kidnappings, acts of extortion, and other forms of violence primarily directed against Philippine security forces. In May 2013, the Armed Forces of the Philippines reported that from 2011 through the first quarter of 2013, 383 people—including 158 civilians—were killed in encounters between CPP/NPA and government forces.

Despite a ceasefire with the Government of the Philippines in December 2014, CPP/NPA continued to carry out attacks. During the first few months of 2015, the CPP/NPA attacked a Dole plantation, killed several soldiers and a policeman in an attempt to seize a police station in Mati City, and ambushed and killed several Philippine soldiers in Kalinga and Compostela Valley.

In 2016, the group was responsible for a number of clashes and attacks against government forces, police, and businesses. In August, multinational banana company Dole-Stanfilco closed its operation in Barobo due to escalating attacks by suspected CPP/NPA members.

Throughout 2016, several attempts were made to establish a ceasefire and peace deal between the CPP/NPA and the Armed Forces of the Philippines. Reported violations from both sides, however, including reports of CPP/NPA's continued recruitment in the Philippines, have stalled any peace efforts. In November, CPP/NPA announced it would not sign a deal until the government formally broke ties with the United States and resisted foreign intervention from other nations, including China and Russia.

Strength: The Philippine government estimates the group has approximately 4,000 members. CPP/NPA also retains a significant amount of support from communities in rural areas of the Philippines.

Location/Area of Operation: The Philippines, including Rural Luzon, Visayas, and parts of northern and eastern Mindanao. There are also CPP/NPA cells in Manila and other metropolitan centers.

Funding and External Aid: The CPP/NPA raises funds through extortion and theft.

Continuity Irish Republican Army

Also known as Continuity Army Council; Continuity IRA; Republican Sinn Fein

Description: Designated as a Foreign Terrorist Organization on July 13, 2004, the Continuity Irish Republican Army (CIRA) is a terrorist splinter group that became operational in 1986 as the clandestine armed wing of Republican Sinn Fein following its split from Sinn Fein. "Continuity" refers to the group's belief that it is carrying on the original goal of the Irish Republican Army (IRA), to force the British out of Northern Ireland. CIRA cooperates with the larger Real IRA (RIRA). In 2012, CIRA released a statement claiming new leadership, after the previous leadership was ousted over allegations it was acting to the detriment of the organization.

Activities: CIRA has been active in Belfast and the border areas of Northern Ireland, where it has carried out bombings, assassinations, kidnappings, hijackings, extortion operations, and robberies. On occasion, it has provided advance warning to police of its attacks. Targets have included the British military, Northern Ireland security forces, and Loyalist paramilitary groups. CIRA did not join the Provisional IRA in the September 2005 decommissioning and remains capable of effective, if sporadic, terrorist attacks.

In January 2013, the group claimed responsibility for firing shots at police officers in Drumbeg, Craigavon County, Northern Ireland. In March 2014, CIRA claimed an attempted bomb attack on the home and vehicle of a Police Service of Northern Ireland (PSNI) officer. In July 2015, PSNI officers narrowly escaped harm when an explosive detonated as police were investigating another concealed device that they later described as an "elaborate trap." In February 2016, CIRA claimed responsibility for a shooting at a boxing event at the Regency Hotel in Dublin that left one dead. Dublin police said the incident was related to a feud between criminal gangs.

Strength: Membership is small, with possibly fewer than 50. Police counterterrorism operations have reduced the group's strength.

Location/Area of Operation: Northern Ireland and the Republic of Ireland

Funding and External Aid: CIRA supports its activities through criminal activities, including smuggling.

Gama'a Al-Islamiyya

Also known as al-Gama'at; Egyptian al-Gama'at al-Islamiyya; GI; Islamic Gama'at; IG; Islamic Group

Description: Gama'a al-Islamiyya (IG) was designated as a Foreign Terrorist Organization on October 8, 1997. Formed in the 1970s, IG was once Egypt's largest terrorist group. In 2011, it formed the Building and Development political party that competed in the 2011 parliamentary elections and won 13 seats. Egypt-based members of IG released from prison prior to the 2011 revolution have renounced terrorism, although some members located overseas have worked with or joined al-Qa'ida (AQ). The external wing, composed mainly of exiled members in several countries, maintained that its primary goal was to replace the Egyptian government with an Islamist state. IG's "spiritual" leader, the "blind Sheikh," Omar Abd al-Rahman, is serving a life sentence in a U.S. prison for his involvement in the 1993 World Trade Center bombing. Supporters of Abd al-Rahman have called for reprisal attacks in the event of his death in prison.

Activities: In the 1990s, IG conducted armed attacks against Egyptian security, other government officials, and Coptic Christians. IG claimed responsibility for the June 1995 attempted assassination of Egyptian president Hosni Mubarak in Addis Ababa, Ethiopia. The group also launched attacks on tourists in Egypt, most notably the 1997 Luxor attack. In 1999, part of the group publicly renounced violence. IG is not known to have committed a terrorist attack in recent years; the group remained dormant in 2016.

Strength: At its peak, IG likely commanded several thousand core members and a similar number of supporters. Security clampdowns following the 1997 attack in Luxor and the 1999 cease-fire, along with post-September 11 security measures and defections to AQ, have likely resulted in a substantial decrease in what is left of the group.

Location/Area of Operation: Unknown

Funding and External Aid: Unknown

Hamas

Also known as the Islamic Resistance Movement; Harakat al-Muqawama al-Islamiya; Izz al-Din al Qassam Battalions; Izz al-Din al Qassam Brigades; Izz al-Din al Qassam Forces; Students of Ayyash; Student of the Engineer; Yahya Ayyash Units; Izz al-Din al-Qassim Brigades; Izz al-Din al-Qassim Forces; Izz al-Din al-Qassim Battalions

Description: Designated as a Foreign Terrorist Organization on October 8, 1997, Hamas was established in 1987 at the onset of the first Palestinian uprising, or Intifada, as an outgrowth of the Palestinian branch of the Muslim Brotherhood. The armed element, the Izz al-Din al-Qassam Brigades, has conducted anti-Israeli attacks, including suicide bombings against civilian targets inside Israel. Hamas also manages a broad, mostly Gaza-based, network of "Dawa" or ministry activities that include charities, schools, clinics, youth camps, fundraising, and political activities. After winning Palestinian Legislative Council elections in January 2006, Hamas gained control of significant Palestinian Authority (PA) ministries in Gaza, including the Ministry of Interior. As of 2016, the group retained control of Gaza.

Activities: Prior to 2005, Hamas conducted numerous anti-Israeli attacks, including suicide bombings, rocket launches, improvised explosive device (IED) attacks, and shootings. U.S. citizens have died and been injured in the group's attacks. In June 2007, after Hamas took control of Gaza from the PA and Fatah, the Gaza borders were closed and Hamas increased its use of tunnels to smuggle weapons into Gaza via the Sinai and maritime routes. Hamas has since focused on solidifying its control of Gaza, including hardening its defenses, building weapons caches, tightening security, and conducting limited operations against Israeli military forces.

Hamas fought a 23-day war with Israel from late December 2008 to January 2009. From November 14–21, 2012, Hamas fought another war with Israel during which it claims to have launched more than 1,400 rockets into Israel. Despite the Egypt-mediated ceasefire between Israel and Hamas in 2012, operatives from Hamas and Palestine Islamic Jihad (PIJ) coordinated and carried out a November bus bombing in Tel Aviv that wounded 29 people. On July 8, 2014, Israel launched Operation Protective Edge in Gaza with the intent of preventing rocket fire into the country; rocket fire had increased following Israeli military operations after Hamas kidnapped and murdered three Israeli teenagers in June 2014, including 16-year-old U.S.-Israeli citizen Naftali Fraenkel. In March 2015, Amnesty International released a report in which it accused Hamas of committing war crimes for launching rockets and mortars into civilian areas in Israel during Operation Protective Edge. In May 2015, Amnesty International published another report declaring Hamas's abduction, torture, and killing of Palestinians during the 2014 Gaza war as further evidence the group had committed war crimes.

In 2016, Hamas continued striking Israel. In April, a Hamas member carried out a suicide attack on a bus in Jerusalem, killing 20. The attack came just two months after Hamas released a music video on its al-Aqsa TV and several social media sites encouraging suicide bombings of Israeli buses. In October, Israeli authorities arrested seven members of a Hamas cell based in the West Bank that was plotting shooting and kidnapping attacks. The group was also held responsible for several rocket attacks launched from Gaza, including a July strike in Sderot that hit a kindergarten and damaged several neighboring buildings.

Throughout 2016, Hamas cooperated with ISIL-Sinai Province, providing its members with funding, training, and organizational support.

Attacks on Israel continue.

Strength: Several thousand Gaza-based operatives.

Location/Area of Operation: Since 2007, Hamas has controlled Gaza and also has a presence in the West Bank. The group retains leaders and facilitators that conduct political, fundraising, and arms-smuggling activities throughout the region. Hamas also has a presence in the Palestinian refugee camps in Lebanon. The group's current leader Khaled Meshaal ruled from exile in Qatar in 2016.

Funding and External Aid: Historically, Hamas has received funding, weapons, and training from Iran. The group also raises funds in the Gulf countries and receives donations from Palestinian expatriates around the world. Hamas also receives donations from its charity organizations. Hamas' supply lines have suffered since the crackdown on smuggling tunnels in the Sinai Peninsula by the Egyptian military.

Haqqani Network

Also known as HQN

Description: Designated as a Foreign Terrorist Organization on September 19, 2012, the Haqqani Network (HQN) was formed in the late 1970s, around the time of the then-Soviet Union's invasion of Afghanistan. HQN's founder Jalaluddin Haqqani established a relationship with Osama bin Laden in the mid-1980s, and joined the Taliban in 1995. After the fall of the Taliban in Afghanistan in 2001, Jalaluddin retreated to Pakistan where, under the leadership of Jalaluddin's son Sirajuddin Haqqani, the group continued to direct and conduct terrorist activity in Afghanistan. In July 2015, Sirajuddin Haqqani was appointed a Deputy leader of the Taliban.

Activities: HQN has planned and carried out a number of significant kidnappings and attacks against U.S. and Coalition Forces in Afghanistan, the Afghan government, and civilian targets. HQN's most notorious attacks include the June 2011 attack on the Intercontinental Hotel in Kabul, which killed 11 civilians and two Afghan policemen; a September 2011 truck bombing in Wardak Province, Afghanistan, which wounded 77 U.S. soldiers; the September 2011 19-hour attack on the U.S. Embassy and International Security Assistance Force headquarters in Kabul, which killed 16 Afghans, including six children; a June 2012 suicide bomb attack against Forward Operating Base Salerno, which killed two U.S. soldiers and wounded more than 100; and a 12-hour siege of the Spozhmai Hotel in Kabul in June 2012, which resulted in the death of at least 18 Afghans, including 14 civilians. HQN was also involved in holding U.S. Army Sergeant Bowe Bergdahl, who was kidnapped in 2009 and remained in captivity until his May 2014 release.

HQN's attacks continued throughout 2016. In January, Afghan officials blamed HQN for a suicide attack in Kabul, which killed seven journalists affiliated with a television station that had been critical of the Taliban. In April, HQN was blamed for an attack in Kabul against a government security agency tasked with providing protection to senior government officials, killing 64 people and injuring more than 300 others in what was the deadliest attack in Kabul in 15 years. Afghan officials also blamed HQN for being involved in a June double suicide attack outside of Kabul against Afghan police cadets and first responders; 30 people were killed. In addition to these and other suspected HQN attacks, Afghan police disrupted multiple HQN plots and many suicide operations in Kabul before they could be carried out.

Strength: HQN is believed to have several hundred core members, but it is estimated that the organization is able to draw upon a pool of upwards of 10,000 fighters. HQN is integrated into—and cooperates closely with—the larger Afghan Taliban and draws strength through cooperation with other terrorist organizations operating in Afghanistan, including al-Qa'ida and Jaish-e Mohammad.

Location/Area of Operation: HQN is active along the Afghanistan-Pakistan border and across much of southeastern Afghanistan, particularly in Loya Paktia (which consists of Paktia, Paktika, and Khost provinces, and includes parts of Logar and Ghazni), and has repeatedly targeted Kabul in its attacks. The group's leadership has historically maintained a power base around Pakistan's tribal areas.

Funding and External Aid: In addition to the support it receives through its connections to other terrorist organizations, HQN receives much of its funds from donors in Pakistan and the Gulf, as well as through criminal activities such as kidnapping, extortion, smuggling, and other licit and illicit business ventures.

Harakat ul-Jihad-i-Islami

Also known as HUJI; Movement of Islamic Holy War; Harkat-ul-Jihad-al Islami; Harkat-al-Jihad-ul Islami; Harkat-ul-Jehad-al-Islami; Harakat ul Jihad-e-Islami; Harakat-ul Jihad Islami

Description: Harakat-ul Jihad Islami (HUJI) was designated as a Foreign Terrorist Organization on August 6, 2010. The group was formed in 1980 in Afghanistan to fight against the former Soviet Union. Following the Soviet withdrawal from Afghanistan in 1989, the organization redirected its efforts to India. HUJI seeks the annexation of the state of Jammu and Kashmir and the expulsion of Coalition Forces from Afghanistan, and has supplied fighters to the Taliban in Afghanistan.

Since 2001, HUJI has refocused its activities on the Afghanistan-Pakistan front and has established several camps in Pakistan. HUJI is mostly composed of Pakistani militants and veterans of the Soviet-Afghan war. In recent years HUJI has experienced a number of internal splits and a portion of the group has aligned with al-Qa'ida (AQ), including training its members in AQ training camps. Mohammad Ilyas Kashmiri, one of HUJI's top leaders (who also served as an AQ military commander and strategist), died on June 3, 2011.

Activities: HUJI has been involved in a number of terrorist attacks. HUJI claimed responsibility for the September 7, 2011 bombing of the New Delhi High Court, which left at least 11 dead and an estimated 76 wounded. The group sent an email to the press stating that the bomb was intended to force India to repeal a death sentence of a HUJI member. In 2015, it was discovered that HUJI had set up bases at Silchar's Assam University with the help of a group of professors and students.

Strength: HUJI has an estimated strength of several hundred members.

Location/Area of Operation: HUJI's area of operation extends throughout South Asia, with its terrorist operations focused primarily in Afghanistan, India, and Pakistan.

Funding and External Aid: Unknown

Harakat ul-Jihad-i-Islami/Bangladesh

Also known as HUJI-B; Harakat ul Jihad e Islami Bangladesh; Harkatul Jihad al Islam; Harkatul Jihad; Harakat ul Jihad al Islami; Harkat ul Jihad al Islami; Harkat-ul-Jehad-al-Islami; Harakat ul Jihad Islami Bangladesh; Islami Dawat-e-Kafela; IDEK

Description: Designated as a Foreign Terrorist Organization on March 5, 2008, Harakat ul-Jihad-i-Islami/Bangladesh (HUJI-B) was formed in April 1992 by a group of former Bangladeshi Afghan veterans wanting to establish Islamist rule in Bangladesh. In October 2005, Bangladeshi authorities banned the group. The leaders of HUJI-B signed the February 1998 *fatwa* sponsored by Osama bin Laden that declared U.S. civilians legitimate targets.

HUJI-B has connections to Pakistani terrorist groups advocating similar objectives, including HUJI, al-Qa'ida, and Lashkar e-Tayyiba (LeT).

Activities: In December 2008, three HUJI-B members were convicted for the May 2004 grenade attack that wounded the British High Commissioner in Sylhet, Bangladesh. In 2011, Bangladeshi authorities formally charged multiple suspects, including HUJI-B leader Mufti Abdul Hannan, with the killing of former Finance Minister Shah AMS Kibria of the Awami League in a grenade attack on January 27, 2005. Although HUJI-B committed no known attacks in 2013, in March 2013, police in Dhaka arrested a group of militants which included HUJI-B members. The group was preparing attacks on public gatherings and prominent individuals; bombs, bomb-making material, and counterfeit currency were found at the time of arrest. In 2014, HUJI-B continued its operations; reports at the time suggested that some HUJI-B members may have traveled to Pakistan to receive military training from LeT. In December 2016, an elite police force in Bangladesh raided a HUJI-B den, arresting four of the group's members and foiling a HUJI-B plot to attack law enforcement officers to loot arms and ammunition. Law enforcement officials confiscated 18 bombs, electrical devices, and sharp weapons from the HUJI-B hideout.

Strength: HUJI-B leaders claim that up to 400 of its members are Afghan war veterans; its total membership is unknown.

Location/Area of Operation: HUJI-B operates primarily in Bangladesh and India; the group trains and has a network of madrassas in Bangladesh.

Funding and External Aid: HUJI-B funding comes from a variety of sources. Several international nongovernmental organizations may have funneled money to HUJI-B and other Bangladeshi terrorist groups.

Harakat ul-Mujahideen

Also known as HUM; Harakat ul-Ansar; HUA; Jamiat ul-Ansar; JUA; al-Faran; al-Hadid; al-Hadith; Harakat ul-Mujahidin; Ansar ul Ummah

Description: Designated as a Foreign Terrorist Organization on October 8, 1997, Harakat ul-Mujahideen (HUM) seeks the annexation of the state of Jammu and Kashmir and the expulsion of Coalition Forces from Afghanistan. In January 2005, HUM's long-time leader Fazlur Rehman Khalil stepped down and was replaced by Dr. Badr Munir. HUM operated terrorist training camps in eastern Afghanistan until Coalition air strikes destroyed them in 2001. In 2003, HUM began using the name Jamiat ul-Ansar; Pakistan banned the group in November 2003.

Activities: HUM has conducted a number of operations against Indian troops and civilian targets in the state of Jammu and Kashmir. It is linked to the Kashmiri terrorist group al-Faran, which kidnapped, and later killed, five Western tourists in the state of Jammu and Kashmir in July 1995. HUM was also responsible for the hijacking of an Indian airliner in December 1999. The attack led to the release of Masood Azhar, an important leader of the former Harakat ul-Ansar (HUA), who was imprisoned by India in 1994; after his release, Azhar founded Jaishe-Mohammed (JeM). India also released former HUA member Ahmed Omar Sheik as a result of the hijacking. Sheik was later convicted of the 2002 abduction and murder of U.S. journalist Daniel Pearl.

The group has conducted numerous attacks targeting Indian interests including the late December 2015 strikes in Handwor and Poonch, which resulted in the deaths of five Indian army personnel. In January 2016, Indian security forces arrested five HUM militants planning assaults on VIPs and security forces. Four of the militants arrested were involved in grenade-throwing incidents in downtown Srinagar and all five were planning an attack on the Srinagar-Baramulla highway on Republic Day.

Strength: HUM has a few hundred armed supporters. After 2000, however, a significant portion of HUM's membership defected to JeM and only a small number of cadres are reported to be active.

Location/Area of Operation: HUM conducts operations primarily in the state of Jammu and Kashmir and Afghanistan. It operates from Muzaffarabad in Azad Jammu and Kashmir, as well as several cities in Pakistan.

Funding and External Aid: HUM collects donations from wealthy and grassroots donors in Pakistan.

Hizballah

Also known as the Party of God; Islamic Jihad; Islamic Jihad Organization; Revolutionary Justice Organization; Organization of the Oppressed on Earth; Islamic Jihad for the Liberation of Palestine; Organization of Right Against Wrong; Ansar Allah; Followers of the Prophet Muhammed

Description: Hizballah was designated as a Foreign Terrorist Organization on October 8, 1997. Formed in 1982 following the Israeli invasion of Lebanon, the Lebanon-based radical Shia group takes its ideological inspiration from the Iranian revolution and the teachings of the late Ayatollah Khomeini. The group generally follows the religious guidance of the Iranian Supreme Leader. Hizballah is closely allied with Iran and the two often work together on shared initiatives, although Hizballah also acts independently. Hizballah shares a close relationship with Syria, and like Iran, provides assistance—including fighters—to Syrian regime forces in the Syrian conflict.

Activities: Hizballah is responsible for multiple large scale terrorist attacks, including the 1983 suicide truck bombings of the U.S. Embassy and U.S. Marine barracks in Beirut; the 1984 attack on the U.S. Embassy annex in Beirut; and the 1985 hijacking of TWA flight 847, during which U.S. Navy diver Robert Stethem was murdered.

Elements of the group were responsible for the kidnapping, detention, and murder of U.S. citizens and other Westerners in Lebanon in the 1980s. Hizballah was implicated, along with Iran, in the 1992 attacks on the Israeli Embassy in Argentina and in the 1994 bombing of the Argentine-Israeli Mutual Association in Buenos Aires. In 2000, Hizballah operatives captured three Israeli soldiers in the Shebaa Farms area and, separately, kidnapped an Israeli noncombatant in Dubai. The noncombatant survived, but on November 1, 2001, Israeli Army Rabbi Israel Weiss pronounced the soldiers dead. The surviving noncombatant and the bodies of the Israeli soldiers were returned to Israel in a prisoner exchange with Hizballah in 2004.

Hizballah is believed to have carried out two attacks against UN Interim Force in Lebanon peacekeepers, an attack in late July 2011 that wounded six French citizens and a second attack, days later, which injured three French soldiers. Also in 2011, four Hizballah members were indicted by the UN-based Special Tribunal for Lebanon, an international tribunal investigating the 2005 assassination of Lebanese prime minister Rafik Hariri. A fifth Hizballah member, Hassan Habib Merhi, was indicted in October 2013.

In January 2012, Thai police detained Hizballah operative Hussein Atris on immigration charges as he was attempting to depart Thailand. Atris was convicted of possessing bomb-making materials by a Thai court in September 2013 and sentenced to two years and eight months in prison. He was released in September 2014 and is believed to reside in Lebanon. In July 2012, a suspected Hizballah operative was detained by Cypriot authorities for allegedly helping plan an attack against Israeli tourists on the island. On March 21, 2013, a Cyprus court found the operative guilty of charges based on his surveillance activities of Israeli tourists

Hizballah was also responsible for the July 2012 attack on a passenger bus carrying 42 Israeli tourists at the Sarafovo Airport in Bulgaria, near the city of Burgas. The explosion killed five Israelis and one Bulgarian, and injured 32 others.

In May 2013, Hizballah publicly admitted to playing a significant role in the ongoing conflict in Syria, rallying support for Syrian president Bashar al-Assad.

In May 2015, Cypriot authorities arrested Hizballah member and Lebanese-Canadian national Hussein Bassam Abdallah after finding 8.2 tons of liquid ammonium nitrate in the basement of a residence in Larnaca. Abdallah was charged by the Republic of Cyprus on five offenses, including participation in a terrorist organization and providing support to a terrorist organization. On June 29, 2015, Abdallah was sentenced to six years in prison; he is currently serving his sentence in Larnaca, Cyprus.

Hizballah's support for Syria's al-Assad regime continues into 2018. There are reportedly about 7,000 Hizballah fighters in Syria; several senior Hizballah military commanders and hundreds of fighters have died in the Syrian conflict.

Strength: Tens of thousands of supporters and members worldwide.

Location/Area of Operation: Hizballah is based in the southern suburbs of Beirut, the Bekaa Valley, and southern Lebanon. As evidenced by Hizballah's activities during the course of 2012 and 2013, the group is capable of operating around the world. Since 2013, Hizballah fighters have assisted Assad regime forces in many areas across Syria.

Funding and External Aid: Iran continues to provide Hizballah with training, weapons, and explosives, as well as political, diplomatic, monetary, and organizational aid; Syria has furnished training, weapons, and diplomatic and political support. Hizballah also receives funding from private donations and profits from legal and illegal businesses. Hizballah receives financial support from Lebanese Shia communities in Africa, Asia, Europe, North America, and South America. Hizballah supporters are often engaged in a range of licit and illicit activities that benefit the group financially. These activities include smuggling contraband goods, passport falsification, trafficking in narcotics, money laundering, and credit card, immigration, and bank fraud.

Indian Mujahedeen

Also known as Indian Mujahidin; Islamic Security Force-Indian Mujahideen (ISF-IM)

Description: The Indian Mujahedeen (IM) was designated as a Foreign Terrorist Organization on September 19, 2011. The India-based terrorist group is responsible for dozens of bomb attacks throughout India since 2005, and has caused the deaths of hundreds of civilians. IM maintains ties to other terrorist entities including Pakistan-based Lashkar e-Tayyiba, Jaish-e-Mohammed, and Harakat ul-Jihad Islami. In 2016, it was reported that the IM was also cooperating with ISIS. IM's stated goal is to carry out terrorist actions against Indians for their oppression of Muslims. IM has also expanded its area of operations into Nepal, which is now the biggest hub for IM operatives.

Activities: IM is known for carrying out multiple coordinated bombings in crowded areas against economic and civilian targets to maximize terror and casualties. In 2008, for example, IM was responsible for 16 synchronized bomb blasts in crowded urban centers. That same year, IM was also responsible for an attack in Delhi that killed 30 people and an attack at a local hospital in Ahmedabad that killed 38 and injured more than 100. In 2010, IM bombed a popular German bakery in Pune, India, frequented by tourists; 17 people were killed and more than 60 people were injured in the attack.

In January 2015, the arrest of three IM militants linked the group to the December 2014 low-intensity blast near a restaurant in Bangalore that killed one woman and injured three others. The arrest also uncovered that the group planned to carry out attacks on Republic Day and had provided explosives to carry out attacks in other parts of the country. In May 2015, IM fugitive commander Muhammad 'Bada' Sajid, one of the perpetrators of the 2008 bombings, died fighting with ISIS forces in Syria.

In 2016, IM was increasingly linked to ISIS. In May, for example, six IM operatives were identified in an ISIS propaganda video threatening attacks on India. A month later, it was reported that an IM cell linked to ISIS was plotting attacks on multiple targets in Hyderabad and had purchased chemicals to make high-grade explosives for the planned operations. In May, the Indian National Intelligence Agency arrested senior IM member Abdul Wahid Siddibapa, who was connected to several IM plots to attack India and known as a recruiter for the organization. He was linked to the 2006 Mumbai blasts, the 2008 attack in Delhi, and a 2010 attack in Bangalore.

Strength: Unknown

Location/Area of Operation: India, Nepal, and Pakistan

Funding and External Aid: Suspected to obtain funding and support from other terrorist organizations, as well as from sources in Pakistan and the Middle East.

Islamic Jihad Union

Also known as Islamic Jihad Group; Islomiy Jihod Ittihodi; al-Djihad al-Islami; Dzhamaat Modzhakhedov; Islamic Jihad Group of Uzbekistan; Jamiat al-Jihad al-Islami; Jamiyat; The Jamaat Mojahedin; The Kazakh Jama'at; The Libyan Society

Description: The Islamic Jihad Union (IJU) is a Sunni violent extremist organization that was designated as a Foreign Terrorist Organization on June 17, 2005. The group splintered from the Islamic Movement of Uzbekistan in the early 2000s and has been based in Pakistan's Federally Administered Tribal Areas. Najmiddin Jalolov founded the organization as the Islamic Jihad Group in March 2002, but the group was renamed Islamic Jihad Union in May 2005. Although IJU remains committed to overthrowing the Government of Uzbekistan, it also has a global agenda, demonstrated by its attacks on international forces in Afghanistan.

Activities: The IJU primarily operates against international forces in Afghanistan and remains a threat to Central Asia. IJU claimed responsibility for attacks in March and April 2004 in Uzbekistan, which targeted police at several roadway checkpoints and at a popular bazaar, killing approximately 47 people, including 33 IJU members, some of whom were suicide bombers. In July 2004, the group carried out near-simultaneous suicide bombings of the Uzbek Prosecutor General's office and the U.S. and Israeli Embassies in Tashkent. In September 2007, German authorities disrupted an IJU plot to attack U.S. military bases and personnel by detaining and prosecuting three IJU operatives, including two German citizens. Foreign terrorist fighters from Germany, Turkey, and elsewhere in Europe continued to travel to the Afghanistan-Pakistan border area to join IJU to fight against U.S. and Coalition Forces. In 2013, two IJU videos showed attacks against an American military base in Afghanistan and an IJU sniper shooting an Afghan soldier at a base in Afghanistan.

According to statements and photos released by the group, IJU participated in the April-September 2015 Taliban siege of Kunduz city. At least 13 police were killed in the attacks and, according to the United Nations, some 848 civilians were killed or wounded. In June 2015, IJU claimed it carried out multiple attacks in eastern Afghanistan with al-Qa'ida and the Taliban, including an attack on an Afghan military base that killed "many apostate soldiers." In July 2015, the IJU stated it had expanded its base of operations and was engaged in activity in southern Afghanistan (in the provinces of Paktia, Paktika, and Nangarhar) and in Badakhshan and Kunduz. On August 20, IJU pledged allegiance to the then newly appointed Taliban leader Mullah Mansour who was killed in a U.S .drone strike in May 2016 and succeeded by Mawalwi Hibatullah Akhundzada.

Strength: 100 to 200 members

Location/Area of Operation: Afghanistan and Pakistan; IJU members are also scattered throughout Central Asia and Europe.

Funding and External Aid: Unknown

Islamic Movement of Uzbekistan

Also known as IMU

Description: Designated as a Foreign Terrorist Organization on September 25, 2000, the Islamic Movement of Uzbekistan (IMU) seeks to overthrow the Uzbek government and establish an Islamic state. For most of the past decade, however, the group has recruited members from other Central Asian states and Europe. Despite its objective to set up an Islamic state in Uzbekistan, the group operates primarily along the Afghanistan-Pakistan border and in northern Afghanistan, where it fights against international forces. Several IMU members are also suspected of having traveled to Syria to fight with violent extremist groups.

The IMU has had a decade-long relationship with al-Qa'ida (AQ), the Taliban, and Tehrik-e Taliban Pakistan (TTP). Top IMU leaders have integrated themselves into the Taliban's shadow government in Afghanistan's northern provinces.

In August 2015, IMU leader Usman Ghazi publicly announced the group's shift of allegiance to ISIS. Numerous IMU members, including possibly Ghazi himself, were subsequently reported to have been killed as a result of hostilities between ISIS and its former Taliban allies.

Activities: Since the beginning of Operation Enduring Freedom, the IMU has predominantly been focused on attacks against international forces in Afghanistan. In late 2009, NATO forces reported an increase in IMU-affiliated foreign terrorist fighters in Afghanistan. In 2010, IMU claimed responsibility for the September 19 ambush that killed 25 Tajik troops in Tajikistan. On October 15, 2011, IMU claimed responsibility for a suicide assault on a U.S.-led Provincial Reconstruction Team in Panjshir Province, Afghanistan, that killed two Afghan civilians and wounded two security guards. On June 8, 2014, IMU claimed responsibility for participating in an attack on Karachi's international airport that resulted in the deaths of at least 39 people.

Throughout 2015, the IMU actively threatened the Afghan government, specifically in the northern part of the country. In April, the group released a video showing IMU members beheading an individual they claimed to be an Afghan soldier and threatened to behead Hazara (a historically persecuted ethnic group in Afghanistan that follows Shia Islam) hostages, in supposed retaliation for the Afghan security forces capture of several female members of IMU. Also in 2015, Uzbek refugee Fazliddin Kurbanov was convicted and sentenced by a U.S. federal court to 25 years in prison for planning a bomb attack in Idaho. Kurbanov had been in contact with members of IMU online, seeking advice on how to make explosives and formulate attack plans. He discussed attacking U.S. military bases and the U.S. Military Academy at West Point. In late 2016, a dual Dutch-Turkish citizen who served as an IMU fundraiser and facilitator was convicted and sentenced in the United States for providing material support to the IMU and acting on the group's behalf in Afghanistan, France, Jordan, the Netherlands, Pakistan, and elsewhere, between 2006 and 2008.

In June 2016, a faction of IMU announced its continued commitment to the Taliban and AQ, marking a split with its leader Ghazi and the rest of the group, which announced its loyalty to ISIS in 2015 and has since cooperated with Islamic State's Khorasan Province.

Strength: 200 to 300 members

Location/Area of Operation: Central Asia, Iran, Afghanistan, Pakistan

Funding and External Aid: The IMU receives support from a large Uzbek diaspora, terrorist organizations, and donors from Europe, Central and South Asia, and the Middle East.

Islamic State of Iraq and Syria (ISIS)

Also known as al-Qa'ida in Iraq; al-Qa'ida Group of Jihad in Iraq; al-Qa'ida Group of Jihad in the Land of the Two Rivers; al-Qa'ida in Mesopotamia; al-Qa'ida in the Land of the Two Rivers; al-Qa'ida of Jihad in Iraq; al-Qa'ida of Jihad Organization in the Land of the Two Rivers; al-Qa'ida of the Jihad in the Land of the Two Rivers; al-Tawhid; Jam'at al-Tawhid Wa'al-Jihad; Tanzeem Qa'idat al Jihad/Bilad al Raafidaini; Tanzim Qa'idat al-Jihad fi Bilad al-Rafidayn; The Monotheism and Jihad Group; The Organization Base of Jihad/Country of the Two Rivers; The Organization Base of Jihad/Mesopotamia; The Organization of al-Jihad's Base in Iraq; The Organization of al-Jihad's Base in the Land of the Two Rivers; The Organization of al-Jihad's Base of Operations in Iraq; The Organization of al-Jihad's Base of Operations in the Land of the Two Rivers; The Organization of Jihad's Base in the Country of the Two Rivers; al-Zarqawi Network; Islamic State in Iraq; Islamic State in Iraq and al-Sham; Islamic State in Iraq and Syria; ad-Dawla al-Islamiyya fi al-'Iraq wa-sh-Sham; Daesh; Dawla al Islamiya; Al-Furqan Establishment for Media Production; Islamic State; ISIL; ISIS

Description: Al-Qa'ida in Iraq (AQI) was designated as a Foreign Terrorist Organization on December 17, 2004. In the 1990s, Jordanian militant Abu Mus'ab al-Zarqawi organized a terrorist group called al-Tawhid wal-Jihad to oppose the presence of U.S. and Western military forces in the Middle East and the West's support for, and the existence of, Israel. In late 2004, he joined al-Qa'ida (AQ) and pledged allegiance to Osama bin Laden, at which point al-Tawhid wal-Jihad became known as al-Qa'ida in Iraq (AQI). Zarqawi traveled to Iraq during Operation Iraqi Freedom and led his group against U.S. and Coalition Forces until his death in June 2006.

In October 2006, AQI publicly renamed itself the Islamic State in Iraq and in 2013 it adopted the moniker ISIS to express its regional ambitions as it expanded its operations to include the Syrian conflict. On May 15, 2014, the Department of State amended the Foreign Terrorist Organization designation of AQI to add several aliases, including ISIS. ISIS is led by Abu Bakr al-Baghdadi, aka Ibrahim Awwad Ibrahim Ali al-Badri, aka Abu Du'a. In June 2014, ISIS leader al-Baghdadi declared an Islamic caliphate.

Activities: As AQI, ISIS conducted numerous high profile attacks, including improvised explosive device attacks against U.S. military personnel and Iraqi infrastructure, videotaped beheadings of U.S. citizens, suicide bombings against both military and civilian targets, and rocket attacks. ISIS perpetrates suicide and mass casualty bombings in Iraq using foreign and Iraqi operatives. In 2014, ISIS was responsible for the majority of deaths of the more than 12,000 Iraqi civilians killed that year. ISIS was heavily involved in the fighting in Syria during 2014, including against other militant opposition groups, and participated in a number of kidnapping incidents against civilians, including aid workers and reporters.

ISIS remained active throughout 2015 and 2016, conducting several large-scale attacks in Iraq and Syria and across the globe. In February 2016, a series of ISIS suicide and car bombs killed at least 129 people in Homs and Damascus, Syria. In March, ISIS carried out a suicide attack at a crowded park in Iskandariya, Iraq at the end of a football match, killing approximately 29 and wounding more than 60 others. Also in March, at least 60 people were killed in an attack claimed by ISIS in Hilla, Iraq when an explosives-laden fuel tanker ran into an Iraqi security checkpoint.

In early May 2016, two suicide car bombs claimed by ISIS killed 32 and wounded another 75 in Samawa, in southern Iraq. In mid-May, ISIS conducted a series of attacks in and around Baghdad, including suicide bombings and a car bombing at a crowded market in Sadr City that killed at least 88 people—most of them women and children. In late May, ISIS killed 12 people when the group attacked a Real Madrid Supporters' Club in Iraq where people had gathered to watch the Champions League final match. In July, ISIS claimed a car bombing at a popular shopping center in Baghdad that killed nearly 300 people, making it the single deadliest bombing in Iraq's capital city since 2003. In October, it was revealed that ISIS was using hundreds to thousands of Iraqi civilians as human shields. The group reportedly rounded up and massacred 284 men and boys before dumping their bodies in a mass grave in northern Mosul, Iraq.

The group since at least 2015 has also been responsible for integrating local children and ones brought by foreign terrorist fighters into its forces and using them as executioners and suicide attackers. ISIS has systematically prepared child soldiers in Iraq and Syria using its education and religious infrastructure as part of its training and recruitment of members.

In Iraq and Syria, ISIS's use of military equipment captured in the course of fighting gave the group greater capabilities in line with a more conventional military force, including the reported use of eastern bloc tanks, artillery, and self-developed unmanned aerial drones. According to estimates from the UN Assistance Mission for Iraq, acts of terrorism, violence, and armed conflict killed more than 6,000 civilians and injured more than 11,300 in 2016.

ISIS continued to commit atrocities against groups in areas under its control, including Yezidis, Christians, Shia Muslims, Sunni Muslims, Kurds, and other groups. In 2015 and 2016, ISIS abducted, systematically raped, and abused thousands of women and children, some as young as eight years of age. Women and children were sold and enslaved, distributed to ISIS fighters as spoils of war, forced into marriage and domestic servitude, or subjected to physical and sexual abuse. ISIS established "markets" where women and children were sold with price tags attached and has published a list of rules on how to treat female slaves once captured. (For further information, refer to *the Trafficking in Persons Report 2016,* http://www.state.gov/j/tip/rls/tiprpt/2015/index.htm.)

ISIS also directs, enables, and inspires individuals to conduct attacks and act on behalf of the group around the world, including in the United States and Europe. In November 2015, ISIS carried out a series of coordinated attacks in Paris, France, including at a rock concert at the Bataclan concert hall, killing approximately 130 people and injuring more than 350 others; 23-year-old U.S. citizen Nohemi Gonzalez was among the dead. In March 2016, ISIS directed two simultaneous attacks in Brussels, Belgium—one at the Zaventem Airport and the other at a metro station. The attacks killed 32 people, including four U.S. citizens, and injured more than 250 people. In June 2016, a gunman who pledged allegiance to ISIS killed 49 individuals and injured 53 others at the Pulse nightclub in Orlando, Florida. In July 2016, ISIS claimed an attack in which a cargo truck drove into a crowd in Nice, France, during Bastille Day celebrations, resulting in 86 deaths, including three U.S. citizens. In December 2016, ISIS claimed responsibility for a truck attack on a crowded Christmas market in Berlin, Germany that killed 12 people and injured 48 others.

Although ISIS has not claimed responsibility, it was likely responsible for several attacks involving chemical-filled munitions in Iraq and Syria during 2016. The United States has been proactively working with our allies to dismantle this chemical weapons capability, as well as deny ISIS and other non-state actors access to chemical, biological, radiological, and nuclear (CBRN)-useable materials and expertise through interdictions and strengthening the ability of regional governments to detect, disrupt, and respond effectively to suspected CBRN activity.

In July 2017, the group lost control of its largest city, Mosul, to the Iraqi army. ISIS continued to lose territory to the various states and other military forces allied against it, until it controlled no meaningful territory by November 2017. U.S. military officials reported in December 2017 that the group retained a mere 2 percent of the territory they had previously held. On 10 December 2017, Iraq's prime minister Haider al-Abadi announced that Iraqi forces had driven the last remnants of Islamic State from the country, three years after the militant group captured about a third of Iraq's territory. President Putin also announced that ISIS had been driven from Syria.

Strength: Estimates suggest between 12,000 and 15,000 members.

Location/Area of Operation: ISIS's operations were predominately in Iraq and Syria, but 2015 and 2016 witnessed the continued creation of external ISIS branches and networks. In 2016, several branches of ISIS were designated as Foreign Terrorist Organizations (FTOs) and Specially Designated Global Terrorists (SDGTs): ISIL-Libya as an FTO and SDGT in 2016; ISIS in Yemen as an SDGT in 2016; and ISIS in Saudi Arabia as an SDGT in 2016. Islamic State's Khorasan Province was designated as an FTO and SDGT in 2015. ISIS branches in the Caucasus Province and Algeria were also designated as SDGTs in 2015. In addition, supporters and associates worldwide inspired by the group's ideology may be operating without direction from ISIS central leadership.

Funding and External Aid: ISIS receives most of its funding from a variety of businesses and criminal activities. Criminal activities include robbing banks, smuggling oil, looting and selling antiquities and other goods, as well as extortion, human trafficking, and kidnapping-for-ransom.

Islamic State's Khorasan Province

Also known as ISIL Khorasan; Islamic State's Khorasan Province; ISIS Wilayat Khorasan; ISIL's South Asia Branch; South Asian chapter of ISIL

Description: Islamic State's Khorasan Province (ISIS-K) was designated as a Foreign Terrorist Organization on January 14, 2016. The group is based in Afghanistan, conducts operations in Afghanistan and Pakistan, and is composed primarily of former members of Tehrik-e Taliban Pakistan and the Afghan Taliban. ISIS-K's senior leadership has pledged allegiance to ISIS leader Abu Bakr al-Baghdadi. Since Baghdadi accepted the pledge in late January 2015, ISIS-K has carried out suicide bombings, small arms attacks, and kidnappings in Afghanistan against civilians and Afghan National Security and Defense Forces. The group has also claimed responsibility for attacks on civilians in Pakistan.

It was reported that ISIS-K leader Hafiz Saeed Khan was killed in July 2016. Khan's former deputy and the former Taliban Commander from Logar Province, Abdul Hasib, took over leadership responsibilities for ISIS-K.

Activities: ISIS-K was responsible for several attacks on civilians in Karachi, Pakistan in May 2015. In January 2016, the group claimed it carried out a strike on the Pakistani consulate in Afghanistan, resulting in the deaths of seven Afghan security personnel. In July, the group conducted a bomb attack at a peaceful protest in Kabul, Afghanistan, that killed approximately 80 people and wounded another 230. In August, ISIS-K claimed it carried out a shooting and suicide bombing at a hospital in Quetta targeting lawyers, which killed 94. In October, ISIS-K killed 18 worshippers at a shrine in Kabul, and in November, carried out a suicide bombing at a mosque in Kabul, killing at least 30. It also claimed responsibility in November 2016 for a suicide bombing at the Shah Noorani Shrine in Baluchistan province, Pakistan, killing more than 50 people.

Strength: Approximately 1,500 to 2,000

Location/Area of Operation: Afghanistan and Pakistan

Funding and External Aid: ISIS-K receives some funding from ISIS. Other funding sources are unknown.

ISIL-Libya

Also known as Islamic State of Iraq and the Levant-Libya; Islamic State of Iraq and the Levant in Libya; Wilayat Barqa; Wilayat Fezzan; Wilayat Tripolitania; Wilayat Tarablus; Wilayat al-Tarabulus

Description: ISIL-Libya was designated as a Foreign Terrorist Organization on May 20, 2016. In 2014, ISIS leader and SDGT Abu Bakr al-Baghdadi dispatched a group of ISIS operatives from Syria to Libya to establish a branch of the terrorist group. The ISIS operatives set up a base in Darnah and in late November 2014, Baghdadi formally established the ISIL-Libya branch after announcing he had accepted oaths of allegiance from fighters in the country.

Activities: ISIL-Libya poses a significant security threat in the region. Since being established in November 2014, the group has carried out multiple attacks in the country and has threatened to expand ISIS's presence into other countries in Africa. In October 2014, several hundred ISIL–Libya fighters established a presence in the eastern Libya city of Darnah. In January 2015, ISIL-Libya claimed responsibility for a suicide attack on a luxury hotel in Tripoli that killed eight people, including a U.S. contractor. That same month, ISIL-Libya also killed 16 people in an attack on a Libyan army checkpoint.

In February 2015, ISIL-Libya released a propaganda video showing the murder of 21 Egyptian Coptic Christians who had been kidnapped in Sirte, Libya, in two separate incidents in December 2014 and January 2015. Also in February, ISIL-Libya claimed responsibility for bomb attacks against a petrol station, a police station, and the home of parliamentary speaker Agila Salah in the town of al-Qubbah. The attacks killed at least 40 people and wounded dozens of others.

ISIL-Libya claimed responsibility for a suicide bombing targeting an army checkpoint in Benghazi in March 2015, which killed five soldiers and two civilians, and it attacked a Tripoli prison in an attempt to free detainees in September 2015, killing three security personnel.

Between 2015 and 2016, ISIL-Libya doubled its presence in the country; in early 2016, reports suggested the group counted as many as 6,000 fighters in its ranks. In 2016, ISIL-Libya expanded operations into Libya's oil crescent, launching attacks on some of the country's largest oil installations; ISIL-Libya burned oil tanks, killing dozens and forcing oil fields to shut down operations. The group also conducted attacks on Libya's medical facilities. In March, the group attacked a Sirte hospital and kidnapped 20 medical workers. The group also attacked medical facilities and carried out other attacks in Benghazi and Tripoli. In November, the group was responsible for a car bomb that exploded in Benghazi, killing three.

In December 2016, Libyan forces drove ISIL-Libya from its stronghold and main base in Sirte into the desert areas and neighboring cities.

Strength: In early 2016, ISIL-Libya was reported to have 6,000 members.

Location/Area of Operation: At the end of 2016, ISIL-Libya was no longer in control of any towns in Libya, but its members continued to operate throughout the eastern, southern, and western regions of the country. The group has also carried out attacks in Tripoli and Benghazi.

Funding and External Aid: ISIL-Libya's funding comes from a variety of sources, including criminal activity, such as smuggling and extortion, and external funding. The group also receives support from ISIS.

ISIL-Sinai Province

Also known as Ansar Bayt al-Maqdis; Ansar Jerusalem; Supporters of Jerusalem; Ansar Bayt al-Maqdes; Ansar Beit al-Maqdis; ISIL Sinai Province;

Islamic State-Sinai Province; Islamic State in the Sinai; Jamaat Ansar Beit al-Maqdis; Jamaat Ansar Beit al-Maqdis fi Sinaa; Sinai Province; Supporters of the Holy Place; The State of Sinai; Wilayat Sinai

Description: Originally designated as a Foreign Terrorist Organization on April 9, 2014, Ansar Bayt al-Maqdis (ABM) rose to prominence in 2011 following the uprisings in Egypt. It is responsible for attacks against Israeli and Egyptian government and security elements, and against tourists in Egypt. In November 2014, ABM officially declared allegiance to the Islamic State of Iraq and Syria (ISIS). In September 2015, the Department of State amended ABM's designation to add the primary name ISIL Sinai Province (ISIL-SP).

Activities: ISIL-SP has claimed responsibility for numerous attacks against Israeli interests, including a July 2012 attack against a Sinai pipeline exporting gas to Israel, an August 2012 rocket strike on the southern Israeli city of Eilat, and a September 2012 attack targeting an Israeli border patrol, which killed one soldier and injured another.

In October 2013, ISIL-SP claimed responsibility for a suicide bombing targeting the South Sinai Security Directorate in el Tor, which killed three people and injured more than 45. In January 2014, ISIL-SP successfully downed an Egyptian military helicopter in a missile attack, killing five soldiers on board. In 2014, the group also claimed responsibility for four attacks involving car bombs and hand grenades in Cairo, which left six people dead and more than 70 wounded.

ISIL-SP has targeted government officials, including the September 2013 attempted assassination of the Egyptian Interior Minister, and the January 2014 assassination of the head of the Interior Minister's technical office. In February 2014, ISIL-SP claimed responsibility for the bombing of a tour bus in the Sinai Peninsula, killing the Egyptian driver and three South Korean tourists; the attack was ISIL-SP's first against foreign tourists.

In October 2014, ISIL-SP claimed a strike on a security checkpoint that killed 33 Egyptian soldiers and wounded 26 others, including civilians. In April 2015, ISIL-SP claimed responsibility for three separate attacks on Egyptian security forces on the same day in Northern Sinai; the strikes killed 12 people, including one civilian. On November 4, ISIL-SP released an audio recording in which it claimed responsibility for the October 31 bombing of a Russian passenger plane carrying 224 people from the Egyptian resort town of Sharm el-Sheikh to St. Petersburg, Russia. All 224 passengers and seven crew members were killed. ISIL-SP also claimed a suicide bombing that killed four Egyptian police officers and wounded at least five civilians in northern Sinai in November 2015.

On August 5, 2015, ISIL-SP claimed responsibility for the July 22 abduction of Croatian citizen Tomislav Salopek, who worked as a topographer for a French energy company. Salopek was kidnapped on October 6 in the western desert, approximately 20 km west of the suburbs of Cairo. ISIL-SP demanded the release of all female Muslims in Egyptian prisons within 48 hours in exchange for Salopek. Salopek was ultimately beheaded and ISIL-SP claimed responsibility for the killing.

Throughout the course of 2016, ISIL-SP carried out numerous attacks in the Sinai, including a January double bombing that killed two Egyptian policemen and two military officers. The group was also responsible for a July wide-scale coordinated attack on several military checkpoints that reportedly killed more than 50 people and a separate July mortar attack on a military checkpoint near the town of Skeikh Zuweid that killed five members of the Egyptian security forces. In November, ISIL-SP assassinated two Egyptian generals in North Sinai, and carried out a car bomb attack on a checkpoint in North Sinai that killed at least 12 Egyptian soldiers.

In February 2017, ISIL-SP fired four rockets from Egyptian territory into Israel and in April it was reported that an Egyptian air strike had killed 19 ISIL-SP fighters including three leaders. In November 2017, about 40 gunmen attacked the al-Rawda mosque killing 311 people and injuring more than 120. ISIL-SP claimed responsibility.

Strength: ISIL-SP is estimated to have several hundred fighters in the Sinai Peninsula and affiliated cells in the Nile Valley.

Location/Area of Operation: ISIL-SP operations are based out of the Sinai Peninsula, but its reach extends to Cairo, the Egyptian Nile Valley, as well as possibly more than 200 militants in Gaza.

Funding and External Aid: Although the source of ISIL-SP's funding is largely unknown, there are indications that it may receive funding from external actors.

Jama'atu Ansarul Muslimina fi Biladis-Sudan (Ansaru)

Also known as Ansaru; Ansarul Muslimina Fi Biladis Sudan; Vanguards for the Protection of Muslims in Black Africa; JAMBS; Jama'atu Ansaril Muslimina Fi Biladis Sudan

Description: Designated as a Foreign Terrorist Organization on November 14, 2013, Jama'atu Ansarul Muslimina fi Biladis-Sudan (Ansaru) publicly splintered from Boko Haram in January 2012. Ansaru's leadership structure is unclear, although it is known Khalid al-Barnawi held a top leadership position within the group until his alleged capture by the Nigerian army in April 2016. Since its inception, Ansaru has targeted civilians, including Westerners, and Nigerian government and security officials. Ansaru purportedly aims to defend Muslims throughout all of Africa by fighting against the Nigerian government and international interests. While Ansaru claims to identify with

Boko Haram's objectives and struggle, it has criticized the group for killing fellow Muslims.

Activities: In November 2012, Ansaru raided a police station in Abuja, killing Nigerian police officers and freeing detained terrorists from prison. In January 2013, Ansaru attacked a convoy of Nigerian peacekeepers on their way to Mali.

Ansaru has also carried out multiple kidnapping operations targeting civilians. In late 2012, Ansaru kidnapped a French engineer claiming the action was justified due to French involvement in Mali. In early 2013, Ansaru kidnapped and subsequently killed seven international construction workers.

Ansaru did not publicly claim any attacks between 2014 and 2016. On April 4, 2016, the Nigerian army announced the capture of Ansaru leader Terrorist Khalid al-Barnawi.

Strength: Total membership is unknown. Given its narrower scope of operations, it is estimated that Ansaru's membership is much smaller than that of Boko Haram.

Location/Area of Operation: Northern Nigeria

Funding and External Aid: Unknown

Jaish-e-Mohammed

Also known as the Army of Mohammed; Mohammed's Army; Tehrik ul-Furqaan; Khuddam-ul-Islam; Khudamul Islam; Kuddam e Islami; Jaish-i-Mohammed

Description: Pakistan-based Jaish-e-Mohammed (JeM) was designated as a Foreign Terrorist Organization on December 26, 2001. JeM was founded in early 2000 by former senior Harakat ul-Mujahideen leader Masood Azhar upon his release from prison in India in exchange for 155 hijacked Indian Airlines hostages. The group aims to annex the state of Jammu and Kashmir to Pakistan and expel international forces from Afghanistan. JeM has openly declared war against the United States.

Activities: JeM has continued to operate openly in parts of Pakistan despite the country's 2002 ban on its activities. Since its founding, JeM has conducted many fatal terrorist attacks in the region. JeM claimed responsibility for several suicide car bombings in the state of Jammu and Kashmir, including an October 2001 suicide attack on the Jammu and Kashmir legislative assembly building in Srinagar that killed more than 30 people. The Indian government has publicly implicated JeM, along with Lashkar e-Tayyiba (LeT), in the December 2001 attack on the Indian Parliament that killed nine people and injured 18 others.

In 2002, Pakistani authorities arrested and convicted a JeM member for the abduction and murder of U.S. journalist Daniel Pearl. Pakistani authorities reportedly suspect that JeM members may have been involved in the 2002 anti-Christian attacks in Islamabad, Murree, and Taxila that killed two U.S. citizens. In December 2003, Pakistan implicated JeM members in two assassination attempts against then-president Pervez Musharraf. In December 2013, JeM threatened to kill Indian politician Narendra Modi if he were elected prime minister. In 2015, JeM launched three small-scale attacks targeting Indian army camps; approximately five to 10 people were killed in the attacks.

In 2016, Indian officials blamed JeM for a January attack on an Indian Air Force base in Pathankot. One civilian and seven Indian security force personnel were killed, and an additional 20 security force personnel were wounded in the strike. Also in January, JeM was suspected of conducting an attack on the Indian consulate in Mazar-e-Sharif, Afghanistan. From February through May 2016, both LeT and JEM were suspected of engaging in at least three firefights with Indian security forces in Kupwara district, Jammu and Kashmir, injuring approximately two Indian personnel. In June 2016, it is unclear whether LeT or JEM were responsible for an ambush on an Indian security force convoy in Pulwama district, Jammu and Kashmir, killing 8 and injuring 20. In December 2016, the Indian National Intelligence Agency announced that JeM leader Azhar and his brother, Rauf Asghar, were directly implicated in the attack. Attacks have continued.

Strength: JeM has at least several hundred armed supporters.

Location/Area of Operation: India, including the state of Jammu and Kashmir; Afghanistan; and Pakistan, particularly southern Punjab.

Funding and External Aid: To avoid asset seizures by the Pakistani government, since 2007 JeM has withdrawn funds from bank accounts and invested in legal businesses, such as commodity trading, real estate, and the production of consumer goods. JeM also collects funds through donation requests in magazines and pamphlets, sometimes using charitable causes to solicit donations.

Jaysh Rijal Al-Tariq Al-Naqshabandi (JRTN)

Also known as Jaysh Rijal al-Tariq al-Naqshabandi; Army of the Men of the Naqshbandi Order; Armed Men of the Naqshabandi Order; Naqshbandi Army; Naqshabandi Army; Men of the Army of al-Naqshbandia Way; Jaysh Rajal al-Tariqah al-Naqshbandia; JRTN; JRN; AMNO.

Description: Jaysh Rijal al-Tariq al-Naqshabandi (JRTN) was designated as a Foreign Terrorist Organization in September 2015. The group first announced insurgency operations against international forces in Iraq in December 2006 in response to the execution of Saddam Hussein. Izzat Ibrahim al-Douri, former vice president of Saddam Hussein's Revolutionary

Council, leads the group, which consists of former Baath Party officials, military personnel, and Sunni nationalists. Al-Douri was once one of the most-wanted men in Iraq by Coalition Forces. JRTN aims to overthrow the government of Iraq, install a new Ba'athist regime, and to end external influence in Baghdad. JRTN believes in Iraqi and Arab secular nationalism and Naqshabandi Sufi Islam ideals.

Activities: Between its founding in 2006 and the 2011 withdrawal of Coalition Forces from Iraq, JRTN claimed, and distributed numerous videos of, attacks on U.S. bases and forces. JRTN is also known to have used vehicle-borne improvised explosive devicess against Iraqi government security forces. JRTN's influence has grown by exploiting the disintegration of other insurgent groups.

In 2014, elements of JRTN joined military forces with ISIS in opposition to the Iraqi government. The group played a major role in the capture of Mosul from Iraqi security forces in 2014. Elements of JRTN continued cooperating with ISIS in 2016, but did not publicly claim any specific attacks.

Strength: The most recent membership numbers available are 1,500 to 5,000, as of 2011. Membership is almost certainly much lower today.

Location/Area of Operation: Primarily Iraq

Funding and External Aid: JRTN historically received funding from former regime members, major tribal figures in Iraq, and external contributions from the Gulf.

Jemaah Anshorut Tauhid

Also known as JAT; Jemmah Ansharut Tauhid; Jem'mah Ansharut Tauhid; Jamaah Ansharut Tauhid; Jama'ah Ansharut Tauhid; Laskar 99

Description: The Department of State designated Jemaah Anshorut Tauhid (JAT) as a Foreign Terrorist Organization on March 13, 2012. Formed in 2008, the Indonesia-based group seeks to establish an Islamic caliphate in Indonesia, and has carried out numerous attacks on Indonesian government personnel, police, military, and civilians. In 2011, Abu Bakar Ba'asyir, the founder and leader of JAT, was convicted and sentenced to 15 years in prison for his role in organizing a militant training camp in Aceh. Ba'asyir is also the co-founder and former leader of Jemaah Islamiya (JI). JAT maintains ties to JI and other terrorist groups in Southeast Asia.

Activities: JAT has conducted multiple attacks targeting civilians and Indonesian officials, resulting in the deaths of numerous Indonesian police and innocent civilians. In October 2012, authorities blamed JAT for torturing and killing two police officers investigating an alleged terrorist camp linked to JAT in Poso. In December 2012, four police officers were killed and two wounded in an attack by suspected local JAT members in Central Sulawesi.

Since Abu Bakar Ba'asyir's pledge of allegiance to ISIS in 2014, many JAT members have joined Indonesia's ISIS-affiliated groups; others have joined al-Qa'ida-affiliated groups.

It is thought that JAT members continue to be involved in ISIS operations in Southeast Asia.

Strength: JAT is estimated to have several thousand supporters and members, although internal disagreements over aligning with ISIS have likely reduced its membership.

Location/Area of Operation: Indonesia

Funding and External Aid: JAT raises funds through membership donations and legitimate business activities. JAT has also conducted cyber hacking, robbed banks, and carried out other illicit activities to fund the purchase of assault weapons, ammunition, explosives, and bomb-making materials.

Jemaah Islamiya

Also known as Jemaa Islamiyah; Jema'a Islamiyah; Jemaa Islamiyya; Jema'a Islamiyya; Jemaa Islamiyyah; Jema'a Islamiyyah; Jemaah Islamiah; Jemaah Islamiyah; Jema'ah Islamiyah; Jemaah Islamiyyah; Jema'ah Islamiyyah; JI

Description: Designated as a Foreign Terrorist Organization on October 23, 2002, Jemaah Islamiya (JI) is a Southeast Asia-based terrorist group co-founded by Abdullah Sungkar and Abu Bakar Ba'asyir. The group seeks to establish an Islamic caliphate spanning Brunei, Indonesia, Malaysia, Singapore, the southern Philippines, and southern Thailand. More than 400 JI operatives have been captured or killed since 2002, including operations chief and al-Qa'ida (AQ) associate Hambali and, in January 2015, bomb-maker Zulfiki bin Hir (aka Marwan). The group has been linked to AQ in the past.

Activities: JI has carried out many attacks in the region. Significant JI attacks include the 2002 Bali bombings, which killed more than 200 people, among them seven U.S. citizens; the August 2003 bombing of the J.W. Marriott Hotel in Jakarta; the September 2004 bombing outside the Australian Embassy in Jakarta; and the October 2005 suicide bombing in Bali, which left 26 dead, including the three suicide bombers.

On July 17, 2009, a JI faction led by Noordin Mohamed Top claimed responsibility for suicide attacks at the J.W. Marriott and Ritz-Carlton hotels in Jakarta that killed seven people and injured more than 50, including seven U.S. citizens.

In November 2014, Indonesian authorities released former JI bomb-maker Taufik Abdul Halim from prison after he served a 12-year sentence for attempting to bomb a Jakarta shopping mall in 2001.

In January 2015, 44 Philippines policemen and three civilians were killed when a police counterterrorism squad was ambushed while conducting a raid in Mamasapano in the southern island of Mindanao. That same month, it was reported that JI-linked violent extremists had attempted to carry out attacks during a papal visit to Manila and Tacloban. In October 2015, two senior JI leaders—Zarkashi and JI military leader Abu Dujana—were released from prison after serving seven years each in Indonesian jails. In December 2015, police in East Java arrested four individuals associated with JI involved in manufacturing firearms for terrorist activities.

There were no reported attacks by JI in 2016. JI is believed to be recruiting and strengthening its network and finances, however. In early 2016, it was also reported that the group was sending members to Syria to train.

Strength: Estimates vary from 500 to several thousand members.

Location/Area of Operation: JI has been focusing on its operations and increasing its presence in Indonesia. The group has also carried out attacks in Malaysia and the Philippines in the past.

Funding and External Aid: JI fundraises through membership donations and criminal and business activities. It has received financial, ideological, and logistical support from Middle Eastern contacts and illegitimate charities and organizations.

Jundallah

Also known as People's Resistance Movement of Iran (PMRI); Jonbesh-i Moqavemat-i-Mardom-i Iran; Popular Resistance Movement of Iran; Soldiers of God; Fedayeen-e-Islam; Former Jundallah of Iran; Jundullah; Jondullah; Jundollah; Jondollah; Jondallah; Army of God (God's Army); Baloch Peoples Resistance Movement (BPRM)

Description: Jundallah was designated as a Foreign Terrorist Organization on November 4, 2010. Since its inception in 2003, Jundallah, which operates primarily in the province of Sistan va Balochistan of Iran, and the Baloch areas of Pakistan and Afghanistan, has engaged in numerous attacks, killing and maiming scores of Iranian civilians and government officials. Jundallah's stated goals are to secure recognition of Balochi cultural, economic, and political rights from the Government of Iran, and to spread awareness of the plight of the Baloch people. The 2010 execution of Jundallah's leader by Iran and the killing and arrest of many Jundallah fighters seriously diminished Jundallah's capacity to operate.

Activities: Jundallah claimed responsibility for an October 2009 suicide bomb attack in the Sistan va Balochistan province that killed more than 40 people and was reportedly the deadliest terrorist attack in Iran since the 1980s. In a statement on its website, Jundallah also claimed the December

15, 2010, suicide bomb attack inside the Iman Hussein Mosque in Chabahar, which killed an estimated 35 to 40 civilians and wounded 60 to 100. In July 2010, Jundallah attacked the Grand Mosque in Zahedan, killing approximately 30 and injuring an estimated 300. Jundallah did not publicly claim any attacks in 2016.

Strength: Unknown

Location/Area of Operation: Jundallah has traditionally operated throughout the Sistan va Balochistan province in southeastern Iran and the greater Balochistan area of Afghanistan and Pakistan.

Funding and External Aid: Unknown

Kahane Chai

Also known as American Friends of the United Yeshiva; American Friends of Yeshivat Rav Meir; Committee for the Safety of the Roads; Dikuy Bogdim; DOV; Forefront of the Idea; Friends of the Jewish Idea Yeshiva; Jewish Legion; Judea Police; Judean Congress; Kach; Kahane; Kahane Lives; Kahane Tzadak; Kahane.org; Kahanetzadak.com; Kfar Tapuah Fund; Koach; Meir's Youth; New Kach Movement; Newkach.org; No'ar Meir; Repression of Traitors; State of Judea; Sword of David; The Committee Against Racism and Discrimination (CARD); The Hatikva Jewish Identity Center; The International Kahane Movement; The Jewish Idea Yeshiva; The Judean Legion; The Judean Voice; The Qomemiyut Movement; The Rabbi Meir David Kahane Memorial Fund; The Voice of Judea; The Way of the Torah; The Yeshiva of the Jewish Idea; Yeshivat Harav Meir

Description: Radical Israeli-American Rabbi Meir Kahane founded Kach—the precursor to Kahane Chai—with the aim of restoring Greater Israel (Israel, the West Bank, and Gaza). Its offshoot, Kahane Chai (translation: "Kahane Lives"), was founded by Meir Kahane's son Binyamin, following his father's 1990 assassination. Both organizations were designated as Foreign Terrorist Organizations on October 8, 1997. The group has attempted to gain seats in the Israeli Knesset over the past several decades but has not won a seat since 1984.

Activities: Kahane Chai has harassed and threatened Arabs, Palestinians, and Israeli government officials, and vowed revenge for the December 2000 death of Binyamin Kahane and his wife. The group is suspected of involvement in a number of low-level attacks since the start of the Second Palestinian Intifada in 2000. Since 2003, Kahane Chai activists have physically intimidated Israeli and Palestinian government officials who favored the dismantlement of Israeli settlements. Kahane Chai was last linked to an attack in 2005, when a member of the terrorist group shot dead four people on a bus in Shfaram, Israel. The group is believed to have been dormant since 2016.

Strength: Since 2005, Kahane Chai's core membership has been estimated to be fewer than 100. The group's membership and support networks were overwhelmingly composed of Israeli citizens that lived mostly in West Bank settlements.

Location/Area of Operation: Israel and West Bank settlements, particularly Qiryat Arba in Hebron

Funding and External Aid: Has received support from sympathizers in the United States and Europe.

Kata'ib Hizballah

Also known as Hizballah Brigades; Hizballah Brigades in Iraq; Hizballah Brigades-Iraq; Kata'ib Hezbollah; Khata'ib Hezbollah; Khata'ib Hizballah; Khattab Hezballah; Hizballah Brigades-Iraq of the Islamic Resistance in Iraq; Islamic Resistance in Iraq; Kata'ib Hizballah Fi al-Iraq; Katibat Abu Fathel al-A'abas; Katibat Zayd Ebin Ali; Katibut Karbalah

Description: Formed in 2006 and designated as a Foreign Terrorist Organization on July 2, 2009, Kata'ib Hizballah (KH) is a radical Shia Islamist group with an anti-Western outlook and violent extremist ideology. Prior to the withdrawal of U.S. troops in 2011, the group conducted attacks against Iraqi, U.S., and Coalition targets in Iraq, and threatened the lives of Iraqi politicians and civilians supporting the legitimate political process in Iraq. The group is notable for its extensive use of media operations and propaganda, including filming and releasing videos of attacks. KH has ideological ties to and receives support from Iran.

Activities: KH has claimed responsibility for numerous terrorist attacks since 2007, including improvised explosive device bombings, rocket-propelled grenade attacks, and sniper operations. In 2007, KH gained notoriety for its attacks against U.S. and Coalition Forces in Iraq. In June 2011, five U.S. soldiers were killed in a rocket attack in Baghdad when KH assailants fired between three and five rockets at U.S. military base Camp Victory. The group remained active in 2015, participating in fighting in Syria in support of the Assad regime, and in Iraq against ISIS. In June and July 2015, the group broadcast its recruitment contact information and an appeal for donations on a pro-Iran channel and on YouTube in an effort to recruit fighters to Syria and Iraq.

In 2016, the group continued to fight ISIS by joining the Iraqi Army in the Battle of Mosul. KH ultimately advanced farther west of Mosul to clear the Syria/Iraq border, while other Iranian–backed Shia militias continued fighting closer to the city.

Strength: Membership is estimated as at least 400 individuals.

Location/Area of Operation: Predominately Iraq-based, but also participates in fighting alongside pro-Assad regime forces in Syria. Traditionally,

KH conducted the majority of its operations in Baghdad, but its operations have since expanded into other parts of Iraq.

Funding and External Aid: KH is heavily dependent on support from Iran.

Kurdistan Workers' Party

Also known as the Kurdistan Freedom and Democracy Congress; the Freedom and Democracy Congress of Kurdistan; KADEK; Partiya Karkeran Kurdistan; the People's Defense Force; Halu Mesru Savunma Kuvveti; Kurdistan People's Congress; People's Congress of Kurdistan; KONGRA–GEL.

Description: Founded by Abdullah Ocalan in 1978 as a Marxist-Leninist separatist organization, the Kurdistan Workers' Party (PKK) was designated as a Foreign Terrorist Organization on October 8, 1997. The group, composed primarily of Turkish Kurds, launched a campaign of violence in 1984. The PKK's original goal was to establish an independent Kurdish state in southeastern Turkey. In recent years, however, the PKK has spoken more about autonomy within a Turkish state that guarantees Kurdish cultural and linguistic rights.

Activities: In the early 1990s, the PKK moved beyond rural-based insurgent activities to include urban terrorism. Anatolia became the scene of significant violence, with some estimates suggesting at least 40,000 casualties. Following his capture in 1999, Ocalan announced a "peace initiative," ordering members to refrain from violence and requesting dialogue with Ankara on Kurdish issues. Ocalan's death sentence was commuted to life imprisonment; he remains the symbolic leader of the group. The PKK foreswore violence until June 2004, when its hardline militant wing took control and renounced the self-imposed ceasefire of the previous five years. Striking over the border from bases within Iraq, the PKK engaged in terrorist attacks in eastern and western Turkey. In 2009, the Turkish government and the PKK resumed peace negotiations, but talks broke down after the PKK carried out an attack in July 2011 that left 13 Turkish soldiers dead. In 2012, the PKK claimed responsibility for multiple car bombings resulting in the deaths of at least 10 people. Primary targets included Turkish government security forces, local Turkish officials, and villagers who oppose the organization in Turkey.

Between January and mid-July 2015, the PKK carried out small-scale armed attacks against Turkey's security forces and military bases. A ceasefire, negotiated between 2012 and 2014, ended in July 2015 when the PKK killed two security personnel. During 2016, the PKK continued its terrorist activity in Turkey. Turkish officials blamed the PKK for the August 17 vehicle-borne improvised explosive device (VBIED) attack at a police station in the province of Van, which killed four and wounded 72. The PKK also claimed an August 26 VBIED strike against Sirnak police headquarters, which killed 11 and wounded more than 70.

Clashes continue between the KPP and Turkish military.

Strength: Approximately 4,000 to 5,000 members; 3,000 to 3,500 of which are located in northern Iraq.

Location/Area of Operation: The group is located primarily in Turkey, Iraq, and Syria.

Funding and External Aid: The PKK receives financial support from the large Kurdish diaspora in Europe and from criminal activity.

Lashkar E-Tayyiba

Also known as al Mansooreen; Al Mansoorian; Army of the Pure; Army of the Pure and Righteous; Army of the Righteous; Lashkar e-Toiba; Lashkar-i-Taiba; Paasban-e-Ahle-Hadis; Paasban-e-Kashmir; Paasban-i-Ahle-Hadith; Pasban-e-Ahle-Hadith; Pasban-e-Kashmir; Jamaat-ud-Dawa; JUD; Jama'at al-Dawa; Jamaat ud-Daawa; Jamaat ul-Dawah; Jamaat-ul-Dawa; Jama'at-i-Dawat; Jamaiat-ud-Dawa; Jama'at-ud-Da'awah; Jama'at-ud-Da'awa; Jamaati-ud-Dawa; Idara Khidmat-e-Khalq; Falah-i-Insaniat Foundation; FiF; Falah-e-Insaniat Foundation; Falah-e-Insaniyat; Falah-i-Insaniyat; Falah Insania; Welfare of Humanity; Humanitarian Welfare Foundation; Human Welfare Foundation; Al-Anfal Trust; Tehrik-e-Hurmat-e-Rasool; Tehrik-e-Tahafuz Qibla Awwal; Al-Muhammadia Students; Al-Muhammadia Students Pakistan; AMS

Description: Designated as a Foreign Terrorist Organization (FTO) on December 26, 2001, Lashkar e-Tayyiba (LeT) is one of the largest and most proficient of the anti-India-focused terrorist groups. LeT formed in the late 1980s as the terrorist wing of the Islamist extremist organization Markaz ud Dawa ul-Irshad, a Pakistan-based Islamic fundamentalist mission organization and charity originally formed to oppose the Soviet presence in Afghanistan. LeT is led by Hafiz Muhammad Saeed and is not connected to any political party. Shortly after LeT was designated as an FTO, Saeed changed the group's name to Jamaat-ud-Dawa (JUD) and launched humanitarian projects to circumvent restrictions. LeT disseminates its message through JUD's media outlets. In addition to the creation of JUD, LeT has repeatedly changed its name in an effort to avoid sanctions; other LeT aliases and front groups include Al-Anfal Trust, Tehrik-e-Hurmat-e-Rasool, and Tehrik-e-Tahafuz Qibla Awwal. In December 2016, the Department of State also amended the LeT FTO designation to add the alias Al Muhammadia Students.

Elements of LeT and Jaish-e-Muhammad (JeM) combine with other groups to mount attacks. The Pakistani government banned LeT in January 2002 and temporarily arrested LeT's leader Hafiz Saeed following the 2008

Mumbai attack. LeT and Saeed continued to spread terrorist rhetoric and virulent hate speech condemning India, Israel, the United States, and other perceived enemies in 2016.

Activities: LeT has conducted operations, including several high profile attacks, against Indian troops and civilian targets in the state of Jammu and Kashmir since 1993. The group has also attacked Coalition Forces in Afghanistan. LeT uses assault rifles, machine guns, mortars, explosives, and rocket-propelled grenades.

LeT was responsible for the November 2008 attacks in Mumbai against luxury hotels, a Jewish center, a train station, and a popular café that killed 166 people—including six U.S. citizens—and injured more than 300. India has charged 38 people in the case; most are at large, however, and thought to be in Pakistan.

In March 2010, Pakistani-American businessman David Headley pled guilty in a U.S. court to charges related to his role in the November 2008 LeT attacks in Mumbai, as well as to charges related to a separate plot to bomb the Danish newspaper *Jyllands-Posten*. In May 2011, Headley was a witness in the trial of Tahawwur Rana, who was charged and prosecuted successfully for providing material support to LeT; he was sentenced to 14 years in prison in January 2013. In December 2015, Headley agreed to testify on behalf of the prosecution via videoconference in the trial of Zabiuddin Ansari in India.

LeT was responsible for the May 23, 2014, attack on the Indian consulate in Herat, Afghanistan. Throughout the course of 2014–2015, LeT terrorists also engaged in repeated gun battles with Indian security forces in Jammu and Kashmir.

LeT was behind a July 2015 attack in Gurdaspur, Punjab, which killed seven people. In August 2015, operatives affiliated with LeT attacked Indian security forces in Udhampur, Jammu and Kashmir. In December 2015, LeT carried out an attack on a paramilitary convoy after it left Srinagar, Kashmir; three militants opened fire on the convoy, injuring one civilian and seven Indian military personnel.

From February through May 2016, LeT was suspected of engaging in at least three firefights with Indian security forces in Kupwara district, Jammu and Kashmir, injuring two Indian personnel. In June 2016, LeT was suspected of conducting an ambush on an Indian security force convoy in Pulwama district, Jammu and Kashmir, killing eight and injuring 20. Some media reports also alleged the group's involvement in the September 2016 attack on an Indian Army camp in Uri, Jammu and Kashmir; 20 soldiers were killed in the attack.

Strength: The precise size of LeT is unknown, but it has several thousand members in Azad Jammu and Kashmir and Pakistani Punjab;

Khyber-Pakhtunkhwa and Punjab Provinces in Pakistan; and in the state of Jammu and Kashmir, India.

Location/Area of Operation: LeT has global connections and a strong operational network throughout South Asia. LeT maintains a number of facilities, including training camps, schools, and medical clinics in Pakistan. LeT is also active in Afghanistan.

Funding and External Aid: LeT collects donations in Pakistan and the Gulf as well as from other donors in the Middle East and Europe—particularly the United Kingdom, where it is a designated terrorist organization. In 2016, LeT front organizations continued to openly fundraise in Pakistan and solicited donations in the Pakistani press.

Lashkar I Jhangvi

Also known as Army of Jhangvi; Lashkar e Jhangvi; Lashkar-i-Jhangvi

Description: Designated as a Foreign Terrorist Organization on January 30, 2003, Lashkar I Jhangvi (LJ) is the terrorist offshoot of the Sunni Deobandi sectarian group Sipah-i-Sahaba Pakistan. LJ carries out attacks in Afghanistan and Pakistan, focusing primarily on anti-Shia operations. The Government of Pakistan banned the group in August 2001 as part of an effort to rein in sectarian violence, causing many LJ members to seek refuge in Afghanistan with the Taliban, with whom the group had existing ties. After the collapse of the Taliban government in Afghanistan, LJ members became active in aiding other terrorists; they provided safe houses, false identities, and protection in Pakistani cities, including Karachi, Peshawar, and Rawalpindi. LJ works closely with the Tehrik-e Taliban Pakistan.

Activities: LJ specializes in armed attacks and bombings and has admitted to numerous killings of Shia religious and community leaders in Pakistan. In January 1999, the group attempted to assassinate Prime Minister Nawaz Sharif and his brother Shahbaz Sharif, Chief Minister of Punjab Province. Media reports linked LJ to attacks on Christian targets in Pakistan, including a March 2002 grenade assault on the Protestant International Church in Islamabad that killed two U.S. citizens.

In January 2014, at least 24 people were killed and 40 others wounded in a bus bombing targeting Shia pilgrims; LJ claimed responsibility for the attack. LJ also claimed responsibility for the December 2015 suicide bombing that targeted a market in the predominantly Shia town of Parachinar, Pakistan, that killed at least 23 people and wounded 50.

In November 2016, two individuals suspected of belonging to LJ were arrested by police in Pakistan for their alleged involvement in 25 cases of targeted killings which included the murder of Pakistani singer Amjad Sabri, as well as army and police personnel. In December, Naeem Bukhari was sentenced to death by a Pakistani military court for his involvement in several attacks as the head of a faction of LJ. Several attacks claimed by Islamic State's Khorasan Province in Pakistan in 2016 were suspected of being conducted in collaboration with LJ affiliates.

Strength: Membership is assessed to be in the low hundreds.

Location/Area of Operation: Primarily in Pakistan's Punjab province, the Federally Administered Tribal Areas, Karachi, and Balochistan.

Funding and External Aid: Funding comes from wealthy donors in Pakistan, as well as the Middle East, particularly Saudi Arabia. The group engages in criminal activity, including extortion, to fund its activities.

Liberation Tigers of Tamil Eelam

Also known as Ellalan Force; Tamil Tigers

Description: Founded in 1976 and designated as a Foreign Terrorist Organization on October 8, 1997, the Liberation Tigers of Tamil Eelam (LTTE) is a Tamil secessionist group in Sri Lanka. Despite its military defeat at the hands of the Sri Lankan government in 2009, the LTTE's international network of sympathizers and financial support has persisted.

Activities: Although the LTTE has been largely inactive since its military defeat in 2009, in the past it was responsible for an integrated battlefield insurgent strategy that targeted key installations and senior Sri Lankan political and military leaders. In early 2009, Sri Lankan forces recaptured the LTTE's key strongholds, including their capital of Kilinochchi. In May 2009, government forces defeated the last LTTE fighting forces, killed LTTE leader Velupillai Prabhakaran and other members of the LTTE leadership and military command, and declared military victory.

There have been no known attacks in Sri Lanka that could verifiably be attributed to the LTTE since the end of the war in 2009, but a total of 13 LTTE supporters, several of whom had allegedly planned attacks against U.S. and Israeli diplomatic facilities in India, were arrested in Malaysia in 2014. Additional LTTE members were arrested in Malaysia and India in 2015, one of whom was accused of exhorting other Sri Lankans to fund and revive the LTTE.

Strength: Exact strength is unknown.

Location/Area of Operation: Sri Lanka and India

Funding and External Aid: LTTE's financial network of support continued after the LTTE's military defeat in 2009; the group used its international contacts and the large Tamil diaspora in North America, Europe, and Asia to procure weapons, communications, funds, and other needed supplies. The group employed charities as fronts to collect and divert funds for its activities.

Mujahidin Shura Council in the Environs of Jerusalem

Also known as MSC; Mujahideen Shura Council in the Environs of Jerusalem; Mujahideen Shura Council; Shura al-Mujahedin Fi Aknaf Bayt al-Maqdis; Majlis Shura al-Mujahidin; Majlis Shura al-Mujahideen; Magles Shoura al-Mujahddin

Description: The Mujahidin Shura Council in the Environs of Jerusalem (MSC) was designated as a Foreign Terrorist Organization on August 19, 2014. The MSC is a consolidation of several Salafi terrorist groups based in Gaza that have claimed responsibility for numerous attacks against Israel since the group's founding in 2012.

Activities: On August 13, 2013, MSC claimed responsibility for a rocket attack targeting the southern Israeli city of Eilat. Previously, MSC claimed responsibility for the March 21, 2013, attack in which Gaza-based militants fired at least five rockets at Sderot, Israel; and the April 17, 2013, attack in which two rockets were fired at Eilat, Israel. There have been no known MSC attacks since 2016.

Strength: MSC is estimated to have several hundred fighters.

Location/Area of Operation: MSC operates in Gaza.

Funding and External Aid: Unknown

Al-Mulathamun Battalion

Also known as al-Mulathamun Brigade; al-Muwaqqi'un bil-Dima; Those Signed in Blood Battalion; Signatories in Blood; Those who Sign in Blood; Witnesses in Blood; Signed-in-Blood Battalion; Masked Men Brigade; Khaled Abu al-Abbas Brigade; al-Mulathamun Masked Ones Brigade; al-Murabitoun; The Sentinels

Description: The al-Mulathamun Battalion (AMB) was designated as a Foreign Terrorist Organization on December 19, 2013. AMB was originally part of al-Qa'ida in the Islamic Maghreb (AQIM), but became a separate organization in late 2012 after its leader, Mokhtar Belmokhtar, split from AQIM. In his first public statement after the split, Belmokhtar threatened to fight against Western interests and announced the creation of the sub-battalion, "Those Who Sign in Blood." In August 2013, AMB and the Mali-based Movement for Unity and Jihad in West Africa (MUJAO) announced that the two organizations would merge under the name "al-Murabitoun." In late 2015, AMB announced a remerger with AQIM.

Activities: AMB's "Those Who Sign in Blood" sub-battalion claimed responsibility for the January 2013 attack against the Tiguentourine gas facility near In Amenas, in southeastern Algeria. More than 800 people were taken hostage during the four-day siege, resulting in the deaths of 39 civilians, including three U.S. citizens. Seven other U.S. citizens escaped.

Before their merger in August 2013, AMB cooperated with MUJAO in twin suicide bombings on a northern Nigerien military base and a French uranium mine in Arlit in May 2013. The coordinated attacks killed at least 20 people, including all of the attackers. AMB's attacks continued in 2014. In February, AMB claimed responsibility for attacking French forces with rockets near the Timbuktu airport.

In March 2015, AMB claimed responsibility for an attack at La Terrasse restaurant in Bamako, Mali. A French national, a Belgian national, and three Malians were killed when a masked gunman fired indiscriminately on the restaurant, and nine others were also wounded in the attack. In April, AMB claimed responsibility for a suicide bombing outside a UN Multidimensional Integrated Stabilization Mission in Mali (MINUSMA) camp in Ansongo, killing three civilians and wounding nine Nigerian peacekeepers. AMB also claimed the August hotel siege in central Mali; 17 people were killed, including four Malian soldiers and nine civilians, five of whom worked for MINUSMA. In November, AMB operatives participated in the strike on the Radisson Blu Hotel in Bamako, Mali, taking more than 170 people hostage—including U.S. citizens. As many as 27 people were killed in the attack; one of those killed was an U.S. international development worker.

AMB also was reportedly involved in the AQIM January 2016 attack on a popular tourist hotel in Burkina Faso that killed nearly 30, including one U.S. citizen. It continues to direct operations from Libya.

Strength: Membership levels of AMB are unknown.

Location/Area of Operation: Algeria, Burkina Faso, Libya, Mali, and Niger

Funding and External Aid: In addition to the support it may receive through its connections to other terrorist organizations in the region, AMB is likely funded through kidnapping-for-ransoms and other criminal activities.

National Liberation Army

Also known as ELN; Ejercito de Liberacion Nacional

Description: The National Liberation Army (ELN) was designated as a Foreign Terrorist Organization on October 8, 1997. The ELN is a Colombian Marxist-Leninist group formed in 1964. It is primarily rural-based, although it has several urban units. The ELN remains focused on attacking economic infrastructure—in particular oil and gas pipelines and electricity pylons—and on extorting foreign and local companies. ELN attacks in 2016 appeared to focus on improving the group's negotiation position in advance of peace talks with the Colombian government.

Activities: In 2016, the ELN continued to target Colombia's infrastructure, particularly oil pipelines. The ELN also launched mortars at police stations and the military, placed explosive devices near roads, and engaged in sniper attacks, roadblocks, and ambushes. In February, the ELN initiated a wave of violent attacks to force a 72-hour economic shutdown in several parts of the country. On May 22, the ELN kidnapped three journalists in Norte de Santander, who were released six days later to a Catholic Church delegation. On October 27, the ELN killed two civilian truck drivers in Arauca.

In September 2017, the ELN announced a truce.

Strength: Approximately 1,500 armed combatants and an unknown number of active supporters.

Location/Area of Operation: Primarily in the rural and mountainous areas of northern, northeastern, and southwestern Colombia, as well as the border regions with Venezuela.

Funding and External Aid: The ELN draws its funding from the illicit narcotics trade and from extortion of oil and gas companies. Additional funds are derived from kidnapping ransoms. There is no known external aid.

Al-Nusrah Front

Also known as Jabhat al-Nusrah; Jabhet al-Nusrah; The Victory Front; al-Nusrah Front for the People of the Levant; al-Nusrah Front in Lebanon; Jabhat al-Nusra li-Ahl al-Sham min Mujahedi al-Sham fi Sahat al-Jihad; Support Front for the People of the Levant; Jabhat Fath al-Sham; Jabhat Fath al Sham; Jabhat Fatah al-Sham; Jabhat Fateh al-Sham; Front for the Conquest of Syria; The Front for liberation of al Sham; Front for the Conquest of Syria/the Levant; Front for the Liberation of the Levant; Conquest of the Levant Front; Fatah al-Sham Front; Fateh Al-Sham Front

Description: Al-Nusrah Front (ANF) is al-Qa'ida's affiliate in Syria and was designated as a Foreign Terrorist Organization on May 15, 2014. It is led by Abu Muhammad al-Jawlani aka al-Julani. The group was formed in late 2011 when then-al-Qa'ida in Iraq (AQI)—now ISIS—leader and SDGT Abu Bakr al-Baghdadi sent al-Jawlani to Syria to organize terrorist cells. In 2013, the group split from AQI and became an independent entity. ANF's stated goal is to oust Syria's Assad regime and replace it with a Sunni Islamic state. The group currently controls a portion of territory in northwest Syria from which it participates in the Syrian conflict. It also operates in the mountainous, mostly uninhabited zones of the undemarcated Lebanese-Syrian border, and maintains a small presence in southern Syria.

Activities: ANF has been active in a number of operations against other factions in the Syrian conflict. In December 2013, ANF abducted 13 nuns from a Christian monastery in Maaloula and held them until March 9, 2014. In 2014, ANF also carried out multiple suicide bomb attacks and kidnappings, including the abduction of UN peacekeepers.

ANF continued fighting in Syria throughout 2015; it participated in fighting against other opposition groups and carried out a number of kidnappings against civilians. In March 2015, ANF claimed an attack on the intelligence headquarters of Syria's air force in Aleppo, killing an estimated 20 members of the security force. In April 2015, ANF reportedly kidnapped, and later released, approximately 300 Kurdish civilians from a checkpoint in Syria. In June 2015, ANF claimed responsibility for the massacre of the Druze village Qalb Lawzeh in Idlib province, Syria, which killed 20. In July 2015, the group claimed responsibility for a suicide bombing of an army outpost in Aleppo, which killed at least 25 soldiers and allied militia.

In 2016, the group carried out attacks in Aleppo and in other parts of Syria controlled by the Syrian army, killing both military officials and civilians. ANF also attacked various Syrian opposition groups.

In July, ANF leader Jawlani announced the group had adopted a new name, Jabhat Fath al-Sham, and would no longer be known as ANF. The Department of State amended the designation of al-Nusrah Front in November 2016 to add additional aliases, including Jabhat Fath al-Sham. The group remained an al-Qa'ida's affiliate in Syria.

Strength: Five to 10 thousand members

Location/Area of Operation: Syria and Lebanon

Funding and External Aid: ANF receives funding from a variety of sources, including kidnapping-for-ransom payments and donations from external Gulf-based donors.

Palestine Islamic Jihad

Also known as PIJ; PIJ-Shaqaqi Faction; PIJ-Shallah Faction; Islamic Jihad of Palestine; Islamic Jihad in Palestine; Abu Ghunaym Squad of the Hizballah Bayt al-Maqdis; Al-Quds Squads; Al-Quds Brigades; Saraya al-Quds; Al-Awdah Brigades

Description: Palestine Islamic Jihad (PIJ) was designated as a Foreign Terrorist Organization on October 8, 1997. Formed by militant Palestinians in Gaza during the 1970s, PIJ is committed to the destruction of Israel through attacks against Israeli military and civilian targets and to the creation of an Islamic state in all of historic Palestine, including present day Israel.

Activities: PIJ terrorists have conducted numerous attacks, including large-scale suicide bombings, against Israeli civilian and military targets. Although U.S. citizens have died in PIJ attacks, the group has not directly targeted U.S. interests. Between 2008 and 2011, PIJ primarily conducted rocket attacks aimed at southern Israeli cities and attacked Israeli targets with explosive devices. In 2012 and 2014, PIJ operatives carried out attacks on Israeli buses in Tel Aviv. In March 2014, PIJ carried out a wave of rocket attacks into Israeli territory; 60 rockets are believed to have reached Israel.

In early 2015, PIJ began rearming and replenishing its ranks. In March of that year, reports suggested that approximately 200 new recruits between the ages of 19 and 22 were undergoing various PIJ training programs lasting anywhere from 36 days to six months. That same month, PIJ revealed its militants were smuggling weapons, including rockets and mortars made inside Gaza, through tunnels in Gaza, in preparation for future attacks against Israel. In May, Israeli forces blamed PIJ for firing a rocket that landed in Gan Yazne, a region close to the Gaza border. The rocket was the first mid-range rocket fired at Israel since the August 2014 ceasefire. In August 2015, four militants, including one PIJ operative, were arrested in conjunction with what Israel's domestic security agency Shin Bet described as a PIJ plot to attack Jewish worshippers at Joseph's Tomb in the West Bank. Later that month, the Israel Defense Forces (IDF) claimed PIJ operatives in Syria fired four rockets at the Golan Heights and Upper Galilee.

Over the course of 2016, PIJ continued to strike Israel, primarily through light arms fire at IDF patrols. In September, Israeli authorities arrested PIJ operative Mahmoud Yusuf Hasin Abu Taha upon his entry into Israel from Gaza, interrupting a PIJ plot to abduct and kill an IDF soldier, and carry out a mass-casualty attack on a reception hall in Beersheba. PIJ also boycotted the Palestinian municipal elections in October, arguing in favor of resistance, rather than elections, as a means to achieve Palestinian goals.

Strength: PIJ has close to 1,000 members.

Location/Area of Operation: Primarily Gaza, with minimal operational presence in the West Bank and Israel. Other leadership elements reside in Lebanon and official representatives are scattered throughout the Middle East.

Funding and External Aid: PIJ receives financial assistance and training primarily from Iran. PIJ has partnered with Iranian- and Syrian-sponsored Hizballah to carry out joint operations.

Palestine Liberation Front—Abu Abbas Faction

Also known as PLF; PLF-Abu Abbas; Palestine Liberation Front

Description: The Palestinian Liberation Front—Abu Abbas Faction (PLF) was designated as a Foreign Terrorist Organization on October 8, 1997. In the late 1970s, the Palestine Liberation Front (PLF) splintered from the Popular Front for the Liberation of Palestine-General Command (PFLP-GC). It later split into pro-Palestinian Liberation Organization (PLO), pro-Syrian and pro-Libyan factions. The pro-PLO faction was led by Muhammad Zaydan (a.k.a. Abu Abbas) and based in Baghdad prior to Operation Iraqi Freedom.

Activities: The PLF was responsible for the 1985 attack on the Italian cruise ship *Achille Lauro* and the murder of U.S. citizen Leon Klinghoffer. The PLF was suspected of supporting terrorism against Israel by other Palestinian groups into the 1990s. In April 2004, Abu Abbas died of natural causes while in U.S. custody in Iraq. After not claiming an attack for 16 years, the PLF claimed responsibility for the March 14, 2008, assault against an Israeli military bus in Huwarah, Israel, and the shooting of an Israeli settler. Following the attacks, a PLF Central Committee member reaffirmed the group's commitment to using "all possible means to restore" its previous glory and to adhering to its role in the Palestinian "struggle" and "resistance" through its military. On February 18, 2010, the PLF claimed responsibility for an improvised explosive device (IED) attack against an Israel Defense Forces patrol, which caused minor injuries to a soldier; another IED was discovered during a search of the area. The group continues to maintain a strong presence in many refugee camps in Gaza, Lebanon, and Syria.

Strength: Estimates have placed membership between 50 and 500.

Location/Area of Operation: PLF leadership and members are based in Gaza, Lebanon, and the West Bank.

Funding and External Aid: Unknown

Popular Front for the Liberation of Palestine

Also known as PFLP; Halhul Gang; Halhul Squad; Palestinian Popular Resistance Forces; PPRF; Red Eagle Gang; Red Eagle Group; Red Eagles; Martyr Abu-Ali Mustafa Battalion

Description: Designated as a Foreign Terrorist Organization on October 8, 1997, the Popular Front for the Liberation of Palestine (PFLP) is a Marxist-Leninist group that was formed in 1967 by George Habash after splitting from the Arab Nationalist Movement. The group earned a reputation for large-scale international attacks in the 1960s and 1970s, including airline hijackings that killed at least 20 U.S. citizens.

Activities: The PFLP increased its operational activity during the Second Intifada. During that time, the group assassinated Israeli Tourism Minister Rehavam Ze'evi in 2001, carried out at least two suicide operations, and launched multiple joint operations with other Palestinian terrorist groups. In 2008 and 2009, the PFLP claimed responsibility for numerous attacks on Israel Defense Forces (IDF) in Gaza.

Between 2008 and 2011, the PLFP claimed responsibility for multiple mortar and rocket attacks fired from Gaza into Israel. In 2011, the group carried out an attack on a group of Israeli citizens. In 2012, the Israeli Security Agency arrested several members of the PFLP for plotting to carry out attacks on IDF checkpoints and planning to conduct kidnappings.

On November 18, 2014, two Palestinians reportedly affiliated with the PFLP entered a Jerusalem synagogue and attacked Israelis with guns, knives, and axes, killing five people—including three U.S. citizens—and injuring 12. In May 2014, Jerusalem police arrested three suspects, one of whom was affiliated with the PFLP, on suspicion of plotting an attack in East Jerusalem; and in December 2014, the PFLP claimed responsibility for several rocket attacks along the Lebanese-Israeli border.

In August 2016, the Abu Ali Mustafa Brigades—the military wing of PFLP—fired a rocket at the Israeli town of Sderot. No casualties or damages were reported.

Strength: Unknown

Location/Area of Operation: Gaza, Israel, Lebanon, Syria, and the West Bank.

Funding and External Aid: Unknown

Popular Front for the Liberation of Palestine—General Command

Also known as PFLP-GC

Description: The Popular Front for the Liberation of Palestine-General Command (PFLP-GC) was designated as a Foreign Terrorist Organization on October 8, 1997. The PFLP-GC split from the Popular Front for the Liberation of Palestine (PFLP) in 1968, claiming it wanted to focus more on resistance and less on politics. Originally, the group was violently opposed to the Arafat-led Palestinian Liberation Organization. Ahmad Jibril, a former captain in the Syrian Army, has led the PFLP-GC since its founding. The PFLP-GC is closely tied to both Syria and Iran.

Activities: The PFLP-GC carried out dozens of attacks in Europe and the Middle East during the 1970s and 1980s. The organization was known for cross-border terrorist attacks into Israel using unusual means, such as hot-air balloons and motorized hang gliders. Since the early 1990s, the group has primarily focused on supporting Hizballah's attacks against Israel, training members of other Palestinian terrorist groups, and smuggling weapons. The PFLP-GC maintained an armed presence in several Palestinian refugee camps, at its own military bases in Lebanon, and along the Lebanon-Syria border. In recent years, the PFLP-GC has been implicated by Lebanese security officials in several rocket attacks against Israel. In 2009, the group was responsible for wounding two civilians in an armed attack in Nahariyya, Northern District, Israel. In 2011, the PFLP-GC Jihad Jibril Brigades targeted Israeli communities in Eshkolot, Southern District, Israel, in a March 20 rocket attack. The attack caused no injuries or damage.

In November 2012, PFLP-GC claimed responsibility for a bus bombing in Tel Aviv that injured 29 people, although four Palestine Islamic Jihad and Hamas operatives were later arrested for being behind the attack. In 2015, the PFLP-GC reportedly began fighting alongside the Assad regime in Syria, while also receiving logistical and military aid from Hizballah and Iran.

Separately, in December 2015, the PFLP-GC took responsibility for rocket fire aimed at Israeli territory. The attack, in which at least three rockets were fired from Lebanon into northern Israel, landed near Shlomi, a small town near the Lebanese border. The group remains an active participant in the Syrian conflict and took part in the battle against ISIS for control of the Yarmouk refugee camp in Damascus.

Strength: Several hundred members

Location/Area of Operation: Political leadership is headquartered in Damascus, with bases in southern Lebanon and a presence in the Palestinian refugee camps in Lebanon and Syria. The group also maintains a small presence in Gaza.

Funding and External Aid: The PFLP-GC receives safe haven and logistical and military support from Syria and financial support from Iran.

Al-Qa'ida

Also known as al-Qa'eda; Qa'idat al-Jihad (The Base for Jihad); formerly Qa'idat Ansar Allah (The Base of the Supporters of God); the Islamic Army; Islamic Salvation Foundation; The Base; The Group for the Preservation of the Holy Sites; The Islamic Army for the Liberation of the Holy Places; the World Islamic Front for Jihad Against Jews and Crusaders; the Osama bin Laden Network; the Osama bin Laden Organization; al-Jihad; the Jihad Group; Egyptian al-Jihad; Egyptian Islamic Jihad; New Jihad

Description: Designated as a Foreign Terrorist Organization on October 8, 1999, al-Qa'ida (AQ) was established in 1988. The group helped finance, recruit, transport, and train violent Sunni Islamist extremists for the Afghan resistance against the former Soviet Union. AQ strives to eliminate Western influence and presence from the Muslim world, topple "apostate" governments of Muslim countries, and establish a pan-Islamic caliphate governed by its own interpretation of Sharia law that would ultimately be at the center of a new international order. These goals remain essentially unchanged since the group's 1996 public declaration of war against the United States. AQ leaders issued a statement in February 1998 under the banner of "The World Islamic Front for Jihad against the Jews and Crusaders," saying it was the duty of all Muslims to kill U.S. citizens—civilian and military—and their allies everywhere. AQ merged with al-Jihad (Egyptian Islamic Jihad) in June 2001. Many AQ leaders have been killed in recent years, including Osama bin Laden in May 2011. AQ's leader Ayman al-Zawahiri remained at-large in 2016.

Activities: AQ and its supporters conducted three bombings targeting U.S. troops in Aden in December 1992, and claimed responsibility for shooting down U.S. helicopters and killing U.S. soldiers in Somalia in 1993. AQ also carried out the August 1998 bombings of the U.S. Embassies in Nairobi and Dar es Salaam, killing up to 300 individuals and injuring more than 5,000. In October 2000, AQ conducted a suicide attack on the USS Cole in the port of Aden, Yemen, with an explosive-laden boat, killing 17 U.S. Navy sailors and injuring 39 others.

On September 11, 2001, 19 AQ members hijacked and crashed four U.S. commercial jets—two into the World Trade Center in New York City, one into the Pentagon, and the last into a field in Shanksville, Pennsylvania. Nearly 3,000 civilians, police, and first responders were killed. The dead included U.S. and foreign citizens from at least 77 countries.

In a December 2011 video, al-Zawahiri claimed AQ was behind the kidnapping of U.S. aid worker Warren Weinstein in Pakistan. Weinstein was held captive until his death in January 2015.

In February 2014, AQ removed ISIS as an AQ-affiliate. In September 2014, al-Zawahiri and other AQ leaders announced the establishment of Pakistan-based al-Qa'ida in the Indian Subcontinent (AQIS). Two days after the announcement, two Pakistani warships were attacked in Karachi; AQIS claimed responsibility for the plot, which included commandeering missile systems to attack nearby U.S. warships. AQIS claimed the orders came from al-Zawahiri.

In September 2015, five senior AQ leaders were released from Iranian custody in exchange for an Iranian diplomat kidnapped in Yemen. Of the five, Saif al Adel and Abu Mohammed al Masri are wanted for the August 1998 U.S. Embassy bombings in Kenya and Tanzania; al Adel and Abu Mohammed al Masri were experienced operational planners.

In January 2016, AQ leader Ayman al-Zawahiri publicly released two audio messages and one seven-page statement, condemning the Government of Saudi Arabia and its role in the Syria conflict, encouraging AQ activity in Southeast Asia—especially Indonesia, Malaysia, and the Philippines, and acknowledging support for its affiliate in Syria, al-Nusrah Front. On October 3, Abu al-Faraj al-Masri, one of AQ's most senior leaders, involved in planning AQ external attacks, was killed in Syria. As of September, AQ leaders and members were active in at least seven provinces in Afghanistan.

AQ is maintaining a low profile and in many areas its influence has been considerably reduced. However, it still has bases and affiliates in many parts of the world and it cannot be discounted.

Strength: In South Asia, AQ's core has been seriously degraded. The death or arrest of dozens of mid- and senior-level AQ operatives, including bin Laden in May 2011, have disrupted communication, financial support, facilitation nodes, and a number of terrorist plots. AQ, however, remains a focal point of "inspiration" for a worldwide network of affiliated groups—al-Qa'ida in the Arabian Peninsula (AQAP), al-Qa'ida in the Islamic Maghreb (AQIM), al-Nusrah Front, al-Shabaab, and AQIS—and other violent Sunni Islamist extremist groups, including the Islamic Movement of Uzbekistan, Islamic Jihad Union, Lashkar i Jhangvi, Harakat ul-Mujahideen, and Jemaah Islamiya. The Tehrik-e Taliban Pakistan and the Haqqani Network also have ties to AQ. In addition, supporters and associates worldwide "inspired" by the group's ideology may operate without direction from AQ central leadership.

Location/Area of Operation: AQ was based in Afghanistan until Coalition Forces removed the Afghan Taliban from power in late 2001. Subsequently, the group's core leadership was largely based in Pakistan's Federally Administered Tribal Areas. AQ affiliates—al-Nusrah Front, AQAP, AQIM, al-Shabaab, and AQIS—operate in Syria and Lebanon, Yemen, the Trans-Sahara, Somalia, and the Indian subcontinent, respectively.

Funding and External Aid: AQ primarily depends on donations from like-minded supporters, as well as from individuals who believe that their money is supporting a humanitarian cause. Some funds are diverted from Islamic charitable organizations.

Al-Qa'ida in the Arabian Peninsula

Also known as al-Qa'ida in the South Arabian Peninsula; al-Qa'ida in Yemen; al-Qa'ida of Jihad Organization in the Arabian Peninsula; al-Qa'ida Organization in the Arabian Peninsula; Tanzim Qa'idat al-Jihad fi Jazirat al-Arab; AQAP; AQY; Ansar al-Shari'a

Description: Al-Qa'ida in the Arabian Peninsula (AQAP) was designated as a Foreign Terrorist Organization on January 19, 2010. In January 2009, the now-deceased leader of al-Qa'ida in Yemen (AQY), Nasir al-Wahishi, publicly announced that Yemeni and Saudi al-Qa'ida (AQ) operatives were working together under the banner of AQAP. The announcement signaled the rebirth of an AQ franchise that previously carried out attacks in Saudi Arabia. AQAP's self-stated goals are to establish a caliphate and Sharia law in the Arabian Peninsula and the wider Middle East.

Activities: AQAP has claimed responsibility for numerous terrorist acts against both internal and foreign targets since its inception in January 2009, including a March 2009 suicide bombing against South Korean tourists in Yemen and the December 25, 2009, attempted attack on Northwest Airlines Flight 253 from Amsterdam to Detroit, Michigan. In October 2010, AQAP claimed responsibility for a foiled plot to send explosive-laden packages to the United States via cargo planes. The parcels were intercepted in the United Kingdom and in the United Arab Emirates.

AQAP, operating under the alias Ansar al-Shari'a (AAS), carried out a May 2012 suicide bombing in Sana'a that killed 96 people. Also in May 2012, press reported that AQAP allegedly planned to detonate a bomb aboard a U.S.-bound airliner using an improvised explosive device.

In September 2014, AQAP launched two rocket attacks against the U.S. Embassy in Sana'a. The second attack injured multiple embassy security guards. Also in 2014, AQAP attempted attacks against the U.S. and British Ambassadors to Yemen, and carried out a strike on the Iranian ambassador's residence in Sana'a that killed one guard and two pedestrians.

In January 2015, brothers Cherif and Said Kouachi attacked the satirical newspaper *Charlie Hebdo* in Paris, France, killing 12 people. One of the brothers, who had traveled to Yemen in 2011 and met with now-deceased Anwar al-Aulaqi, claimed the attack on behalf of AQAP. AQAP later formally claimed responsibility.

Also in 2015, AQAP took advantage of Yemen's deteriorating political and economic environment after the Yemeni government was overthrown by Houthi rebels in January. The United States and several other countries closed their embassies in February amid the violence. In April, AQAP stormed the city of Mukalla, seizing control of government buildings, releasing terrorists from prison, and stealing millions from the central bank. From 2015 into 2016, AQAP consolidated its control over Mukalla and expanded its reach through large portions of Yemen's south.

In early 2016, AQAP swept through southern Yemen, gaining control of al-Hawta, Azzan, and Habban in Lahij Governorate, and Mahfad and Ahwar in Abyan Governorate. By February 2016, AQAP controlled most of Yemen's southeastern coast. Although the group lost control of Mukalla in April, when the port city was retaken by forces backed by the Saudi-led Coalition, these territorial losses are not believed to have significantly degraded AQAP's capabilities, although they did deprive the group of an important source of income.

AQAP also attempted to carry out multiple attacks targeting Yemeni government and security forces. In June 2016, Yemeni authorities foiled an attack targeting the local administrative headquarters in Mukalla after discovering a car laden with explosives before the operative could carry out the attack. In July, two car bombs targeting security checkpoints outside Mukalla killed at least nine Yemeni soldiers and wounded many others.

Strength: AQAP is estimated to have up to four thousand members.

Location/Area of Operation: Yemen

Funding and External Aid: AQAP's funding has historically come from theft, robberies, and kidnap-for-ransom operations; and donations from like-minded supporters. After seizing Mukallah in April 2015, the group had access to additional sources of revenue, including the millions it stole from the central bank. This access continued until Mukallah was retaken by government forces in April 2016.

Al-Qa'ida in the Indian Subcontinent (AQIS)

Also known as al-Qaeda in the Indian Subcontinent; Qaedat al-Jihad in the Indian Subcontinent

Description: On July 1, 2016, the Department of State designated al-Qa'ida in the Indian Subcontinent (AQIS) as a Foreign Terrorist Organization (FTO). AQIS is an affiliate of al-Qa'ida that focuses on terrorist activity in Bangladesh, India, and Pakistan. Its leader is Asim Umar, a former member of the FTO Harakat ul-Mujahidin.

Activities: AQIS claimed responsibility for the September 6, 2014, attack on a naval dockyard in Karachi, in which militants attempted to hijack a Pakistani Navy frigate. AQIS has also claimed attacks against activists and writers in Bangladesh, including U.S. citizen Avijit Roy, U.S. Embassy local employee Xulhaz Mannan, and Bangladeshi nationals Oyasiqur Rahman Babu, Ahmed Rajib Haideer, and A. K. M. Shafiul Islam.

Strength: Total membership is unknown.

Location/Area of Operations: Primarily Afghanistan and Pakistan, with elements operating in Bangladesh and India.

Funding and External Aid: AQIS likely receives funding from al-Qa'ida senior leadership and engages in general criminal activity, kidnapping, and extortion.

Al-Qa'ida in the Islamic Maghreb

Also known as AQIM; GSPC; Le Groupe Salafiste Pour la Predication et le Combat; Salafist Group for Preaching and Combat; Salafist Group for Call and Combat; Tanzim al-Qa'ida fi Bilad al-Maghrib al-Islamiya

Description: The Salafist Group for Call and Combat (GSPC) was designated as a Foreign Terrorist Organization on March 27, 2002. The Department of State amended the GSPC designation on February 20, 2008, after the GSPC officially joined with al-Qa'ida (AQ) in September 2006 and became al-Qa'ida in the Islamic Maghreb (AQIM). Although AQIM remains largely a regionally focused terrorist group, it has adopted a more anti-Western rhetoric and ideology, and has aspirations of overthrowing "apostate" African regimes and creating an Islamic state. Abdelmalek Droukdel, aka Abu Mus'ab Abd al-Wadoud, is the group's leader.

Activities: Following AQIM's 2007 bombing of the UN headquarters building and an Algerian government building in Algiers, which killed 60

people, AQIM's northern leadership was largely contained to the mountainous region of northeastern Algeria, while the group's southern battalions focused mostly on its kidnapping-for-ransom efforts. In 2011 and 2012, however, AQIM took advantage of the deteriorating security situation across Libya, Mali, and Tunisia to plan and conduct expanded operations. Militants with ties to AQIM were involved in the September 11, 2012, attack on U.S. facilities in Benghazi that killed then-U.S. Ambassador to Libya J. Christopher Stevens and three other Embassy staff members. In April 2014, AQIM killed 14 Algerian soldiers in an ambush on a convoy in mountains to the east of Algiers, one of the deadliest attacks on the Algerian military in several years. AQIM also claimed responsibility for the May 27, 2014, attack on the house of Tunisia's interior minister, Lotfi Ben Jeddou.

In January 2015, AQIM claimed responsibility for an attack on a UN vehicle in Kidal, which wounded seven peacekeepers. Also in 2015, AQIM twice attacked UN convoys near Timbuktu with small arms and rocket-propelled grenades; three peacekeepers were killed in a May attack and six in a July attack. In November 2015, AQIM, in cooperation with other terrorist groups, attacked the Radisson Blu Hotel in Bamako, Mali, taking more than 170 people hostage, including U.S. citizens. As many as 27 people were killed in the attack; one of those killed was an U.S. international development worker.

In January 2016, AQIM carried out an attack on a hotel in Burkina Faso, resulting in 28 deaths and another 56 injuries. In March, AQIM claimed responsibility for a strike on a popular tourist beach resort in Cote d'Ivoire; at least 16 people were killed and another 33 were injured.

AQIM has also continued to conduct kidnapping-for-ransom operations. Its targets have historically been Western citizens from governments or third parties that have established a pattern of making concessions in the form of ransom payments for the release of individuals in custody. In November 2014, AQIM released a video of two Western hostages, a Dutch and a French national, who were later released in December 2014.

In June 2015, AQIM published a video featuring one Swedish and one South African hostage who continued to be held captive since they were kidnapped in Timbuktu in 2011.

Strength: AQIM has several hundred fighters operating in Algeria and the Sahel, including remote regions of northern Mali and southwest Libya. Since the French intervention in northern Mali, AQIM's safe haven in northern Mali is less tenable for the organization and elements have moved to remote regions of northern Mali or to southwestern Libya. AQIM is attempting to reorganize in the wake of setbacks inflicted upon it by the combined French and African forces.

Location/Area of Operation: Southern and Eastern Algeria (including isolated parts of the Kabylie region), Burkina Faso, Cote D'Ivoire, Libya, northern Mali, Niger, and Tunisia.

Funding and External Aid: AQIM members engage in kidnapping-for-ransom and criminal activities to finance their operations. AQIM also successfully fundraises globally, and received limited financial and logistical assistance from supporters residing in Western Europe.

Real IRA

Also known as RIRA; Real Irish Republican Army; 32 County Sovereignty Committee; 32 County Sovereignty Movement; Irish Republican Prisoners Welfare Association; Real Oglaigh Na hEireann

Description: The Real Irish Republican Army (RIRA) was designated as a Foreign Terrorist Organization on May 16, 2001. The group was formed in 1997 as the clandestine armed wing of the 32 County Sovereignty Movement, a "political pressure group" dedicated to removing British forces from Northern Ireland and unifying Ireland. The RIRA has historically sought to disrupt the Northern Ireland peace process and did not participate in the September 2005 weapons decommissioning. In September 1997, the 32 County Sovereignty Movement opposed Sinn Fein's adoption of the Mitchell principles of democracy and nonviolence. Despite internal rifts and calls by some jailed members, including the group's founder Michael "Mickey" McKevitt, for a cease-fire and disbandment, the RIRA has pledged additional violence and continues to conduct attacks.

Activities: Many RIRA members are former Provisional Irish Republican Army members who left the organization after that group renewed its cease-fire in 1997. These members brought an extensive experience in terrorist tactics and bomb-making to the group. Targets have included civilians (the most notorious example is the Omagh bombing in August 1998), British security forces, and police officers in Northern Ireland. The Independent Monitoring Commission, which oversees the peace process, assessed that RIRA members were likely responsible for the majority of the shootings and assaults that occurred post-Irish Republican Army (IRA)-decommissioning in Northern Ireland.

In May 2015, Irish police carried out 20 searches aimed at known dissident republicans across Ireland. Six individuals with links to RIRA and Continuity Irish Republican Army (CIRA) were arrested after police discovered explosive devices and components. In spring 2016, the RIRA bombed the van of an Irish prison officer in East Belfast; the officer died from complications following the attack. Dublin police also linked the RIRA to an explosives cache they found in Dublin in April.

Strength: According to the Irish government, the RIRA has approximately 100 active members. The organization may receive limited support from IRA hardliners and Republican sympathizers who are dissatisfied with the IRA's continuing ceasefire and with Sinn Fein's involvement in the peace process.

Location/Area of Operation: Great Britain including Northern Ireland and the Republic of Ireland

Funding and External Aid: The RIRA is suspected of receiving funds from sympathizers in the United States and of attempting to buy weapons from U.S. gun dealers. The RIRA reportedly purchased sophisticated weapons from the Balkans and occasionally collaborated with the CIRA.

Revolutionary Armed Forces of Colombia

Also known as FARC; Fuerzas Armadas Revolucionarias de Colombia

Description: Designated as a Foreign Terrorist Organization on October 8, 1997, the Revolutionary Armed Forces of Colombia (FARC) is Latin America's oldest, largest, and best-equipped terrorist organization. The FARC, founded in the 1960s, is responsible for large numbers of kidnappings-for-ransom in Colombia and, in past years, has held as many as 700 hostages. A Colombian military offensive targeting key FARC units and leaders degraded its capacity and halved the FARC's numbers to approximately 7,000. After four years of negotiation in Havana, Cuba, a peace agreement was crafted which, after an unsuccessful plebiscite and further negotiation, was approved by Colombia's Congress on November 30, 2016, which put in motion a six-month disarmament, demobilization, and reintegration process. In accordance with the peace agreement, the FARC began to demobilize in early December under UN supervision.

Activities: Over the years, the FARC has perpetrated a large number of high profile terrorist acts, including the 1999 murder of three U.S. missionaries working in Colombia, and multiple kidnappings and assassinations of Colombian government officials and civilians. In July 2008, the Colombian military conducted a dramatic rescue of 15 high-value FARC hostages including U.S. Department of Defense contractors Marc Gonsalves, Keith Stansell, and Thomas Howe, who were held captive for more than five years, along with former Colombian presidential candidate Ingrid Betancourt.

In August 2016, they agreed to end their 52-year war with the government as part of a peace process started in 2012.

Strength: Approximately 7,000 members, with several thousand additional supporters.

Location/Area of Operation: FARC leaders and combatants are located in Colombia and have been known to use neighboring countries for weapons sourcing and logistical planning.

Funding and External Aid: In 2016, the FARC was primarily funded by extortion and the international drug trade.

Revolutionary People's Liberation Party/Front

Also known as DHKP/C; Dev Sol; Dev Sol Armed Revolutionary Units; Dev Sol Silahli Devrimci Birlikleri; Dev Sol SDB; Devrimci Halk Kurtulus Partisi-Cephesi; Devrimci Sol; Revolutionary Left

Description: Designated as a Foreign Terrorist Organization on October 8, 1997, the Revolutionary People's Liberation Party/Front (DHKP/C) was originally formed in 1978 as Devrimci Sol, or Dev Sol, a splinter faction of Dev Genc (Revolutionary Youth). It was renamed in 1994 after factional infighting. "Party" refers to the group's political activities, while "Front" is a reference to the group's militant operations. The group advocates a Marxist-Leninist ideology and opposes the United States, the North Atlantic Treaty Organization (NATO), and Turkish establishments. It strives to establish a socialist state and to abolish Turkish prisons.

Activities: Since the late 1980s, the group has primarily targeted current and retired Turkish security and military officials. In 1990, the group began to conduct attacks against foreign interests, including against U.S. military and diplomatic personnel and facilities. The DHKP/C assassinated two U.S. military contractors and wounded a U.S. Air Force officer in the 1990s, and bombed more than 20 U.S. and NATO military, diplomatic, commercial, and cultural facilities. DHKP/C added suicide bombings to its repertoire in 2001, with attacks against Turkish police in January and September of that year. Since the end of 2001, DHKP/C has typically used improvised explosive devices (IEDs) against official Turkish and U.S. targets of opportunity.

The DHKP/C was dealt a major ideological blow when its leader, Dursun Karatas died in August 2008. After the loss of its leader, the DHKP/C reorganized in 2009 and was reportedly in competition with the Kurdistan Workers' Party for influence in Turkey and with the Turkish diaspora in Europe.

The group was responsible for a number of high-profile attacks in 2012 that included a suicide bombing of a police station in Istanbul. This tactic continued in 2013 when, on February 1, a DHKP/C operative exploded a suicide vest inside the employee entrance to the U.S. Embassy in Ankara. The explosion killed a Turkish guard and seriously wounded a visiting Turkish journalist. In March 2013, three members of the group attacked the Ministry of Justice and the Ankara headquarters of the Turkish Justice and Development political party using grenades and rocket launchers. The DHKP/C claimed a similar attack in September 2013, when two members of the group fired rockets at the Turkish National Police headquarters and a police guesthouse.

In March 2014, the DHKP/C also claimed a shooting at a protest that left one man dead.

In 2015, the DHKP/C claimed responsibility for a suicide bombing that killed a police officer and wounded another. In March, Turkish prosecutor Mehmet Selim Kiraz was taken hostage and killed from multiple gunshot wounds by the DHKP/C after police attempted to rescue him. In August, two women opened fire on the U.S. Consulate in Istanbul; one woman was identified as a member of the DHKP/C. On March 3, 2016, two DHKP/C militants attacked a police vehicle in Istanbul, wounding two officers.

Strength: Membership includes an estimated several dozen members inside Turkey, with a support network throughout Europe.

Location/Area of Operation: Turkey, primarily in Adana, Ankara, Istanbul, and Izmir. Other members live and plan operations in European countries.

Funding and External Aid: The DHKP/C finances its activities chiefly through donations and extortion. The group raises funds primarily in Europe.

Revolutionary Struggle

Also known as RS; Epanastatikos Aghonas; EA

Description: Designated as a Foreign Terrorist Organization on May 18, 2009, Revolutionary Struggle (RS) is a radical leftist group with Marxist ideology that has conducted attacks against both Greek and U.S. targets in Greece. RS emerged in 2003 following the arrests of members of the Greek leftist groups 17 November and Revolutionary People's Struggle.

Activities: RS first gained notoriety when it claimed responsibility for the September 5, 2003, bombings at the Athens Courthouse during the trials of 17 November members. From 2004 to 2006, RS carried out a number of improvised explosive device attacks, including a March 2004 attack outside of a Citibank office in Athens. RS claimed responsibility for the January 12, 2007, rocket-propelled grenade (RPG) attack on the U.S. Embassy in Athens, which damaged the building; and the March 2009 bombing of a Citibank branch in Athens. In September 2009, RS also claimed responsibility for a car bomb attack on the Athens Stock Exchange, which caused widespread damage and injured a passerby.

The Greek government has made significant strides in curtailing RS' terrorist activity. On April 10, 2010, Greek police arrested six suspected RS members, including purported leader Nikos Maziotis, who later escaped. In addition to the arrests, the Greek raid resulted in the seizure of a RPG launcher, possibly the one used against the U.S. Embassy in Athens in the 2007 attack. On April 3, 2013, five members of RS were convicted by an Athens appeals court, three of them receiving maximum prison sentences. Maziotis and another accused RS conspirator, Paula Roupa, were convicted in absentia. Prior to Maziotis's recapture (following a gunfight with police on July 16, 2014), RS conducted a bomb attack outside a Bank of Greece office in Athens in April 2014; the blast caused extensive damage to surrounding structures but no casualties.

In March 2016, a Greek court sentenced Maziotis to life in prison plus 129 years and a fine of 20,000 Euros (US $21,750). Roupa was sentenced in absentia to 11 years in prison on misdemeanor charges; she will stand trial on felony charges upon capture. Greek authorities suspect she was involved in a September 2016 bank robbery in Malesina, Greece.

Strength: Unknown

Location/Area of Operation: Athens, Greece

Funding and External Aid: Unknown, but the group most likely supports itself through criminal activities, including bank robbery.

Al-Shabaab

Also known as The Harakat Shabaab al-Mujahidin; al-Shabab; Shabaab; the Youth; Mujahidin al-Shabaab Movement; Mujahideen Youth Movement; Mujahidin Youth Movement

Description: Al-Shabaab was designated as a Foreign Terrorist Organization on March 18, 2008. Al-Shabaab was the militant wing of the former Somali Islamic Courts Council that took over parts of southern Somalia during the second half of 2006. Since the end of 2006, al-Shabaab and associated militias have undertaken a violent insurgency using guerrilla warfare and terrorist tactics against the transitional governments of Somalia. In 2016, the group continued to fight to discredit and destabilize the Federal Government of Somalia.

Al-Shabaab is an official al-Qa'ida (AQ) affiliate and has ties to other AQ affiliates, including al-Qa'ida in the Arabian Peninsula and al-Qa'ida in the Islamic Maghreb. The group's leader is Ahmed Diriye aka Ahmed Umar aka Abu Ubaidah.

Al-Shabaab is composed of a mixture of Somali recruits and foreign terrorist fighters. Since 2011, al-Shabaab has seen its military capacity reduced due to the efforts of the African Union Mission in Somalia (AMISOM) and Somali forces, and clashes—some violent—within the group itself. Despite al-Shabaab's loss of urban centers since 2012, the group was able to maintain its hold on large sections of rural areas in south-central Somalia in 2016 and conducted multiple attacks in Somalia and Kenya.

Activities: Al-Shabaab has used intimidation and violence to exploit divisions in Somalia and undermine the country's federal government, recruit new fighters, and kill activists working to bring about peace through political dialogue and reconciliation. The group has claimed responsibility for several high profile bombings and shootings throughout Somalia targeting AMISOM troops and Somali officials. Al-Shabaab fighters and others claiming allegiance to the group have assassinated numerous civil society figures, government officials, journalists, international aid workers, and members of nongovernmental organizations.

Al-Shabaab was responsible for the July 11, 2010, suicide bombings in Kampala, Uganda—its first attack outside of Somalia. The attack, which took place during the World Cup, killed nearly 76 people, including one U.S. citizen. In September 2013, al-Shabaab again expanded its area of operations when it staged a significant attack against the Westgate Mall in Nairobi, Kenya. The siege resulted in the deaths of at least 65 civilians, including foreign nationals from 13 countries and six soldiers and police officers; hundreds of others were injured. In April 2015, al-Shabaab carried out a raid with small arms and grenades on Kenya's Garissa University College that left nearly 150 people dead.

In February 2016, al-Shabaab attempted to down Daallo Airlines Flight 321 with 74 passengers on board. Only the suicide bomber was killed and the plane made an emergency landing in Mogadishu shortly after take-off due to an explosion that created a large hole in the fuselage near the wing. Al-Shabaab killed 10 people in May when members of the group set off a car bomb outside a hotel in Mogadishu. In August, al-Shabaab claimed a double suicide bombing in Galkayo, Somalia, that killed at least 20 people. Al-Shabaab carried out a series of raids in northeast Kenya in October, including one attack that killed at least 12 people at a guesthouse in Mandera, and in November, the group claimed responsibility for a car bombing targeting an army convoy near Parliament in Mogadishu that killed at least two soldiers and injured another five.

In October 2017, more than 500 people were killed by two bomb explosions in Mogadishu.

Strength: Al-Shabaab is estimated to have several thousand members.

Location/Area of Operation: Al-Shabaab has lost full control of significant areas of territory. In September 2012, the group lost control of Kismayo, a vital port it used to obtain supplies and funding through taxes. In October 2014, al-Shabaab lost another strategic port in Baraawe to AMISOM and Somali troops. In 2015, an AMISOM offensive forced the group out of two of its strongholds of Dinsoor and Bardaheere and killed up to 100 personnel. Despite these losses, throughout 2016, al-Shabaab continued to control large swaths of rural areas in the middle and lower Juba regions, as well as the Bakol, Bay, and Shabelle regions. The group also maintained its presence in northern Somalia along the Golis Mountains and within Puntland's larger urban areas, and launched several attacks against targets in the border regions of Kenya.

Funding and External Aid: While al-Shabaab has seen its income diminish due to the loss of the strategic port cities of Baraawe, Kismayo, and Merka, it has received enough income to launch multiple attacks per week in Mogadishu, launch complex attacks against AMISOM bases, and expand operations against civilian aviation targets. Al-Shabaab obtained funds through illegal charcoal production and exports, and taxation of local populations and businesses.

Because al-Shabaab is a multi-clan entity, it reportedly receives donations from individuals in the Somali diaspora; the donations are not always intended to support terrorism, however, but also to support family members.

Shining Path

Also known as SL; Sendero Luminoso; Ejercito Guerrillero Popular (People's Guerrilla Army); EGP; Ejercito Popular de Liberacion (People's Liberation Army); EPL; Partido Comunista del Peru (Communist Party of Peru); PCP; Partido Comunista del Peru en el Sendero Luminoso de Jose Carlos Mariategui (Communist Party of Peru on the Shining Path of Jose Carlos Mariategui); Socorro Popular del Peru (People's Aid of Peru); SPP

Description: The Shining Path (Sendero Luminoso or SL) was designated as a Foreign Terrorist Organization on October 8, 1997. The Peru-based terrorist organization was formed in the late 1960s by former university professor Abimael Guzman, whose teachings created the foundation of SL's militant Maoist doctrine. In the 1980s, SL was one of the most ruthless terrorist groups in the Western Hemisphere. In 1992, the Peruvian government captured Guzman who remains, along with key accomplices, in prison serving a life sentence. The group is now led by brothers Victor and Jorge Quispe Palomino along with SL leader Tarcela Loya Vilchez. Under their direction, the group aims to overthrow the Peruvian government and names the United States a principal enemy.

SL's area of activity and influence is mostly confined to the remote special military emergency zone known as the Apurimac, Ene, and Mantaro River Valley (VRAEM). SL is involved in all logistical aspects of drug trafficking in the VRAEM area.

Activities: SL committed 13 terrorist attacks in 2015, in comparison to 20 terrorist acts in 2014 and 49 in 2013. In 2016, SL terrorist attacks declined. On April 9, the group struck a six-vehicle military caravan transporting election materials ahead of the country's April 10 election; eight soldiers and two civilian contractors were killed when SL members armed with long-range rifles and grenades attacked the caravan.

In March 2017, three police officewrs were killed by Shining Path snipers in the Ene Apurimac Valley.

Strength: Estimates of SL's strength vary, but experts assess SL to number between 250 and 300 combatants.

Location/Area of Operation: Peru, with most activity taking place in rural areas, specifically the Apurimac, Ene, and Montaro River Valley of central Peru.

Funding and External Aid: SL is primarily funded by the illicit narcotics trade.

Terhik-e Taliban Pakistan

Also known as Pakistani Taliban; Tehreek-e-Taliban; Tehrik-e-Taliban; Tehrik-i-Taliban Pakistan; TTP

Description: Designated as a Foreign Terrorist Organization on September 1, 2010, Tehrik-e Taliban Pakistan (TTP) is a Pakistan-based terrorist organization formed in 2007 to oppose Pakistani military efforts in the Federally Administered Tribal Areas. Previously disparate tribal militants agreed to cooperate and eventually coalesced into TTP under the leadership of now deceased leader Baitullah Mehsud. TTP was led by Hakimullah Mehsud from August 2009 until his death in November 2013. Since Hakimullah Mehsud's death, TTP has been led by Mullah Fazlullah, former leader of TTP's chapter in the Swat area of Pakistan. Since Fazlullah became leader, TTP has been fraught with violent infighting. A rival faction led by Khan Said, a former TTP commander who previously clashed with Hakimullah, publicly split from TTP in May 2014, but has since rejoined. Separately, TTP entered into peace talks with the Pakistani government in early 2014, but the talks collapsed in June of that year. In October 2014, the chief spokesman and five regional commanders defected from TTP and publicly pledged allegiance to ISIS.

TTP aims to overthrow the Government of Pakistan by waging a terrorist campaign against the Pakistani military and state, as well as North Atlantic Treaty Organization forces in Afghanistan. TTP uses the tribal belt along the Afghanistan-Pakistan border to train and deploy its operatives, and has ties to al-Qa'ida (AQ). TTP draws ideological guidance from AQ, while AQ relies on TTP for safe haven in the Pashtun areas along the Afghanistan-Pakistani border. This arrangement gives TTP access to both AQ's global terrorist network and the operational expertise of its members.

Activities: TTP has carried out and claimed responsibility for numerous terrorist acts against Pakistani and U.S. interests, including a December 2009 suicide attack on a U.S. military base in Khowst, Afghanistan, which killed seven U.S. citizens; and an April 2010 suicide bombing against the U.S. Consulate in Peshawar, Pakistan, which killed six Pakistani citizens. TTP is suspected of being involved in the 2007 assassination of former Pakistani prime minister Benazir Bhutto. TTP directed and facilitated the failed attempt by Faisal Shahzad to detonate an explosive device in New York City's Times Square on May 1, 2010.

Between 2011 and 2015, TTP continued to carry out attacks against the Government of Pakistan and civilian targets, as well as against U.S. targets in Pakistan. A series of suicide bombings and strikes in 2011 targeted civilians, Pakistani government and military targets, and an American consulate convoy. In 2012, TTP carried out attacks against a mosque, a police checkpoint, a Pakistani Air Force base, and a bus carrying Shia Muslims. In 2013, TTP attacked churches, the home of a government minister in Khyber-Pakhtunkhwa Province, and a Shia neighborhood in Karachi, Pakistan. TTP's attacks in 2013 killed and wounded hundreds of civilians and Pakistani government and law enforcement officials. In 2014, TTP targeted military and police convoys, bazaars, buses, and schools including two consecutive attacks against Karachi's international airport and a siege on a primary school in Peshawar, Pakistan that killed 145 people, 132 of whom were children. Throughout 2015, TTP focused many of its small-scale attacks on Pakistani government and law enforcement officials by targeting convoys, government buildings, motorcades, and police checkpoints. The group also bombed a Shia mosque near Peshawar and carried out suicide bombings at two churches in Lahore.

In 2016, the group continued carrying out attacks, claiming responsibility for a December attack that left the Deputy Superintendent of the police counterterrorism department dead and his son injured in an attack on their vehicle in Peshawar. Sporadic attacks continue.

Strength: Several thousand

Location/Area of Operation: Pakistan and Afghanistan

Funding and External Aid: TTP is believed to raise most of its funds through kidnapping ransoms, criminal activity, and extortion.

Part II

A HISTORY OF TERRORISM

Chapter 5

America's Birth

A Heritage of Conflict and Terror

There seems to be a popular myth in America that "terrorism" is something new. But anyone familiar with the history of the New World, as America was once known, knows that the confluence of European and Native American cultures spawned a heritage of conflict and terror from the start.

Early in the summer of 1002, Leif Erickson fitted a stout ship with a crew of twenty-six men and sailed southwestward from his home in Greenland. The son of Eric the Red—who discovered and colonized Greenland—Leif sought to follow his father's example and seek out unknown lands himself.

The dauntless crew braved rough seas, icebergs, and strong winds as they sailed on southward to find a land that was "charmed by a soft climate" where they sought a harbor, finding one at the mouth of a river where the vessel was swept by the tide into a bay. The waters were filled with salmon, and herds of wild deer roamed the woods.

A young German who was Leif's servant was missing one day. They searched for him in all directions. They found him deep in the forest where he was "full of joy" because he had discovered grapes, delicious and abundant, such as grew in his own country—so Leif named the country Vineland. Shortly after Leif returned home to Brattahlid his father died and Leif assumed the leadership of the Greenland colony.

Thorwald, Leif's younger brother, then bought the good ship and, with thirty companions, sailed for Vineland. They passed the two winters there, occupying the huts built by Leif and his companions, and subsisting as they had done, upon fish. During the spring and summers, they explored. They sailed into a large inlet during the early autumn of the third year. They saw thick wooded highlands on each side.

"Here," said Thorwald, "is a goodly place—here I will make my abode." They found there some natives—dark-skinned people, of small stature, like the Eskimos of Greenland. They were in canoes, and appeared timid and harmless. The Northmen caught them and cruelly put them to death, excepting one who escaped to the hills and aroused his tribesmen. The angry Natives returned silently in their canoes and surprised Thorwald and his company. A sharp fight ensued. Arrows flew thick and fast. Thorwald was mortally wounded, but his companions escaped unhurt. The Natives fled to the wooded hills, and Thorwald's companions buried the body of their chief on the promontory where he intended to settle, with a cross at its head and another at its feet. The survivors passed the winter in Vineland, in mortal fear of the enraged Natives, and in the spring they returned to Greenland.

And so began an episode that would play out time and time again in America. The Native peoples would more often than not welcome the explorers with warm smiles and hospitality. Too often they would be repaid with enslavement, abuse, illness, and death. And their memories of mistreatment were passed from generation to generation.

THE SPANIARDS

> *Impelled by his own religious zeal and prompted by the priests in his train, Cortez at once proceeded to further humiliate, horrify and exasperate the subdued people, by making a clean sweep, with the besom of destruction, over the idols and temples of the empire. In the great square in Mexico, the conqueror and his followers, with their garments stained with the blood of their fellow-creatures, devoutly sang the Te Deum, and prostrating themselves before the image of the Blessed Virgin which they had set up, they reverently thanked God for permitting them to be the humble instruments in annihilating image-worship and in staying the horrid rites of human sacrifice. Such was the spirit and temper of the age in which they lived. So was introduced Christianity into Mexico.*
>
> (Excerpt from *Our Country*, published in 1800)

Pedro Menéndez de Avilés commanded a great treasure-fleet of galleons on their voyage from Mexico to Spain in 1561. One of the vessels containing his son and several relatives and friends disappeared during the voyage. After Menéndez delivered the fleet to the Spanish crown, he asked permission to go back in search of the lost vessel, but was refused. Finally, after two or three years delay, his request was finally granted, but only on condition that he explore and colonize Florida. He fitted out an expedition for the purpose at his own expense.

When Menéndez was about to sail, orders came to him from King Philip that he was to exterminate all protestants he might find there. Philip heard that French Huguenots fleeing persecution in France were hiding in the forests beyond the Atlantic. The king's zeal to please the Catholic Church was such that he gave the order to Menéndez to slaughter the "heretics" wherever they were found. To refuse would mean torture and death in the dungeons of the Spanish Inquisition for Menéndez.

Because the king regarded the voyage as a holy enterprise, he added ships and treasure to support the venture. Soldiers and seamen flocked to the standard of Menéndez in great numbers, and he sailed with a fleet of eleven ships (one of them a galleon of nine hundred tons) with over twenty-six hundred persons, consisting, besides the soldiers and sailors, of adventurers and priests. Only five of the ships survived the crossing.

Stopping briefly at present-day St. Augustine, Menéndez went ashore with great pomp and circumstance to plant the Spanish flag, claiming Florida for God and king. He then set sail to strike at the feeble French Fort Caroline near the mouth of the St. Johns River.

The Huguenot fort was defended by only a handful of soldiers. The Spaniards closed with the fort before their presence was suspected. They attacked without mercy. No person was spared. They were slaughtered in their beds, screaming prayers for mercy, and in terrified flight. A few escaped to the woods, but without food and with scanty clothing. Many later perished. A few survivors made their way to two small French ships, in which they sailed for Europe. According to Chaplain Mendoza, 142 Huguenots were slain, while the Spaniards suffered no losses. The women and children were butchered. A few men were hanged upon trees with a sign posted above inscribed: NOT AS FRENCHMEN, BUT AS LUTHERANS.

Chevalier Dominic de Gourges, a gentleman of Gascony, a member of an eminent family and a devoted Roman Catholic was outraged. His hatred of the Spanish fueled by his experience as a prisoner of war who was enslaved aboard a Spanish galleon which was further stoked by the slaughter of his countrymen and the indifference expressed by his French king. De Gourges sold his property, borrowed money from his friends, and fitted out three small vessels, manned by one hundred soldiers (many of them gentlemen volunteers), and eighty mariners prepared with cross-bows and picks to act as soldiers.

Arriving at a small river north of the St. Johns in 1568, de Gourges met with a leader whose tribe suffered terrible punishment by the Spaniards because of the tribe's friendly relations with the French. The cacique formed an alliance with the French nobleman to war on the Spanish. Over the next few days the allies attacked two Spanish forts south of Fort Caroline (Matheo), then turned north and slaughtered everyone they found except for a few prisoners that they reserved for a special event. Almost 400 Spaniards perished during the conflict. Those few remaining were lined up beneath the same trees upon which the French Huguenots were hanged. De Gourges addressed them as they were strung up, then he posted a wooden sign with the inscription reading: NOT AS SPANIARDS AND MARINERS, BUT AS TRAITORS, ROBBERS, AND MURDERERS.

So much for the Golden Rule—outrage was expressed throughout Europe—but it would be centuries before the sins committed in the name of Christianity would be resolved.

Menéndez remained undaunted. He ordered a fort and a settlement be established at the tip of Port Royal in 1566 where Rene Ribaut's Charlesfort lay in ruins. Dubbed Santa Elena, the settlers found the local Yamassee tribe to be friendly and supportive, supplying corn, fish, and wild game.

Pedro Menéndez de Avilés, adelantado of Florida, died in 1574 while on a mission to Spain. During Menéndez' absence, Don Diego de Velasco, one of Pedro Menéndez' two sons-in-law and lieutenant governor, served as interim governor in his absence. Don Diego continued in that position following Menéndez' death. Menéndez' daughter, Catalina, then inherited the title of adelantado of Florida, and ultimately her husband, Hernando de Miranda, was appointed governor. Miranda, however, did not actually arrive at Santa Elena until February 1576.

During the years that Velasco served as interim governor, he had several run-ins with settlers and mistreated the Natives residing in the vicinity now known as Parris Island, SC. This poor relationship with the Natives led to a series of attacks on Santa Elena. The loss of thirty soldiers in these attacks ultimately forced the temporary abandonment of both the fort and the town at Santa Elena in late summer of 1576. One story has it that the soldiers' wives revolted, took the Spanish ship's captain prisoner, and forced him to set sail for Spain. This early instance of women's suffrage occurred as the once friendly Yamassees rained fire arrows down on the Spaniards' thatched huts. As the soldiers and settlers waited to cross the bar to depart Port Royal Sound, they were able to see the town and fort burning to the ground.

THE ENGLISH

On Roanoke Island the Englishmen were entertained, with a refined hospitality, by the mother of King Wingina (who was absent); and wherever they went, friendship was the rule. To the feelings of the strangers, everything on the islands and on the main was charming. Nature was then garnished in all her summer wealth, and to the eyes of the Englishmen her beauties there were marvelous. Magnificent trees

> *were draped with luxuriant vines clustered with growing grapes, and the forest swarmed with birds of sweetest songs and beautiful plumage. After gathering what information they could about the neighboring country, Barlow and Amidas departed for England,(August 1585) with their company, attended by Manteo and Wanchese, two dusky lords of the woods and waters.*
>
> (Excerpt from *Our Country*, published in 1800)

When the first English settlers ventured to America's shores, there were an estimated one million Natives living throughout what is now the United States. In general, they were peaceable and religious within their own communities and hospitable to strangers. However, they were capable of being very vicious when crossed or threatened. Their culture of stoicism and their ability to live in harmony with nature made them excellent warriors, despite their limited resources—and they were vicious enemies.

By the time that Queen Elizabeth acted upon Sir Walter Raleigh's advice to colonize the New World, atrocities inflicted against the native people by earlier exploring parties and settlers had infused hostility and distrust among the tribes. Realizing that they were no match for soldiers in armor and breastplates with "firing sticks" capable of killing from great distances, the natives relied on cunning and guile to gain advantage. Fearing the inevitability of these foreigners permanently settling on their lands and destroying their hunting grounds, the tribes formed alliances to resist the Europeans.

The conflict between the Native and European cultures could not be avoided, given the choices made for leadership of the new colonies. Too few were enlightened men such as Thomas Harriot, a man of keen observation who was Sir Walter Raleigh's accountant and mathematical mentor, who looked upon everything with the eye of a Christian philosopher. He perceived that the best way to establish viable colonies was to treat the Natives kindly as friends and neighbors. He deprecated the conduct of Sir Richard Grenville whom Raleigh put in charge of the Roanoke colonization venture, and tried to quench the fires of revenge that the leader's cruelty kindled.

Harriot befriended Chief Wingina, sharing his bible and his religion with the chief. When the chief became ill, he dismissed his medicine men and conjurers and called for Harriot to attend to him and pray for him. When Wingina recovered, many of his tribe embraced Christianity.

Grenville, however, responded with terror and intimidation. Upon returning with supplies for the small band of settlers at Roanoke, he sent Manteo, one of the natives who served as interpreter, to precede the party to the mainland to announce their arrival, Grenville soon followed him with Harriot and others. For eight days they explored the country, and were hospitably entertained everywhere. While being hosted at one Native village, a silver cup was stolen from the English and when it was not immediately restored on demand, Grenville ordered the whole town to be burned and the standing corn around it destroyed. A flame of indignation, furious and destructive, was kindled in the Native mind, which could not be quenched. Unaware of the consequences of his act, Grenville left the colonists and returned to England with his ships.

No one knows for sure what happened to the ill-fated colony. But many believe that no sooner than Grenville's sails disappeared over the horizon that the Roanoke settlement was raided. As survivors fled northward they apparently were met by Powhatan's warriors who dispatched all who remained.

> *The reason why worry kills more people than work is that more people worry than work.*
>
> —Robert Frost

Many early attempts at setting up colonies failed simply because the settlers lacked the skills needed to ensure survival in the raw environment of this virgin land. Ironically, had the Natives not been so compassionate and sensitive to the newcomers' needs, most of the settlements would have failed because the colonist were more interested in finding gold and silver than in working the land.

Early Jamestown was no exception. Captain John Smith was a strong, intelligent leader, but half of the settlers who arrived in 1607 were described as "gentlemen" who considered themselves too refined for common work—the other half of the settlers were artisans, craftsmen, soldiers, and laborers—including a tailor, a barber, and two surgeons among them. Smith ordered them all to work a minimum of six hours per day. But while Smith was absent, the gentlemen not only refused to work, they also failed to take care of basic health needs. After 8 months, only 38 of the original 120 pioneers were still alive. New settlers arrived who were no better equipped than the first. Many died of starvation during the "starving time" of 1608–1609. Yet all about them lay fertile ground that only lacked planting of seed and its tilling and harvesting to sustain their needs.

When Smith first met Wahunsunacock—more commonly known as Powhatan, leader of the Powhatan Confederation of Algonquian tribes numbering over 14,000—he told him that the English did not plan on staying in Virginia. As more settlers arrived, however, the chief considered wiping out the small colony. He led an assault against the Jamestown settlers shortly after Smith departed for England to treat severe burns resulting from an explosion. Smith returned and made peace with Powhatan and the settlement began to thrive.

When his favorite daughter, Pocahontas embraced Christianity, changed her name to Rebecca and married John Rolfe, Powhatan pledged that his confederation would let the settlers live in peace—but he remained suspicious.

When his daughter and her son Tom sailed to England with John Rolfe in 1617, Powhatan sent his adviser, Tomocome, along to count how many Englishmen there were in the far land. The population of London was about 300,000 then. When Tomocome returned with his report, Powhatan became despondent at the loss of his daughter, who died of disease, and the realization that he and his people would never be able to stop the steady emigration of the English into their lands.

Chief Powhatan died in 1618 and leadership passed to his younger brother, Opechancanough, who believed that the English intended to seize the lands of his empire and exterminate his race. Opechancanough planned an attack on all English settlements throughout Virginia. The settlers lived primarily on farms spread around the countryside. They were lulled into believing that the Powhatans were not a threat—even to the point of inviting them into their homes for meals and teaching them how to hunt with firearms—which was strictly forbidden by the Jamestown Council.

Many Powhatans visited their English "friends" the evening of March 30, 1622. The settlers invited them to spend the night and share a meal with them the next day. At noon the next day, thousands of Natives painted for war rushed out from the dense woods to attack the remote settlements. Those Natives sharing a meal suddenly drew tomahawks and knives to slaughter their hosts and all their families. In the space of an hour three hundred and fifty men, women, and children were slain. Only the settlers living in Jamestown escaped the mayhem thanks to Chanco, a Christian Native who heard of the conspiracy the evening before the massacre. He hastened to Jamestown to warn a friend of impending danger. The alarm spread, but far too late to reach the more remote settlements. The people at Jamestown repulsed their assailants, and so averted the blow that might have extinguished the colony. Those at a distance, who survived the carnage, beat back the attacks and then fled to Jamestown. In the course of a few days, eighty inhabited plantations were reduced to eight. But a large part of the colony was saved.

The settlers retaliated quickly, waging all-out war on the Powhatan people. After the Indian uprising in 1622, the colonists gave up attempts to Christianize and live peacefully with the Powhatans. For over a decade, the English systematically razed villages, seizing or destroying crops, killing men and women, and capturing children. The English presence expanded rapidly as the Powhatan Empire declined. In 1644, Opechancanough rallied his people for one final attempt to force the English off their land. Hundreds of colonists were killed, but the English captured and shot Opechancanough. Treaties were made with his successor, severely restricting the Powhatan people's territory and confining them to small reservations. By 1669, the Powhatan population in the area dropped to about 1,800—almost one tenth of its size just 60 years before. By 1722, many of the tribes comprising the Chief Powhatan's former empire were reported extinct.

A QUEST FOR INDEPENDENCE

Government is not reason. Government is not eloquence. It is force. And, like fire, it is a dangerous servant and a fearful master.

—George Washington

As new settlers poured into the Atlantic seaboard colonies, the Native Americans were pushed beyond the Allegheny Mountains. The French employed them as allies during the French-Indian War. In fact, throughout the war, both the British and the French provided their Native allies with steel scalping knives specifically designed for that purpose. In frustration, the British became desperate and applied such horrible tactics as inflicting biological warfare on the Natives.

Toward the end of the war, Sir Jeffrey Amherst, commander in chief of the British forces in America, wrote to one of his underlings, a colonel who was then struggling against a particularly volatile Native American revolt on the western frontier, "Could it not be contrived to send smallpox among these disaffected tribes of Indians? We must use every stratagem in our power to reduce them." Evidently the colonel complied as evidence shows the offending tribes suffered from the ravages of smallpox after the colonel presented the Natives with gifts of infected blankets.

Eventually, diplomacy won out and the British signed the Treaty of Easton with the Ohio River Valley American Indians, which promised that if they chose not to fight for the French, the British would not settle the lands west of the Allegheny Mountains after the war.

However, conflicts between the colonialists and the Natives continued at great expense to the British Crown. King George II wanted the colonists to pay some of that expense with higher taxes. He issues the **Proclamation of 1763** prohibiting colonists from moving westward over the Appalachian Mountains.

Not only did this and other proclamations anger the colonists, but it frustrated their hopes of gaining new fertile land beyond the mountains and gave unfair advantage to Crown hunters and traders who monopolized the fur trade with the Native Americans.

Many colonial leaders fought with the British during the French and Indian War. They witnessed the raw beauty and bounty of the Western lands beyond the mountains. They saw it as their birthright and resented being restrained by a king thousands of miles away. The colonials declared their independence and prepared for war.

> *In the beginning of a change, the patriot is a scarce man, brave, hated, and scorned. When his cause succeeds however, the timid join him, for then it costs nothing to be a patriot.*
>
> —Mark Twain

Terror permeated every aspect of the American Revolution. This was civil war at its most intimate—neighbors against neighbors—family against family. In the beginning the loyalists (also called Tories) outnumbered the "Rebels." In fact, by 1779, there were more Americans fighting with the British than with Washington! There were no less than 21 regiments of loyalists (estimated to total 6,500 to 8,000 men each or between 150,000 and 160,000 men) serving in the British army. Washington reported his field army numbered 3,468. About a third of Americans opposed the Revolution.

Back in July 1776, there were an estimated 2.5 million people living in the colonies. English farmers settled in New England villages; Dutch, German, Swedish, Scotch-Irish, and English farmers settled on isolated Middle Colony farms; English and some French farmers settled on plantations in tidewater and on isolated Southern Colony farms in Piedmont; Spanish immigrants, mostly lower middle-class and indentured servants, settled the Southwest and California.

The loyalist majorities were primarily located in Pennsylvania, New York, and the southern colonies of South Carolina and Georgia. Throughout the colonies, they represented a large minority of the elites in their respective communities.

On October 6, 1775, Congress passed a resolution calling for the arrest of all loyalists who "are dangerous to the liberties of America." Many of the loyalists fled north to other British colonies in what would become Canada. Patriots arrested some of those who remained behind. Some were tarred and feathered, and others whipped, beaten, and abused by mobs of angry revolutionists. Lynchings occurred frequently on both sides of the conflict.

Even before the first shots of rebellion were fired at Concord Bridge the "long hunters" were blazing trails into the wilderness to the West. Many of these adventurers were Scotts and Irish immigrants with little love for the British. Many married Native American women—some of their offspring became leaders within their tribes. Others served as scouts and soldiers in the revolutionary militia that engaged the British and their Native allies operating in the frontier. Still others, such as Daniel Boone and Simon Kenton, began to lead settlers into this "dark and bloody ground" called Kentucky to settle.

The Native tribes resisted this incursion into what they considered to be their hunting grounds. Tecumseh, a minor Shawnee war chief, became the leading architect of a plan to unite all Native American tribes against inevitable domination by the whites flooding into their lands from the east. To this end, Tecumseh aligned his warriors with the British and launched a wave of terror throughout what would later become the states of Ohio, Kentucky, Indiana, and Tennessee. No pioneer settlement was safe. Captives were routinely scalped, burned at the stake, and their bodies mutilated.

Daniel Boone gained the respect of the Native tribes because of his courage. But even that failed to save his family from the horror of their oldest son, James, being skinned alive, nor his younger brother Israel of being slaughtered at the battle of Blue Licks (the last battle of the revolution) and having his body chopped to bits.

Following the revolution, thousands more settlers flooded across the mountains. Revolutionary War veterans were granted land in lieu of pay for serving. These grants were sold or given to younger sons and sometimes exchanged to satisfy debts.

As settlers cleared timber for farms and slaughtered buffalo wholesale for their valuable fur, the wildlife so central to Native American subsistence disappeared. In less than two generations huge herds were decimated. The tribes had no choice but to enter reservations or move west.

AMERICA'S FIRST WAR ON TERROR

Millions for defense, not a penny for tribute.

—Thomas Jefferson, 1800

As far back as the mid-1600s, marauding pirates sponsored by a loose confederation of North African states tied together by a tacit fealty to the Turkish Sultan in Constantinople. For almost 200 years, they robbed, raped, ransomed, and sent countless thousands of souls into slavery or worse with impunity. Victims reported being fed near-starvation rations, beaten regularly, and put to work breaking rocks on chain gangs, or scraping barnacles off ship hulls. Some of them were imprisoned for twelve years or more, waiting for their countrymen to save them—which at times came too late for those hostages, who died in captivity. Even the fledgling United States knuckled under following the revolution by paying $80,000 annually in tribute—despite the objections of then US Minister to France Thomas Jefferson.

In his autobiography, Jefferson wrote that in 1785 and 1786 he unsuccessfully "endeavored to form an association of the powers subject to habitual depredation from them. I accordingly prepared, and proposed to their ministers at Paris, for consultation with their governments, articles of a special confederation." Jefferson argued that "the object of the convention shall be to compel the piratical States to perpetual peace." Jefferson prepared a detailed plan for the interested states. "Portugal, Naples, the two Sicily's, Venice, Malta, Denmark and Sweden were favorably disposed to such an association," Jefferson remembered, but there were "apprehensions" that England and France would follow their own paths, "and so it fell through."

So much for Jefferson's efforts at forming the first coalition to oppose terror!

It's not surprising then, that when Jefferson became president in 1801, he was prepared to go it alone against the "barbarians." He refused to accede to Tripoli's demands for an immediate payment of $225,000 and an annual payment of $25,000.

The pasha of Tripoli then declared war on the United States. Although as secretary of state and vice president Jefferson had opposed developing an American navy capable of anything more than coastal defense, as president he dispatched a squadron of naval vessels to the Mediterranean. As he declared in his first annual message to Congress: "To this state of general peace with which we have been blessed, one only exception exists."

The American show of force quickly awed Tunis and Algiers into breaking their alliance with Tripoli. The humiliating loss of the frigate *Philadelphia* and the capture of her captain and crew in Tripoli in 1803, criticism from his political opponents, and even opposition within his own cabinet did not deter Jefferson from his chosen course during four years of war. The aggressive action of Commodore Edward Preble (1803–1804) forced Morocco out of the fight and his five bombardments of Tripoli restored some order to the Mediterranean.

However, it was not until 1805 that a treaty brought an end to hostilities following a successful attack on Derna, Tripoli, by US Navy Captain William Eaton commanding a rag-tagged force of about 50 Christian and about 500 Muslim mercenaries. The attack, spearheaded by Marine First Lieutenant Presley O'Bannon, commanding a detachment of seven Marines, supported by the American naval fleet under Commodore John Rogers, threatened to capture Tripoli and install the brother of Tripoli's pasha on the throne. Negotiated by Tobias Lear, former secretary to President Washington and now consul general in Algiers, the treaty of 1805 still required the United States to pay a ransom of $60,000 for each of the sailors held by the Dey of Algiers, and so it went without Senatorial consent until April 1806.

Nevertheless, Jefferson was able to report in his sixth annual message to Congress in December 1806 that in addition to the successful completion of the Lewis and Clark expedition, "the states on the coast of Barbary seem generally disposed at present to respect our peace and friendship."

In fact, it was not until the second war with Algiers, in 1815, that naval victories by Commodores William Bainbridge and Stephen Decatur led to treaties ending all tribute payments by the United States. However, European nations continued annual payments until the 1830s.

THE AMERICAN CIVIL WAR

Scorched Earth Tactics and Brother against Brother

> *I myself have seen in Missouri, Kentucky, Tennessee, and Mississippi, hundreds and thousands of women and children fleeing from your armies and desperadoes, hungry and with bleeding feet. . . . Now that war comes to you, you feel very different. You deprecate its horrors . . . . But these comparisons are idle. I want peace, and believe it can only be reached through union and war, and I will ever conduct war with a view to perfect an early success.*
>
> —General William Tecumseh Sherman, Letter to the Mayor and Councilmen of Atlanta, September 1864

On November 6, 1860, Abraham Lincoln, after declaring "government cannot endure permanently half slave, half free," is elected president—the first Republican—receiving 180 of 303 possible electoral votes and 40 percent of the popular vote. Southern Americans are outraged and on December 20, 1860, South Carolina secedes from the Union, followed within two months by Mississippi, Florida, Alabama, Georgia, Louisiana, and Texas. The fabric of our great Democracy is rent apart.

The Confederate States of America is formed with Kentuckian Jefferson Davis, a West Point graduate and former US Army officer and senator from Mississippi, as president. Earlier, Davis sought to find a compromise, declaring himself opposed to secession "as long as the hope of a peaceable remedy remained."

Abraham Lincoln was sworn in as 16th president of the United States of America on March 4, 1861. At 4:30 a.m. on April 12, 1861, Confederates under Gen. Pierre Beauregard opened fire with fifty cannons upon Fort Sumter in Charleston, South Carolina. The Civil War begins.

The war was savage! Some of America's best and brightest military men—many classmates at West Point and sons of distinguished patriots who fought in the America Revolution—ended up on opposite sides of the conflict. Slavery certainly was a prominent issue, but for many, loyalties remained stronger to their home states than to the Republic.

Without doubt, the war must be considered as one of the most horrific episodes of our history. Families were rendered apart, physically, psychologically, politically, and spiritually. Civilians suffered greatly as they were caught between the onslaughts of two great, determined armies. *This was terror to the nth degree!*

Although the Confederacy fought valiantly, they were outnumbered and out supplied by the industrialized North. Gettysburg, PA, became the South's high water mark. Thousands of Southern men were mowed down like sheaves of grain as they charged across open fields to attack fortified Union positions. Retreat became their only option.

Gen. Ulysses S. Grant, determined to break the Confederacy's back, began an aggressive strategy of "total war." He launched attacks on the South's infrastructure in 1964. Blockades restricted resupply. Railheads were captured and rails destroyed. Gen. William T. Sherman's relentless string of battles from Shiloh, to Nashville, Stone River, Chattanooga, Rocky Face Ridge, and Resaca forced the Southern armies backwards until they finally were almost eyeball to eyeball on the ridges surrounding Atlanta. On July 20, 1864, the battle of Atlanta was engaged.

The two armies maneuvered against each other. Both sides lost general officers to sharpshooters. For a time, it appeared that the Confederates might win the day—but Sherman's quick initiative to train an artillery barrage against a breech in his lines forced Gen. John Bell Hood to withdraw. Casualties were a total of 12,140—3,641 for the Union and 8,499 for the CSA.

Sherman then imposed a siege on the city until August 25. He entered the city on September 2. Two days later, Sherman issued an order for the inhabitants to leave the town within five days so the city might be prepared for military purposes. He deemed the measure humane, under the circumstances, for he expected the Confederates to attack him. When Hood pleaded with Sherman to withdraw his order, he replied, "God will judge me in good time, and He will pronounce whether it be more humane to fight with a town full of women and the families of a brave people at our backs, or to remove them in time to places of safety among their own friends." In a few days Atlanta was thoroughly evacuated of civilians.

After setting fires to Atlanta's warehouses and railroad facilities, Gen. Sherman, began his "March to the Sea" with 62,000 men on November 15, 1864. President Lincoln on advice from Grant approved the idea. "I can make Georgia howl!" Sherman boasts. He is determined to destroy the South's will to further prosecute the war.

Hood's Rebel Army of 23,000 is crushed on December 15/16, 1864, at Nashville by 55,000 Federals, including Black troops under Gen. George H. Thomas. The Confederate Army of Tennessee ceases as an effective fighting force.

Sherman reached Savannah, Georgia, on December 21, 1864, leaving behind a 300-mile long path of destruction 60 miles wide all the way from Atlanta. Sherman then telegraphs President Abraham Lincoln, offering him Savannah as a Christmas present. Meanwhile, Grant pressed the attacks in Virginia, forcing Gen. Robert E. Lee to withdraw from Richmond. Lee's nine-day retreat ended at Appomattox Court House where he surrendered his army at on April 9, 1865.

More American soldiers died during the US Civil War than in all the other wars Americans have fought in combined. Three million uniformed men entered the four-year struggle. A total of over 600,000 soldiers died (20%). Slightly more Union soldiers than Confederates paid the supreme sacrifice—with more than 400,000 deaths resulting from disease—many of them while suffering inhumane conditions in military prison camps. More than 400,000 were wounded—resulting in one in three of every man engaged becoming a casualty.

Slash and burn tactics on both sides deprived civilians of their homes, and forced them to suffer the loss of their crops, livestock, and any means of livelihood. But the real terror came later.

Union soldiers assigned to protect the innocent too often exceeded their orders as they sought revenge of their own. The federal government was forced to pass the *posse comitatus* law that restrained federal troops from being used for law enforcement duties.

But the terror was not one-sided. As slaves were freed and the vanquished Southerners returned to what was left of their homes, self-proclaimed vigilantes organized as night riders in silk sheets who sought to right perceived wrongs with beatings, torching of homes, and the lynching of innocents. Domestic terrorism would plague Americans for generations to come.

Chapter 6

Domestic Terrorism

> Our adversaries will keep evolving, national security and criminal threats will further converge, and old jurisdictional boundaries will become less and less relevant. If we are to address these trends successfully, we must be willing and able to evolve ourselves. The FBI must continue to build our intelligence capabilities, including a strong intelligence workforce. We must continue hiring and training personnel with technical expertise and foreign language skills. We must continue to seek new ways to share information and collaborate with partners in the Intelligence and Law Enforcement Communities. Above all, we must be agile, and encourage creativity, innovation, and strategic thinking. If we do all of these things, I am confident that we will out-network, out-think, and ultimately defeat our adversaries.
>
> —Robert S. Mueller III, Director, Federal Bureau of Investigation, Before the Senate Committee on Intelligence, February 16, 2005

Since the terrorist attacks of September 11, 2001 (9/11), domestic terrorists—*people who commit crimes within the homeland and draw inspiration from US-based extremist ideologies and movements*—have not received as much attention from federal law enforcement as their violent jihadist counterparts. This was not necessarily always the case. The Federal Bureau of Investigation (FBI) reported in 1999 that "during the past 30 years, the vast majority—but not all—of the deadly terrorist attacks occurring in the United States have been perpetrated by domestic extremists."

The information in this chapter draws heavily from FBI reports, the Southern Poverty Law Center's monitoring of extremist and hate groups, and Congressional committees, especially the August 2017 report on Domestic

Terrorism by the Congressional Research Service, a public policy research arm of the United States Congress.

The US government reacted to 9/11 by greatly enhancing its counterterrorism efforts. However, domestic terrorists broadly fit into the counterterrorism landscape, a terrain that since 9/11 has been largely shaped in response to terrorists inspired by *foreign* ideologies.

Today (perhaps in part because of the government's focus on international terrorist ideologies), it is difficult to evaluate the scope of domestic terrorist activity. For example, federal agencies employ varying terminology and definitions to describe it. Possibly contributing to domestic terrorism's secondary status as a threat at the federal level, a large number of those labeled as domestic terrorists do not necessarily use traditional terrorist tactics such as bombings or airplane hijackings. Additionally, many domestic terrorists do not intend to physically harm people but rather rely on alternative tactics such as theft, trespassing, destruction of property, and burdening US courts with retaliatory legal filings.

While plots and attacks by foreign-inspired homegrown violent jihadists have earned more media attention, domestic terrorists have been busy as well. It is worth noting that in terms of casualties on US soil, an act of domestic terrorism is second only to the events of 9/11. Timothy McVeigh's bombing of the Alfred P. Murrah Federal Building in Oklahoma City on April 19, 1995, claimed 168 lives and injured more than 500 others. Some estimates suggest that domestic terrorists are responsible for carrying out dozens of incidents since 9/11.

Much like their jihadist counterparts, domestic terrorists are often Internet savvy and use the medium as a resource for their operations. Terrorists are typically driven by particular ideologies. In this respect, domestic terrorists are a widely divergent lot, drawing from a broad array of philosophies and worldviews. These individuals can be motivated to commit crimes in the name of ideas such as animal rights, white supremacy, and abortion, for example. However, the expression of these worldviews (minus the commission of crimes) involves constitutionally protected activity.

Aware of the lines between constitutionally protected speech and criminality, domestic terrorists often rope themselves off from ideological (aboveground) elements that openly and often legally espouse similar beliefs. In essence, the practitioners who commit violent acts are distinct from the propagandists who theorize and craft worldviews that could be interpreted to support these acts.

Thus, in decentralized fashion, terrorist lone actors (lone wolves) or isolated small groups (cells) generally operate autonomously and in secret, all the while drawing ideological sustenance—*not direction*—from propagandists operating in the free market of ideas.

Domestic terrorists may not be the top federal counterterrorism priority, but they feature prominently among the concerns of some law enforcement officers. For example, in 2011, Los Angeles Deputy Police Chief Michael P. Downing included "black separatists, white supremacist/sovereign citizen extremists, and animal rights terrorists" among his chief counterterrorism concerns.

A 2014 national survey of state and local law enforcement officers found that sovereign citizens were "the top concern" among terrorist threats.

The violence related to protests in Charlottesville, Virginia, on August 12, 2017, also has raised the issue of domestic terrorism, particularly related to public discussions regarding a widely reported incident involving James Alex Fields, who according to witnesses drove his car into a group of people protesting a rally featuring white supremacists in Charlottesville on August 12. Fields allegedly killed one person and injured nineteen others in the incident. The Department of Justice (DOJ) opened a civil rights investigation into the incident, presumably pursuing possible hate crime charges. Additionally, Attorney General Jeff Sessions publicly stated that terrorism investigators were involved in investigating the incident, ostensibly exploring the possibility of characterizing it as an act of domestic terrorism rather than a hate crime.

ENFORCEMENT—JOINT TERRORISM TASK FORCES

Aside from the FBI, other federal agencies such as the Bureau of Alcohol Tobacco Firearms and Explosives (ATF) and the Internal Revenue Service (IRS) play a role in enforcement efforts to counter domestic terrorism. These agencies—as well as state and local law enforcement representatives—typically cooperate within the framework of Joint Terrorism Task Forces (JTTFs), multi-agency investigative units led by DOJ and the FBI across the country.

JTTFs are teams of police officers, federal agents, analysts, linguists, SWAT experts, and other specialists who investigate terrorism and terrorism-related crimes. Seventy-one of the more than 100 JTTFs operated by DOJ and the FBI were created since 9/11. About 4,000 federal, state, and local law enforcement officers and agents—more than four times the pre-9/11 total—work in them. These officers and agents come from more than 500 state and local agencies and 50 federal agencies.

The FBI considers JTTFs "the nation's front line on terrorism." They "investigate acts of terrorism that affect the U.S., its interests, property and citizens, including those employed by the U.S. and military personnel overseas." As this suggests, their operations are highly tactical and can involve developing human sources (informants) as well as gathering intelligence to thwart terrorist plots. JTTFs also offer an important conduit for the sharing of intelligence developed from FBI-led counterterrorism investigations with outside agencies and state and local law enforcement.

WHAT IS DOMESTIC TERRORISM?

In the most general statutory terms, a domestic terrorist engages in terrorist activity that occurs in the homeland. The FBI (the Bureau) has lead responsibility for terrorism investigations at the federal level.

The FBI generally relies on two fundamental sources to define domestic terrorism. First, the Code of Federal Regulations characterizes "terrorism" as including "the unlawful use of force and violence against persons or property to intimidate or coerce a government, the civilian population, or any segment thereof, in furtherance of political or social objectives."

Second, 18 U.S.C. Section 2331(5) more narrowly defines "domestic terrorism" and differentiates it from international terrorism and other criminal activity. This definition comes from Section 802 of the USA PATRIOT Act (P.L. 107–56). According to 18 U.S.C. Section 2331(5), domestic terrorism occurs primarily within US territorial jurisdiction, and involves (A)—acts dangerous to human life that are a violation of the criminal laws of the United States or of any State; (B) appear to be intended:

(i) to intimidate or coerce a civilian population;
(ii) to influence the policy of a government by intimidation or coercion; or
(iii) to affect the conduct of a government by mass destruction, assassination, or kidnapping.

The definitions cited above are too broad to capture what the FBI specifically investigates as "domestic terrorism." Besides the statutory definitions regarding the crime of domestic terrorism, the FBI has historically emphasized particular qualities inherent to the *actors* who engage in domestic terrorism. According to the Bureau, domestic *terrorists* do not simply operate in the homeland, but they also *lack foreign direction*. In fact, the Bureau's practical, shorthand definition of domestic terrorism is "Americans attacking Americans based on US-based extremist ideologies." The Department of Homeland Security (DHS) follows this construction.

On the surface, the FBI's shorthand definition for domestic terrorism appears straightforward. However, there is inherent ambiguity to it. Namely, some of the "U.S.-based extremist ideologies" driving what the Bureau views as domestic terrorism have international roots and active followings abroad. The ideologies supporting eco-extremism and animal rights extremism (discussed below) readily come to mind, and people have long committed crimes in their names outside the United States. At least in part, their origins lay in the United Kingdom. Nazism—with its German origins and foreign believers—is an element within domestic white supremacist extremism. Anarchism, the philosophy followed by anarchist extremists, also has long-standing European roots. The racist skinhead movement traces its origins abroad—to the United Kingdom—as well. These examples illustrate the FBI's challenge when it emphasizes US-based ideologies in its framing of domestic terrorism.

A few more issues make it hard to grasp the breadth of domestic terrorist activity in the United States. First, counting the number of terrorist prosecutions or plots in general has been difficult in the post-9/11 period. Second, there may be some ambiguity in the investigative process regarding exactly when criminal activity becomes domestic terrorism. Third, the federal government appears to use the terms "terrorist" and "extremist" interchangeably when referring to domestic terrorism. It is unclear why this is the case. Finally, and most importantly, which specific *groups* are and should be considered domestic terrorist organizations? The US government does not provide a public answer to this question. Rather, the federal government defines the issue in terms of "threats," not groups.

While statutory and practical federal definitions exist for "domestic terrorism," there is little clear sense of the scope of the domestic terrorist threat based on publicly available US government information. Most broadly, it has been said that in much of the post-9/11 period, the federal courts and DOJ may have applied different parameters when sorting, counting, and categorizing *all types* of terrorist prosecutions—let alone domestic terrorism cases.

A 2009 study (critiqued by DOJ) found that the US federal district courts, DOJ's National Security Division, and federal prosecutors rely on different criteria to determine whether or not specific cases involve terrorism at all. A bit more narrowly, in many instances, individuals considered to be domestic terrorists by federal law enforcement may be charged under non-terrorism statutes, making it difficult to grasp from the public record exactly how extensive this threat is. Regarding the prosecution of domestic terrorism cases, no separate federal crime of "domestic terrorism" exists.

Also, DOJ has noted that "although we do have at least one specialized [federal] statute aimed at animal enterprise terrorism, domestic terrorism cases often involve firearms, arson or explosive offenses; crimes relating to fraud; and threats and hoaxes." In some instances, the crimes committed by people the FBI describes as domestic terrorism suspects do not violate federal law. When this occurs, the Bureau "support[s] its local partners any way it can—sharing intelligence, offering forensic assistance, conducting behavioral analysis, etc." Thus, individuals considered domestic terrorists by federal law enforcement may not necessarily be federally charged as terrorists.

To further cloud matters, another category of criminal activity, hate crime, may appear to involve ideological issues. However, as described by one federal official, a "hate crime" "generally involve[s] acts of personal malice directed at individuals" and is missing the broader motivations driving acts of domestic terrorism. For investigators, distinguishing between "personal malice" and ideologically motivated actions may be difficult in specific cases. This suggests that sorting domestic terrorism from hate crimes depends on the degree of a suspect's intent. Did the suspect articulate an ideology belonging to an extremist movement? The grey area between domestic terrorism and hate crime hints that in some instances, suspects with links to domestic terrorist movements or ideologies supporting domestic terrorism may be charged with hate crimes.

It is unclear to what extent this influences how the government understands the threat posed by extremist movements that hold racist beliefs. If some individuals of this ilk commit crimes against police or judges, for example, is the government more apt to label this activity as terrorism while individuals sharing these same racist motivations but targeting ordinary citizens based on race, religion, disability, ethnic origin, or sexual orientation are charged with hate crimes?

On June 17, 2015, Dylann Roof shot and killed nine African Americans in the Emanuel African Methodist Episcopal Church in Charleston, South Carolina. Three churchgoers survived. DOJ pursued a case involving federal hate crimes and firearms charges. Roof was sentenced to death by a federal jury. According to DOJ, before the shooting incident, Roof had posted on a website a manuscript and photos "expressing his racist beliefs." Additionally, DOJ argued that he "wanted to increase racial tensions across the Nation, and sought retribution for perceived wrongs he believed African-Americans had committed against white people." Although it seems that Roof's alleged motives had ideological underpinnings in white supremacy, then-Attorney General Loretta E. Lynch noted in 2015 that DOJ pursued federal hate crimes charges because "we think that this is exactly the type of case that the federal hate crimes statutes were, in fact, conceived to cover."

The FBI's public description of the case of confessed would-be bomber Kevin Harpham is an example of how difficult it may be to characterize acts as domestic terrorism. Initially, the FBI viewed the case as domestic terrorism. In 2011, Harpham, motivated by white supremacist ideology, left a bomb—which never detonated—along the route of a parade in Spokane, Washington, honoring Dr. Martin Luther King Jr. The FBI's Northwest Joint Terrorism Task Force led the investigation. In prepared public remarks framing the "current state of the terrorism threat" from April 2011, the then-FBI Assistant Director for the Counterterrorism Division noted that Harpham's case was one of "several recent domestic terrorism incidents

[that] demonstrate the scope of the threat." Harpham eventually pled guilty to committing a federal hate crime and attempting to use a weapon of mass destruction. Thereafter, the Bureau described the case as the successful prevention of a "horrific hate crime."

EXTREMISM

The FBI's public formulation of "extremism" suggests two components. First, extremism involves hewing to particular ideologies. Second, it also includes criminal activity to advance these ideologies. Thus, according to this construction, an anarchist believes in a particular ideology—anarchism. An "anarchist extremist" is an anarchist who adopts criminal tactics.

The FBI and DHS have popularized the phrase "homegrown violent extremist" (HVE). It separates domestic terrorists from US-based terrorists *motivated by the ideologies of foreign terrorist organizations.* According to DHS and the FBI, a HVE is "a person of any citizenship who has lived and/or operated primarily in the United States or its territories who advocates, is engaged in, or is preparing to engage in ideologically-motivated terrorist activities (including providing support to terrorism) in furtherance of political or social objectives promoted by a foreign terrorist organization, but is acting independently of direction by a foreign terrorist organization." According to the FBI and DHS, an HVE is not a domestic terrorist—they are two distinct categories of terrorist actors.

DOJ and the FBI do not list domestic terrorist organizations publicly and officially. While not naming specific *groups*, DOJ and the FBI have openly delineated domestic terrorist *threats*. DOJ has identified domestic terrorism threats to include criminal activity by animal rights extremists, ecoterrorists, anarchists, antigovernment extremists such as "sovereign citizens" and unauthorized militias, black separatists, white supremacists, and abortion extremists.

ANIMAL RIGHTS EXTREMISTS AND ENVIRONMENTAL EXTREMISTS

The term "animal rights extremism" covers criminal acts committed in the name of animal rights. Environmental extremism—most often referred to as "ecoterrorism"—includes criminal acts committed in the name of the environment. These terms are not applied to groups or individuals involved with environmental movements or animal welfare protection/rights activism within the "confines of civil society and the rule of law."

Many of the crimes committed by both animal rights extremists and ecoterrorists are perpetrated by independent small cells or individuals who harass and intimidate their victims. These cells or lone actors engage in crimes such as vandalism, theft, the destruction of property, and arson. Most animal rights extremists and ecoterrorists also eschew physical violence directly targeting people or animals. Regardless, crimes committed by ecoterrorists and animal rights extremists have caused millions of dollars in property damage, and some have involved the intimidation and harassment of victims. These two types of extremism are often discussed together, because the two broader radical *movements* from which they draw their philosophical underpinnings have similar beliefs and overlapping membership.

The two *movements*—the Animal Liberation Front (ALF) and the Earth Liberation Front (ELF)—have the greatest reach among animal rights extremists and ecoterrorists. The ALF and the ELF are too diffuse to be called groups. Neither the ALF nor the ELF maintains formal rosters or leadership structures, for example. However, each communicates a sense of shared identity and attracts people who commit crimes in its name. They achieve this via "above-ground" wings.

Largely using websites, ALF and ELF supporters publish literature highlighting movement philosophies, tactics, and accounts (press releases) of recent movement-related criminal activity. Much of this involves protected speech and occurs in the public realm. Press releases allow "underground" extremists to publicly claim responsibility for criminal activity in the name of either movements while maintaining secrecy regarding the details of their operations. The ALF and the ELF do not work alone. Members of other entities such as Stop Huntingdon Animal Cruelty (SHAC) have committed crimes in the name of animal rights, for example.

Additional factors tangle our understanding of the ALF and the ELF. People can simultaneously participate in both. This may partly be true because the movements are so amorphous. The two movements also share similar agendas, and in 1993 they declared solidarity. All of this can play out confusingly in the real world. For example, an individual can commit a crime and claim responsibility for it online in the name of *both* the ALF and the ELF.

One case especially highlights intersections between the ALF and the ELF. In the late 1990s and early 2000s, the FBI uncovered a network that, according to DOJ, committed violent acts in the name of both the ALF and the ELF. The group included about twenty individuals and called itself "the Family." It was reportedly responsible for at least twenty-five criminal incidents totaling approximately $48 million in damages and disbanded at some point in 2001, due to law enforcement pressure on the group. The Family was responsible for an arson attack in 1998 at the Vail Ski Resort. Eight simultaneous fires damaged radio towers, ski lift towers, restaurants, and the ski patrol office at the Colorado site and totaled over $24 million in losses.

The ALF's moral code includes the belief that animals possess basic inalienable rights such as life, liberty, and the pursuit of happiness, and this suggests that animals cannot be owned. According to the ALF, the US legal system—which describes animals as property—is corrupt, and there exists a "higher law than that created by and for the corporate-state complex, a moral law that transcends the corrupt and biased statutes of the US political system." Simply put, the rights of one species do not trump the rights of others. To suggest otherwise is to be prejudiced, according to animal rights adherents.

For the ALF and other animal rights supporters, the favoring of one species, particularly humans, over others has a name: speciesism. For the ALF, speciesism is a "discriminatory belief system as ethically flawed and philosophically unfounded as sexism or racism, but far more murderous and consequential in its implications." Thus, the movement couches the theft or illegal release of animals used in research or for economic gain as "liberation." The ALF views the destruction of laboratory infrastructure or tools as the elimination of items used to enslave species who have the same rights as humans. Intimidation of scientists and employees of businesses tied to animal research or testing is rationalized as confrontation with "oppressors" or those who, in the eyes of movement adherents, abuse and murder animals.

Ecoterrorists are motivated by a mélange of environmental philosophies. There is no single formula for what constitutes the ideological makeup of an ELF follower, but several concepts likely play key roles in the movement. These are biocentrism, deep ecology, social ecology, and green anarchism. Biocentrism argues for the equality of all organisms. Deep ecology suggests that all species are part of "the larger super-organism that is nature." It criticizes industrialization and views modern human impact on the earth as negative and hearkens back to small communities centered on subsistence agriculture.

Social ecology suggests that hierarchical human society leads to social inequalities and environmental harm.

Green anarchism ascribes environmental harm to civilization and domestication and embraces the notion of "rewilding" or rejecting civilization and returning to a hunter-gatherer state to preserve one's natural surroundings.

ANARCHIST EXTREMISTS

According to the FBI, anarchist extremists commit crimes in the name of anarchist ideals. These ideals include belief that individual autonomy and collective equality are fundamental and necessary for a functional, civilized society. Anarchism resists the existing hierarchical structure of society that gives some people authority and control over others. According to anarchists, authority imbues power, and power always is used in illegitimate and self-serving ways by those who have it.

Anarchist extremists as well as anarchists engaging in constitutionally protected activity can oppose government, business, or social interests that they view as dangerous. As this suggests, anarchists advocate some form of revolution that realigns authority in the societies they desire to transform. However, adherents cannot agree to a single means for attaining revolutionary change.

As to be expected, anarchist activity is decentralized. In fact, a basic, temporary organizational structure—the affinity group—likely plays a larger role in shaping the work of US anarchists than any formal long-lasting entities or networks. Affinity groups are "autonomous militant unit[s] generally made up of between five-to-twenty individuals who share a sense of the causes worth defending and the types of actions they prefer to engage in. The decision-making process is anarchist, that is to say, egalitarian, participatory, deliberative, and consensual."

An affinity group often consists of a circle of friends. The friends coalesce around a specific objective and break apart when they achieve their desired ends. Individual groups can band together in "clusters" and clusters can coordinate their efforts, if need be. The efforts can be legal or illegal, violent or nonviolent, covert or open. These structures have a long history among anarchists, but other movements use them as well. Also, anarchists can engage in what they call "black bloc" tactics. These involve secretive planning for public—often criminal—activity in which participants, typically dressed in black, act en-masse.

Adding to the sprawling nature of the anarchist movement, some adherents also participate in the ALF and the ELF. These three movements share general philosophical tenets such as opposition to globalization and capitalism.

In the aftermath of the violence related to protests in Charlottesville, Virginia, on August 12, 2017, there has been media attention devoted to the confrontational tactics of antifascist protesters known as "Antifa." Antifa beliefs can dovetail with the broad anti-government and anti-capitalist views that are part of anarchism. One observer has described Antifa as "a radical pan-leftist politics of social revolution applied to fighting the far right. Its adherents are predominantly communists, socialists and anarchists who reject turning to the police or the state to halt the advance of white supremacy." Antifa beliefs appear to inspire autonomous groups to track the activities of neo-Nazis, and most "anti-fascist organizing is nonviolent." However, some Antifa protesters are willing to violently clash with people publicly advocating what Antifa protesters would see as fascist views, particularly white supremacists. Antifa protesters characterize such violence targeting people as defensive.

The FBI has described anarchist extremists as typically being "event driven," meaning they show up at political conventions, economic and financial summits, environmental meetings, and the like. They usually target symbols of Western civilization that they perceive to be the root causes of all societal ills—that is, financial corporations, government institutions, multinational companies, and law enforcement agencies. They damage and vandalize property, riot, set fires, and perpetrate small-scale bombings.

Law enforcement is also concerned about anarchist extremists who may be willing to use improvised explosives devices or improvised incendiary devices. Anarchist extremists in the United States have been involved in illegal activity during mass protests surrounding events such as the 1999 World Trade Organization Ministerial Conference in Seattle, Washington.

Anarchist extremists reportedly committed crimes during the 2008 Republican National Convention in St. Paul, Minnesota. To coordinate their protests during the convention, some anarchists formed what they called the "RNC Welcoming Committee" (RNCWC). In September 2007, the RNCWC developed a plan to broadly organize the activities of affinity groups intending to disrupt the convention. Law enforcement infiltrated and undermined these efforts, arresting 800 people, including eight involved with the RNCWC. Initially, in Minnesota state court, the eight "had been charged with felonies: first-degree damage to property and second-degree conspiracy to riot. Prosecutors added a more serious charge of conspiracy to riot in furtherance of terrorism, which was later dismissed." Five of the eight pled guilty to gross misdemeanor charges in 2010. The others had all of the charges they faced dismissed.

On April 30, 2012, five men who reputedly had anarchist sympathies were arrested for purportedly scheming to blow up a bridge near Cleveland, Ohio. One was convicted on charges related to the plot. The four others pled guilty. The plot was apparently timed to coincide with peaceful protest activity arranged by Occupy Cleveland, an offshoot of the Occupy Wall Street movement. Occupy Cleveland representatives stated that the would-be bombers "were in no way representing or acting on behalf of Occupy Cleveland." An FBI sting operation led to the arrests. Reportedly, the group relied on an undercover FBI agent to supply them with two inert bombs that the conspirators believed were functional.

Criminal acts involving anarchist extremists do not have to be event-driven. For example, Eric G. King pled guilty to using explosive devices to commit arson in a failed 2014 attempt to ignite a fire at the Kansas City, Missouri, office of US Representative Emanuel Cleaver II. The incident occurred when the office was unoccupied, but King had posted violent commentary regarding police to social media.

WHITE SUPREMACIST EXTREMISTS

The term "white supremacist extremism" (WSE) describes people or groups who commit criminal acts in the name of white supremacist ideology. In May 2017, the FBI and DHS released a joint intelligence bulletin stating that white supremacists "were responsible for 49 homicides in 26 attacks from 2000 to 2016—more than any other domestic extremist movement."

At its core, white supremacist ideology purports that the white race ranks above all others. WSE draws on the constitutionally protected activities of a broad swath of racist hate-oriented groups active in the United States ranging from the Ku Klux Klan to racist skinheads. Some of these groups have elaborate organizational structures, dues-paying memberships, and media wings. Additionally, many individuals espouse extremist beliefs without having formal membership in any specific organization.

A large proportion of white supremacists dualistically divide the world between whites and all other peoples who are seen as enemies. Particular animus is directed toward Jews and African Americans. In fact, a common racist and revisionist historical refrain is that the civil rights movement succeeded only because Jews orchestrated it behind the scenes. Scholars indicate that white supremacists believe in racial separation and that society discriminates against them. To them, whites have lost "ground to other groups and—extreme measures are required to reverse the trend."

All of this has been encapsulated in a slogan known as the "Fourteen Words": "We must secure the existence of our race and a future for white children." This was coined by David Lane, a member of a violent terrorist group active in the 1980s. The Fourteen Words have been described as "the most popular white supremacist slogan in the world."

Neo-Nazism and its obsession with Adolph Hitler and Nazi Germany is also a prominent component of WSE in the United States. The father of American neo-Nazism, George Lincoln Rockwell, became publicly active in the late 1950s. According to one scholar, Rockwell laid down three concepts that have shaped neo-Nazism ever since. For his followers, he reconfigured the racial notion of "white," broadening it beyond "Aryan" to include people of Southern and Eastern European descent. Additionally, Rockwell denied the Holocaust. He also encouraged tying neo-Nazism to religion, and some of his followers took up the obscure creed of Christian Identity.

Aside from racial superiority, a dualistic view of the world, and neo-Nazism, at least two other broad concepts shape white supremacy in the United States. They are the inevitability of violent conflict, and a belief that conspiracies hostile to white supremacy shape the existing world. It can be said that WSE broadly shares these concepts with the militia movement (discussed below).

The FBI has stated that white supremacists "commonly anticipate" waging war against their opponents. For example, the inevitability of RAHOWA—an acronym for "racial holy war"—is a central tenet of the neo-Nazi Creativity Movement, which has its roots in the Church of the Creator, a racist group founded by Ben Klassen in 1973. Klassen, who committed suicide in 1993, argued that whites had no choice but to wage war against non-whites. Likewise, some white supremacists use racism to interpret apocalyptic imagery from Norse mythology embodied in Odinism. Most Odinists are not racists, however.

Conspiracism has been defined as "the idea that most major historic events have been shaped by vast, long-term, secret conspiracies that benefit elite groups and individuals." Conspiracy theories are not the province of a particular movement or group. Regardless, conspiracy theories can particularly shape the outlooks and actions of white supremacist extremists. Media sources have stated that Richard Poplawski—convicted of shooting and killing three Pittsburgh police officers in April 2009—believed that a Zionist conspiracy controlled government and major corporations in the United States.

As in Poplawski's example, anti-Semitism plays a prominent role in the racist conspiracies of many white supremacists. Many anti-Semites—as well as anti-government extremists—believe in something they call the Zionist-Occupied Government (ZOG). ZOG refers to the federal government, which adherents contend is "controlled or manipulated by international Jewish interests." On its website, one WSE group sold versions of a video game titled "ZOG's Nightmare." Gameplay involves shooting nonwhites while being chased by a police agency controlled by Jews. According to adherents, ZOG is said to control the media, arts, religion, science, and education.

In the 1980s and 1990s, a small number of figures dominated white supremacist circles. They were intimately linked to their own relatively cohesive organizations. By the early 2000s, these groups fragmented as they lost their leaders.

Two particularly well-known white supremacist figures died in the early 2000s. William Pierce, head of the National Alliance, died in 2002. Richard Butler, leader of Aryan Nations, died in 2004. Both Pierce and Butler articulated clear ideologies that attracted followers and drew upon resources such as rural headquarters/compounds to sustain their organizations. By the early 2000s, the National Alliance even had a substantial revenue stream estimated at $1 million annually generated from a publishing company and record labels it owned as well as dues. The deaths of Butler and Pierce exacerbated the downfall of both organizations. The decline of these groups also resulted from a number of other forces, such as infighting among members and pressure from law enforcement and watchdog groups. Other groups have emerged since to promote white supremacist ideas. Several movements espousing such views participated in the August 2017 rally in Charlottesville, VA, that led to violence mentioned earlier.

The following are the two of the most prominent white supremacist movements.:

National Socialist Movement

One of the longest-standing white supremacist organizations active in the United States, the National Socialist Movement (NSM) has benefitted from the decline of other groups as well as new leadership in the form of Jeff Schoep. The NSM also capitalized on the expansion of the Internet in the early 2000s. The group, which emerged in 1974, is a descendant of the American Nazi Party, and until the 1990s and early 2000s "it operated only on the fringes of the neo-Nazi movement."

As of 2008, the group had around 500 members and close associates throughout the United States. The NSM is flexible about membership, allowing its members to also participate in other white supremacist organizations.

Individuals allegedly tied to the NSM at some point in their lives have run afoul of the law.

- In Minnesota in April 2012, Joseph Benjamin Thomas was indicted on drug-related charges, and Samuel James Johnson was indicted on weapons-related charges. Purportedly the two were tied to NSM—at one point Johnson had allegedly served as its leader in Minnesota. The duo had reportedly formed their own white supremacist group, gathering weapons and ammunition and planning to attack the government and other targets. In June 2012, Johnson pled guilty to "one count of being a felon in possession of firearms." In July 2012, Thomas pled guilty to "possession with intent to distribute more than 50 grams of high-purity methamphetamine."
- William White, a onetime member of the NSM and founder of his own white supremacist organization, has faced charges in several criminal cases. In September 2014, he was found guilty on charges related to a December 2013 indictment that included "five counts of making threats in aid of extortion over the Internet and one count of the unlawful use of identification information in furtherance of those offenses." He threatened a Florida judge, a state attorney, and an FBI agent, with kidnapping, torture, rape, and murder. According to DOJ, White included the families of these individuals in his threats.

The officials that White threatened had been involved in prosecuting suspects tied to the American Front, a white supremacist organization in Florida. White apparently hoped that his threats would somehow secure the release of the American Front suspects. In January 2011, White was convicted of soliciting violence online against the jury foreman in *U.S. v. Matthew Hale*. In April 2011, a federal judge reversed White's conviction. Upon appeal, the conviction was reinstated. In an unrelated case, in December 2009, White was convicted of four counts of communicating threats in interstate commerce and one count of witness intimidation. One of the convictions for communicating threats in interstate commerce was later reversed.

The witness intimidation charges involved White reportedly attempting to "delay or prevent the testimony" of African Americans in a discrimination case. According to publicly available information, in 2005 and 2006, White was involved with NSM, for a time serving as its national spokesman. His activity with NSM ceased after he had a falling out with Schoep.

Racist Skinheads

In the United States, racist skinheads have a legacy stretching back to the 1980s. However, skinhead culture originated in the United Kingdom in the late 1960s and today has a global reach. Since the early 2000s, the movement in the United States has been characterized by a proliferation of regional groups or crews rather than a united core organization. In law enforcement circles, racist skinheads have a reputation for violence. This is reinforced by hate filled white power music and literature.

They foster their reputed toughness through their appearance (shaved heads or close-cropped hair, white power tattoos) and dress (bomber jackets, "braces" (suspenders), steel-toed boots).

Skinheads emerged as a non-racist movement among British working-class youth in the late 1960s. These early skinheads rejected the hippie lifestyle and embraced elements of Jamaican culture, particularly reggae and ska music. As immigration from South Asia to the United Kingdom grew, some white British skinheads embraced racism and neo-Nazism. This racist skinhead variant of the subculture materialized in the US Midwest and in Texas in the early 1980s.

In the mid-1990s, many US-based racist skinhead groups allied with one another to form the Hammerskin Nation (HSN). HSN eventually developed chapters throughout the United States and in Europe. It had its own annual meeting/concert called Hammerfest, ran a record label, and had a publishing company. In the early 2000s, other groups such as the Outlaw Hammerskins, Hoosier State Hammerskins, and Ohio State Skinheads challenged HSN for preeminence. These groups saw HSN as "elitist."

In January 2010, the FBI released a bulletin that, among other things, emphasized that some racist skinheads formed the most violent segment of WSE adherents. This supported the findings in a 2008 FBI assessment. Between 2007 and 2009, skinheads were involved in thirty-six of the fifty-three violent incidents the FBI identified in the United States as being tied to WSE proponents. The Bureau has stated that "violence is an integral part of the racist skinhead subculture."

Elements within the fractious movement even target one another. These criminal acts are typically unrehearsed and opportunistic, targeting nonwhites and other religious and social minorities. At least one exception involved greater levels of planning. One man was convicted and two others pled guilty in a Connecticut case that involved the illegal sale of firearms and homemade grenades. The scheme included multiple meetings between late 2008 and early 2010 to negotiate the transactions, prepare the firearms, and assemble the grenades. The trio was tied to a skinhead group known as Battalion 14 (originally called the Connecticut White Wolves). They sold the weapons to a convicted felon working as an FBI cooperating witness. The informant posed as a member of the Imperial Klans of America, a Ku Klux Klan organization. Two others in the case, including the leader of Battalion 14 and a man not tied to the group, were acquitted of charges.

ANTI-GOVERNMENT EXTREMISTS

As mentioned above, DOJ considers both unauthorized militias and sovereign citizens as antigovernment extremists. *Neither militia membership nor advocacy of sovereign citizen tenets makes one a terrorist or a criminal.* However, in some instances both militia members and sovereign citizens have committed crimes driven in part by their ideologies.

MILITIA EXTREMISTS

The militia movement became prominent in the 1990s as a collection of armed, paramilitary groups formed to stave off what they perceived as intrusions of an invasive government. Central to this is a fear of firearm confiscation by a federal government thought to be out of control. Some adherents also believe in anti-Semitic and racist ideologies.

Regardless, most militia members engage in constitutionally protected activity. Militia groups typically coalesce around a specific leader. Groups can run training compounds where they rehearse paramilitary tactics, practice their survival skills, and receive weapons instruction and lessons in movement ideology. Some militia groups also maintain websites for recruitment and fundraising.

Extremists within the movement who run afoul of law enforcement "tend to stockpile illegal weapons and ammunition, trying illegally to get their hands on fully automatic firearms or attempting to convert weapons to fully automatic. They also try to buy or manufacture improvised explosive devices."

Segments of the militia movement believe that the US government either is run by some hidden conspiracy or is an overreaching sham. Some see a "New World Order" controlling US institutions such as the media and the federal government. They contend that this is partly fostered by international organizations such as the United Nations. From this perspective, these organizations sap American sovereignty. Some militia supporters believe that agents of an unauthentic "Shadow Government" are interested in seizing lawfully owned firearms as part of a plan to undermine democracy. Importantly, others in the militia movement hold that the federal government has overstepped its constitutional bounds.

Some militia members assert that they have "the right to organize, purchase and use firearms, and enforce the law against agents of the government who behave unconstitutionally."

A small minority of Americans who held anti-government fears formed militias largely in response to two incidents in the early 1990s. These were confrontations between federal law enforcement and private citizens at Ruby Ridge, ID, and at a site near Waco, TX. Both involved warrants related to firearms violations.

- In August 1992, Randy Weaver and his family were engaged in an eleven-day standoff with federal law enforcement agents. Randy Weaver had failed to appear in court on firearms-related charges in 1991. Subsequently, an unsuccessful operation to arrest Weaver led to the death of his fourteen-year-old son and a US Marshal. It also precipitated the standoff. During the standoff, Weaver and a friend were shot and wounded. An FBI sniper also shot and killed Weaver's wife, Vicki. Weaver was eventually found guilty of failing to appear in court on the gun charges that played a role in the standoff. In October 1993, he was sentenced to 18 months in jail and a $10,000 fine. In 1995, Weaver received a $3.1 million settlement in a wrongful death suit filed against the US government. The events at Ruby Ridge helped precipitate the militia movement, whose members tend to view Randy Weaver as a hero and demonize the federal government.
- The militia movement also emerged because of the fifty-one-day standoff between federal law enforcement and a religious sect named the Branch Davidians near Waco. On February 28, 1993, an unsuccessful attempt by Bureau of Alcohol Tobacco Firearms and Explosives (ATF) agents to

arrest the sect's leader, David Koresh, initiated the events near Waco. He was wanted on suspicion of federal firearms and explosives violations. Four ATF agents and six Branch Davidians died in a gunfight during the operation. Protracted discussions followed between federal negotiators and Koresh. These failed. On April 19, federal agents assaulted the Davidian compound, which caught on fire. At least seventy-five Branch Davidians perished in the assault.

If the incidents involving the Weavers and the Branch Davidians helped form the militia movement, Timothy McVeigh's bombing of the Alfred P. Murrah Federal Building in Oklahoma City on April 19, 1995, helped usher in a temporary decline. In the bombing's aftermath, militia groups received greater law enforcement scrutiny. The bombing claimed 168 lives, and until 9/11 was the largest single act of terrorism on US soil. The militia movement declined after the bombing.

Although McVeigh's bombing cannot fully account for a dip in militia activity, it affected the movement by causing some groups to temper their rhetoric while others grew more extreme, and militias became more marginalized.

The militia movement has experienced resurgence in the last decade, likely driven by growing antipathy toward the federal government. A few individuals still draw inspiration from McVeigh. Jerry Drake Varnell was arrested in August 2017 after he allegedly intended to damage or destroy a bank in downtown Oklahoma City by reportedly attempting to detonate what he thought was a bomb in a van. But the inert explosive device was based on materials provided by an undercover FBI agent as part of an investigation that nabbed Varnell, who allegedly was inspired by the Timothy McVeigh's 1995 bombing. Varnell purportedly hewed to militia extremist beliefs and, according to investigators, expressed an interest in founding a small militia.

Two widely reported incidents in recent years have attracted militia and other anti-government extremists.

- In 2014, multiple individuals led by Cliven Bundy engaged in an armed standoff with police officials at Bundy's ranch in Bunkerville, NV, preventing the execution of a court order related to a dispute involving the grazing of cattle on federal lands. One individual involved was sentenced to sixty-eight years in prison for a variety of crimes related to his involvement in the standoff.
- For forty-one days in late 2015 and early 2016, Ammon and Ryan Bundy (sons of Cliven Bundy) led numerous individuals in armed occupation of federal property at the Malheur National Wildlife Refuge in Oregon.

Although the ringleaders of the broader occupation—including the Bundy brothers—were acquitted in federal court of many of the most serious charges levied against them, a violent incident occurred during the occupation. Oregon State Police officers fatally shot LaVoy Finicum. The shooting occurred after Finicum reportedly drove a pickup truck at high speed toward a police roadblock, reputedly endangering officers before careening into a nearby snowbank. According to law enforcement officials, Finicum subsequently left the vehicle, refusing to comply with police commands, and reached for a gun in his pocket before he was shot to death. In the fallout related to the Malheur National Wildlife Refuge takeover, Cliven Bundy was arrested on charges tied to the 2014 Bunkerville standoff. Several other examples highlight how some militia adherents have reportedly engaged in criminal activity since 9/11.

- In November 2011, the FBI arrested four retirees, Samuel J. Crump, Ray H. Adams, Dan Roberts, and Frederick W. Thomas, who allegedly formed a fringe militia group and planned violent attacks on government officials. The group, based in northern Georgia, purportedly had ties to an unnamed militia organization. According to DOJ, the quartet "discussed multiple criminal activities, ranging from murder; theft; manufacturing and using toxic agents; and assassinations in an effort to undermine federal and state government and to advance their interests." Between June and November 2011, Roberts and Thomas met with an FBI undercover agent "to negotiate the purchase of material for the plot: a silencer for a rifle and conversion parts to make a fully automatic rifle, as well as explosives." In October, plotters reportedly discussed making ricin, a deadly poison derived from castor beans. In April 2012, Roberts and Thomas pled guilty to conspiring to obtain an unregistered explosive device and silencer. Crump and Adams were found guilty of "conspiring to make ricin to be used as a weapon in January 2014. Also, they were found guilty of one count each of possessing a biological toxin for use as a weapon." Adams was acquitted of "attempting to develop, produce and possess a biological toxin."
- In June 2012, three individuals were found guilty in Anchorage, AK, of conspiracy and firearms charges related to a scheme purportedly led by Francis "Schaeffer" Cox. He and his followers allegedly plotted "a potential retaliatory response to any attempt by law enforcement to arrest Cox, who had an outstanding bench warrant for not attending a trial over a misdemeanor weapons charge." They were members of the Alaska Peacemaker's Militia based in Fairbanks, AK, and also held sovereign citizen beliefs. The plotters supposedly codenamed their plan "241 (two for one)," because they reputedly intended to kill two government officials for every militia member killed in the operation.

SOVEREIGN CITIZEN MOVEMENT

The FBI defines the sovereign citizen movement as "anti-government," involving people "who believe that even though they physically reside in this country, they are separate or 'sovereign' from the United States. As a result, they do not accept any government authority, including courts, taxing entities, motor vehicle departments, or law enforcement." However, simply holding these views is not a criminal act, and numerous movement adherents solely exercise their beliefs via constitutionally protected activities.

The ideas behind the movement originated during the 1970s with a group known as the Posse Comitatus and enjoyed some popularity in extremist circles during the 1980s and 1990s. Early on, the movement featured white supremacist elements, but this has not kept some African Americans from subscribing to its ideals in recent years. In the 1990s, the movement attracted 250,000 followers and was marked by the FBI's standoff with a group known as the Montana Freemen that lasted 81 days. Recent estimates suggest a membership of around 300,000.

For the most part, the sovereign citizen movement is diffuse and includes few organized groups. The FBI suggests that sovereigns "operate as individuals without established leadership and only come together in loosely affiliated groups to train, help each other with paperwork [critical to some of their schemes], or socialize and talk about their ideology."

The movement involves leaders described as "gurus" who proselytize online, in print publications, or via in-person seminars. These gurus rouse followers into believing a conspiracy theory in which the legitimate federal government has been replaced by a government designed to take away the rights of ordinary citizens. This shares the same broad interplay between concepts of legitimate and illegitimate rule seen in the New World Order and WSE theories about ZOG. Gurus can also promote illegal techniques that individuals can use to supposedly cut their ties to the federal government or avoid its reach, particularly when it comes to taxation.

Sovereign citizens reject the legitimacy of much of the US legal system. Many believe that the Fourteenth Amendment "shifted the nation from its original common-law roots with states' rights to a federal corporation that legally enslaved everyone." According to movement members, the amendment ushered in an illegitimate federal government by supposedly abrogating individual rights and replacing them with a system that "granted privileges through contracts such as marriage and driver's licenses, gun permits, and property codes."

By ignoring all sorts of laws, avoiding taxes, disregarding permit requirements, and destroying government-issued identification documents, some sovereign citizens have tried to cut formal ties with what they perceive as an illegitimate regime. Sovereigns have filed court documents stating that they are not US citizens. They have also created bogus financial documents to harass or defraud their enemies. Sovereign citizens have in some instances created fictitious entities and used fake currency, passports, license plates, and driver licenses.

In 2009, a federal jury found three men guilty of "conspiring to use and sell fraudulent diplomatic credentials and license plates that they believed allowed their customers to subject to being stopped, detained, or arrested by law enforcement personnel."

In 2003, Ronald K. Delorme developed the Pembina Nation Little Shell Band of North America into a sovereign citizen group. It is a sham Native American tribe that anyone can join to try and avoid taxes and government-imposed costs, such as auto registration fees. For example, in June 2010, a sheriff's deputy in Florida pulled over John McCombs when the law enforcement official noticed a Pembina Nation Little Shell license plate on the motorcycle McCombs was driving. McCombs presented a fraudulent letter of diplomatic immunity and an invalid Pembina Nation Little Shell vehicle registration.

Some sovereign citizen fraud appears to be motivated by economic opportunism rather than ideology. This includes "pyramid schemes, other investment schemes, bogus trust scams, real estate fraud, and various types of tax frauds as well as more esoteric scams—ranging from immigration fraud to malpractice insurance fraud."

In November 2011, husband and wife Monty and Patricia Ervin were convicted in federal court of conspiring to defraud the United States as well as three counts of tax evasion. In addition, the federal jury convicted Patricia of structuring transactions to avoid bank reporting requirements. The couple allegedly had not filed federal income tax returns between 2000 and 2008, denied their US citizenship, and dubbed themselves "sovereign" when the IRS investigated. The Ervins earned more than $9 million from investment properties they owned. A group of self-proclaimed sovereign citizens in North Georgia was indicted in March 2011 for using sovereign schemes to allegedly steal millions of dollars' worth of real estate.

Some avowed sovereign citizens have been involved in violent altercations with law enforcement officers:

- In June 2016, Gavin Long shot and killed three police officers and wounded three others in Baton Rouge, LA, before police killed him. He had articulated sovereign citizen views online.
- In May 2010, two self-professed sovereign citizens were involved in a violent confrontation with West Memphis, TN, police officers. During a traffic stop, Joe Kane fired an AK-47 assault rifle and killed two officers. Kane

and his father Jerry fled the scene. Law enforcement sighted their vehicle in a nearby parking lot ninety minutes later. The duo died in the ensuing shootout, which also wounded two more officers. The FBI had investigated Jerry Kane five years before the murders because he was allegedly traversing the United States peddling what the FBI termed a debt elimination scheme.

In August 2013, authorities in Las Vegas, Nevada, arrested two reputed Sovereigns, David Allen Brutsche and Devon Campbell Newman, after a local investigation uncovered the duo's purported schemes to kidnap and kill police officers. The Las Vegas Metropolitan Police Department ran an undercover investigation to nab the two after they encountered Brutsche in what has been characterized as a series of vehicle stops during which Brutsche would espouse his Sovereign Citizen beliefs that he wasn't bound by the law Metro officers were enforcing. Brutsche pled guilty to felony kidnapping conspiracy in February 2014 after prosecutors abandoned the two most serious charges—conspiracy to murder and attempted armed kidnapping.

BLACK SEPARATIST EXTREMISTS

DOJ includes black separatism in its list of movements that potentially spawn domestic terrorists. However, most black separatists solely engage in constitutionally protected behavior. Since 9/11, there has been little public discussion of federal investigations involving black separatist extremists. One group exhibiting what can be described as black separatist views, the New Black Panther Party for Self Defense (NBPP), has received national attention over several incidents.

The NBPP emerged in the early 1990s, and it is not tied to the Black Panthers from the 1960s. Watchdog groups, like the Southern Poverty Law Center, have described the NBPP as "a virulently racist and anti-Semitic organization whose leaders have encouraged violence against whites, Jews, and law enforcement officers," as well as "the largest organized anti-Semitic and racist black militant group in America."

The NBPP, which denies that it is a hate group, engages in "high-profile" rhetoric at rallies or demonstrations intended to encourage confrontation with authorities. The group's actions occur "on behalf of the poor or disadvantaged, involving the ready display of firearms." As an example of the rhetoric the group uses, an NBPP representative characterized the March 2011 shooting death of a drug suspect in Jacksonville, FL, as "a violent act of terrorism" committed by police. Soon after the shooting, the Jacksonville Sheriff's Office said that the confrontation involved undercover officers serving a search warrant at an apartment. Officers claimed that inside the apartment, the victim—an alleged drug dealer with a criminal record—was holding a firearm.

In an infamous recent incident, Micah Johnson shot and killed five police officers in Dallas, Texas, in July 2015. He reportedly "liked" groups on Facebook tied to black separatism and may have been involved at some point with the NBPP in Houston. Johnson purportedly told police that he wasn't affiliated with any groups at the time of the shooting. He died in the altercation with police.

In 2008, the Philadelphia, PA, chapter of the NBPP was involved in a case that generated public controversy. A 2009 civil suit filed by DOJ claimed that two NBPP members wearing the group's paramilitary uniforms loitered around the entrance to a 2008 federal general election polling station in Philadelphia. One of the NBPP members allegedly carried a nightstick. According to DOJ, some poll watchers feared for their safety because of this activity. Philadelphia police officers responding to claims of voter intimidation removed the nightstick-wielding NBPP member and allowed the other to remain (the latter was a certified poll watcher). Police asked people at the polling station whether they had been threatened by the two individuals. All those questioned replied that they had not. However, at least one individual claimed that the presence of the two NBPP members had been intimidating.

The NBPP disavowed the actions of its two members. In May 2009, DOJ voluntarily dismissed claims against defendants in the case, and a July 2009 letter from ten Members of Congress to DOJ's Inspector General questioned the decision to do so. DOJ's Office of Professional Responsibility (OPR) investigated, and in March 2011, OPR issued a report which argued that DOJ officials did not act inappropriately regarding the matter.

ABORTION EXTREMISTS

The vast majority of activists who either favor or oppose abortion engage in constitutionally protected activity. However, abortion extremism involves crimes committed based on such beliefs. Over the past two decades, most abortion-related violence appears to have targeted abortion providers. Eighty-four instances of "extreme violence" targeting abortion providers and clinics occurred in the United States from 1997 through 2015, according to the National Abortion Federation—a group that supports abortion rights and tracks criminal activity intended to limit access to abortion services. These cases involved shootings, bombings, arson incidents, and acid attacks.

Since 1993, eight clinic workers have been murdered by anti-abortion extremists in the United States. Because of a wave of violence focused on abortion providers in the 1980s and early 1990s, Congress passed and President Clinton signed into law the Freedom of Access to Clinic Entrances Act (FACE Act) (18 U.S.C. §248) in 1994. As with other types of domestic terrorism investigations, it is unclear exactly which incidents of violence perpetrated against abortion providers the FBI considers terrorist acts.

Two violent incidents have been prominent in recent years. In 2015, Robert Dear Jr. allegedly killed three people and wounded nine others in a shooting at a Planned Parenthood facility in Colorado Springs, CO. The 2009 murder of George Tiller, an abortion provider, received significant public attention. On January 29, 2010, Scott Roeder was convicted of first-degree murder and two counts of aggravated assault for killing Tiller. Roeder shot Tiller while the latter was at church on May 31, 2009. Roeder was sentenced to life in prison with no possibility of parole for fifty years.

In January 2012, Bobby Joe Rogers was charged in the firebombing of a Pensacola, FL, abortion clinic on New Year's Day 2012. The bombing destroyed the clinic, which had been targeted in the past. In February 2012, a federal grand jury indicted him on two counts—arson and damaging a reproductive health facility. He pled guilty to the charges in July 2012.

One underground network that supports attacks on abortion clinics is the Army of God (AOG). The loosely structured organization openly promotes anti-abortion violence. However, its members deny that they are terrorists. They also deny that attacks against clinics and abortion providers constitute violent activity, because they see it as "Godly work." AOG first made headlines with the 1982 kidnapping of a doctor and his wife, both of whom ran an abortion clinic in Illinois. Three individuals who claimed membership in AOG were responsible. The group disseminates a manual that "is a 'how to' for abortion clinic violence. It details methods for blockading entrances, attacking with butyric acid, arson, bomb making, and other illegal activities. The manual contains anti-abortion language as well as anti-government and antigay/lesbian language. The manual begins with a declaration of war on the abortion industry."

Eric Rudolph, who in the late 1990s bombed an abortion clinic near Atlanta, Georgia, and one in Birmingham, Alabama, published his writings on the AOG website.

ANIMAL RIGHTS EXTREMISTS AND ECOTERRORISM

US law enforcement, some business groups, and some scientists—among others—have stressed that animal rights extremists (and ecoterrorists) are a security and law enforcement concern. In 2008, the FBI stated that animal rights extremists and ecoterrorists together posed a serious domestic terrorism threat for several reasons, including the number of crimes attributed to animal rights extremists and ecoterrorists (between 1,800 and 2,000 incidents accounting for more than $110 million in damages from 1979 to early 2009), the broad pool of victims (such as large pharmaceutical corporations, scientific laboratories, ski resorts, automobile dealerships, individual researchers, and lumber companies), and the movement's rhetoric and destructive tactics. In March 2012, the FBI suggested that the threat from ecoterrorists may be declining.

Animal Enterprise Terrorism Act (P.L. 109-374)

The Animal Enterprise Terrorism Act (P.L. 109-374; AETA) expanded the federal government's legal authority to combat animal rights extremists who engage in criminal activity. Signed into law in November 2006, it amended the 1992 Animal Enterprise Protection Act (P.L. 102-346; AEPA). Namely, the AETA amends the federal criminal code to revise criminal prohibitions against damaging or interfering with the operations of an animal enterprise to include intentional damage or loss to any real or personal property and intentional threats of death or serious bodily injury against individuals (or their family members, spouses, or intimate partners) who are involved with animal enterprises.

The AETA expanded the AEPA to include both successful and attempted conspiracies. It also prohibits intentionally placing a person in "reasonable fear" of death or serious bodily injury while damaging or interfering in the operations of an animal enterprise. The AETA revised and increased monetary and criminal penalties. It also stipulates that it does not prohibit First Amendment-protected activity.

HOW SAFE IS AMERICA?

Domestic terrorist attacks have come nowhere near the devastation of 9/11. However, it is worth noting that (as mentioned above) Timothy McVeigh's bombing of the Alfred P. Murrah Federal Building in Oklahoma City on April 19, 1995, claimed 168 lives and injured more than 500 others. It ranks as the second-deadliest terrorist attack on US soil, behind only the devastation wrought by Al-Qaeda on 9/11.

Domestic terrorists feature prominently among the concerns of some law enforcement officers. For example, in 2011, Los Angeles Deputy Police

Chief Michael P. Downing described violent Islamists such as Al-Qaeda, Hezbollah, and Hamas as Los Angeles' main terrorist threats "along with three other terrorist categories: black separatists, white supremacist/sovereign citizen extremists, and animal rights terrorists."

A 2014 national survey of state and local law enforcement officers found that among terrorist threats, sovereign citizens were "the top concern." In one 2008 study, state police agencies "overwhelmingly reported" dangerous domestic extremist groups present in their jurisdictions.

Of course, most followers of extremist viewpoints pose no threat: most of them are not going to do anything but bore their relatives and friends with ridiculous papers and treatises. However, four broad themes speak to the threat posed by domestic terrorists. First, domestic terrorists likely have been responsible for numerous incidents since 9/11. Second, a large number of those labeled as domestic terrorists do not necessarily use major terrorist tactics such as bombings or airplane hijackings. Third, domestic terrorists—much like their violent jihadist analogues—are often Internet savvy and use the medium as a resource for their operations. Fourth, domestic terrorism can be seen as a somewhat decentralized threat often involving lone wolves and movements operating under the model of leaderless resistance.

In August 2017, the New America Foundation reported that people who can be characterized as domestic terrorists have killed seventy-five people in the United States since 9/11, whereas jihadist terrorists have killed ninety-five.

"NONVIOLENT" STRATEGIES

While some terrorists engage in violence, others commit much different crimes that do not physically harm people. This latter group differs from their homegrown violent jihadist counterparts, who are often bent on killing or harming people. Two types of activities that avoid visiting violence upon people but are commonly associated with subjects of domestic terrorism investigations stand out. First, many animal rights extremists, ecoterrorists, and anarchist extremists believe in "direct action." This typically involves what movement members would characterize as *nonviolent but criminal* protest or resistance activities furthering the movement's ideology.

While direct action has a long legacy among anarchists, in recent years the ALF and the ELF have played a large role in articulating its meaning. Second, "paper terrorism" is a term used to describe some of the *nonviolent* criminal activity committed by sovereign citizens involving the filing of fraudulent documents in the hopes of harassing enemies or bilking state or federal tax authorities.

DIRECT ACTION

Anarchist extremists, animal liberation extremists, and environmental extremists refer to much of their operational activity as "direct action." This term has a long history, and it can be used to describe legitimate protest such as letter writing campaigns or work stoppages. However, it also includes criminal activities such as sabotage and arson.

ALF and ELF members understand that criminality and direct action are one and the same. *The Animal Liberation Primer*, a movement resource, highlights criminality in the actions of supporters: "Anyone working in the ALF is a criminal. You have to begin to think like a criminal." ALF and ELF members also generally view direct action as nonviolent and heroic. Using politically charged language, the ALF allegedly styles itself along the lines of the Underground Railroad, freedom fighters in Nazi Germany, anti-Apartheid protestors, US civil rights activists, and Palestinian groups opposing Israel. The ELF views constitutionally protected protest as "state sanctioned" and eschews such activity. The ELF, much like the ALF, also wraps itself in the mantle of reformers and describes itself as inheriting the spirit of Luddites, abolitionists, suffragists, and even the American revolutionary-era Boston Tea Party.

The ALF frames its version of direct action as "economic sabotage" or "ethical vandalism." The ALF supports the destruction of property and intimidation of individuals and businesses considered by the movement to be involved in the exploitation of animals. Cells and individuals linked to the ALF also engage in trespassing and theft, or what they perceive as "live liberations" or "rescuing" animals from "the horrors of exploitation" and human use by stealing them from places such as legitimate research facilities or farms. Economic sabotage can be virtual. The North American Animal Liberation Press Office (NAALPO) has carried claims of cyber hacking incidents in the name of animal rights. NAALPO is one of the web-based vehicles used by ALF supporters to publicize criminal activities claimed on behalf of the movement.

Like the ALF, the ELF's discussions of direct action also revolve around economic sabotage. The ELF rejects legal protest tactics partly for what it views as pragmatic reasons—"because they have been proven not to work, especially on their own." Economic sabotage in the name of environmentalism has a long history, perhaps stretching back to the 1950s, and has been called "monkeywrenching," a term taken from a 1975 novel, *The Monkey Wrench Gang* by Edward Abbey. The book depicts such activity.

A guidebook that describes monkeywrenching offers what can be interpreted as a call to arms for would-be extremists: It is time for women and men, individually and in small groups to act heroically in the defense of the wild, to put a monkey wrench into the gears of the machine that is destroying natural diversity. Though illegal, this strategic monkeywrenching can be safe, easy, and—most important—effective. The guidebook also defines monkeywrenching as nonviolent by stressing that it should never target people or other forms of life.

ARSON AND EXPLOSIVE DEVICES

Federal officials are especially concerned about the use of incendiary devices and explosives by animal rights extremists and ecoterrorists. In congressional testimony from 2005, then-ATF Deputy Assistant Director Carson Carroll stated that the "most worrisome" trend regarding animal rights extremists and ecoterrorists was their "willingness to resort to incendiary and explosive devices."

In January 2012, NAALPO issued a communiqué in which "unnamed activists" claimed responsibility for setting fires that damaged fourteen tractor trailer rigs at the Harris Ranch, a cattle feedlot in Coalinga, CA. The perpetrators used containers of accelerant, kerosene-soaked rope, and digital timers to set the blazes. According to the communiqué, the fires apparently embodied a reaction to "the horrors and injustices of factory farming."

THE INTERNET AND DOMESTIC TERRORISTS

In the counterterrorism world, there has been much concern regarding violent jihadist use of the Internet and social media. However, domestic terrorists also are computer savvy and active online. A web presence may help extremist groups—some relatively small, with rosters in the 100s or fewer—educate their existing membership and forge a group identity. Also, in many instances they can use websites to focus on outsiders to propagandize, socialize, and recruit new adherents.

A few domestic terrorists also have exploited the web to harm their targets. White supremacists have long been using computer technology to communicate and interact. As one study has suggested, white supremacists "were among the very early users of the electronic communication network that eventually evolved into the Internet." Among a variety of findings, the study indicated that white supremacist extremist websites were possibly an effective recruiting tool that the groups exploited. Membership forms are available on some sites. Others exhibit multimedia material, and some actually retail items such as music and video games.

The Internet allows individuals and groups to connect with one another and to disseminate ideology. It also enables groups to manage how others perceive them. Many white supremacist sites claim that their sponsoring groups are nonviolent and not even racist.

Some white supremacists may be unwilling to affirm their views in public spaces such as work, school, or in street demonstrations. To them, the virtual realm is an important antidote. A 2006 study in the Journal of Political and Military Sociology suggested that "free spaces" in both the real and virtual worlds—where conflict with nonbelievers will be minimized—are important for adherents. In them they can "meet, articulate, and support their views." Supremacists can turn to virtual free spaces to receive indoctrination into movement culture, key narratives outlining movement grievances, adopt ideologies, and "talk of violence against 'racial enemies.'"

A 2016 study found that Americans espousing white supremacist ideals on the social-media platform Twitter outnumber the supporters of the foreign terrorist organization known as the Islamic State by many measures: on Twitter, ISIS's preferred social platform, American white nationalist movements have seen their followers grow by more than 600 percent since 2012. Today, they outperform ISIS in nearly every social metric, from follower counts to tweets per day.

Some domestic terrorists also engage in cyberattacks. According to DOJ, an animal rights extremist cell (SHAC USA, Inc.) active between 2001 and 2004 listed online the personal information—names, addresses, phone numbers—of workers at a firm it was targeting. (The business uses animals in its research.) The extremist cell likely devised the list to help focus the activities of the group's online followers. In some cases, the published information included the names of spouses and children of employees, license plate numbers, churches attended by the employees, as well as the schools their children attended. The websites used by the extremist cell also posted suggestions for action by supporters—including what it described as the "top 20 terror tactics." Supporters across the United States vandalized victims' homes and automobiles and engaged in cyberattacks against the research firms and other companies tied to it, among other activities.

Domestic terrorism can be described as a decentralized threat. Domestic terrorism suspects generally operate on their own or in small, independent cells. In other words, they do not necessarily belong to organizations with cohesive, well-articulated leadership structures or cadres. However, independently acting domestic terrorism suspects are not necessarily isolated, adrift, and cut off from any outside contact or influence. Some take ideological cues from broader movements or groups espousing extremist ideas. These groups or movements publicly disavow violent criminal behavior and engage in constitutionally protected activities. This dynamic—the interplay between aboveground groups or movements proffering extremist dogma or ideology (protected speech) that is then consumed and acted upon by independent underground groups or cells who commit crimes—is a critical feature of domestic terrorism.

LONE WOLVES

Lone wolves have committed crimes in the names of a number of domestic terrorism movements. For example, according to the FBI, when it comes to violence attributed to WSE, lone wolves play a prominent role. Lone wolves filter in and out of WSE groups. They can either get dismissed from these groups because of their "violent tendencies" or voluntarily leave because they find the organizations too passive.

There is little research on the lone wolf phenomenon and no universally accepted definition of the term. Lone wolves can hew to broader ideological causes and use them to justify their actions. This suggests that lone wolves potentially adopt the ideas of broader terrorist movements while not claiming formal membership in them. Defining exactly what "formal membership" constitutes leads to debate regarding whether or not some individuals acted as lone wolves or part of larger movements. For example:

- On January 29, 2010, Scott Roeder (also discussed above) was convicted of first degree murder and two counts of aggravated assault for killing abortion provider George Tiller. Roeder allegedly had "connections with militant abortion foes but few formal ties with known groups." Some supporters of abortion rights consider his contacts among anti-abortion adherents as evidence of possible conspiracy. Meanwhile, some anti-abortion activists have stressed that Roeder was a lone wolf. He remains the only person convicted in Tiller's murder.
- Described as a neo-Nazi and white supremacist, James von Brunn reportedly shot and killed a security guard at the US Holocaust Memorial Museum in Washington, DC, in June 2009. In January 2010, the eighty-nine-year-old von Brunn died in federal prison before he could be tried.

Because lone wolves are not plugged into terrorist organizations, distinguishing them from individuals who commit hate crimes can also be difficult. In these cases, as mentioned above, the FBI likely attempts to determine whether the motives involved were personal (hate crime) and not focused on broader ideologies (domestic terrorism).

Because they operate alone, lone wolves pose a particular challenge to law enforcement. Lone wolves do not participate in terrorist networks or training camps that can be infiltrated or whose communications can be traced. They do not rehearse their schemes or practice their criminal skills with conspirators who can potentially act as cooperating witnesses.

Lone wolves, do, however, have weaknesses. Their lack of tradecraft may make it harder for lone wolves to engage in large-scale attacks. Likewise, lone wolves do not necessarily experience the reinforcement of a closely knit terrorist social network. They cannot rely on others to assist them in any type of complicated plot.

Regardless, apparent lone wolf attacks can be lethal. For example, in March 2017, James Harris Jackson, allegedly acting alone, used a two-foot-long sword to stab to death Timothy Caughman, an African American individual who Jackson is said to have encountered while trolling New York City for victims. According to New York District Attorney Cyrus Vance Jr., Jackson "prowled the streets of New York for three days in search of a black person to assassinate in order to launch a campaign of terrorism. Jackson, who reportedly espoused white supremacist beliefs, faces state-level murder and terrorism charges.

Richard Poplawski shot and killed three Pittsburgh police officers in April 2009. He has been described as a "white supremacist" lone wolf and had posted anti-government messages on racist websites. He was sentenced to death.

Omar Mateen was the lone wolf attacker who killed forty-nine people and injured more than fifty others when he attacked the Pulse nightclub in Orlando in the early hours of Sunday, June 12, 2016. This was both a hate crime and an act of domestic terrorism. The venue was a gay nightclub and the attack remains the deadliest act of violence against the LGBT community in the United States. Also, at one point Mateen told a crisis negotiator by phone, "I pledge my allegiance to Abu Bakr al-Baghdadi of the Islamic State." The FBI has concluded that Mateen was not formally linked to the Islamic State and had no direct ties with it but had been radicalized.

DEALING WITH A HERITAGE OF HATE

The Southern Poverty Law Center points out that somewhere in America someone commits a hate crime every hour. Each day at least eight blacks, three whites, three gays, three Jews, and one Latino become hate crime victims—and every week a cross is burned somewhere in the United States. Their website is one of the best resources to find out more about extremist groups and tracks more than 1,600 groups operating across the country at www.splcenter.org.

"Hate in America is a dreadful, daily constant. The dragging death of a black man in Jasper, Texas; the crucifixion of a gay man in Laramie, Wyo.; and post-9.11 hate crimes against hundreds of Arab Americans, Muslim Americans and Sikhs are not 'isolated incidents.' They are eruptions of a nation's intolerance. Bias is a human condition, and American history is rife with prejudice against groups and individuals because of their race, religion, disability, sexual orientation or other differences. The 20th century saw major progress in outlawing discrimination, and most Americans today support integrated schools and neighborhoods. But stereotypes and unequal treatment persist, an atmosphere often exploited by hate groups."

Is America unique in having this legacy of prejudice and hate? No—but it is our diversity and our openness that permits bias, prejudice, and hate to flourish for a time. We unknowingly welcome bigots and fanatics to our shores as we welcome down-trodden and oppressed immigrants from the world to share in our freedoms and opportunities.

Ironically, some ethnic groups who suffered from prejudice elsewhere choose to inflict prejudice on others when they arrive in America. Biases of every kind arrived with our early settlers, pioneers, and émigrés—and each new wave seems to suffer in their own way until they are assimilated or accepted—the latest being Muslims who suffer the negative fallout generated by the actions of militant Islamists bent on murder and mayhem to advance their distorted view of how the world should be.

America encourages tolerance and understanding of all religious and political views. But tolerance is a two-edged sword. It may encourage diversity—but it may also permit racism, prejudice, and bias to thrive for a season.

Unfortunately, America is not safe—either from enemies foreign or domestic!

Chapter 7

International Terrorism

Terrorism in the modern sense emerged around the mid-nineteenth century. During the French Revolution in the 1780s and 1790s, the harshest period of the rule of the Committee of Public Safety was known as "The Terror." This is the first known use of the term "terrorism" although this is not strictly applicable to the modern use of the word "terrorism."

Ironically, the word "assassin" is more than 1,000 years old and comes from a Shi'ite Muslim sect called the Nizari Isma'ilis, who were also known as the Hashashans because they used hashish when they went out to kill. They were fighting on two fronts—against the Sunni Muslims and against the armies of the Crusaders.

The Hashashans, or Brotherhood of the Assassins as they become known, launched a reign of terror including the slaughter of women and children. How little has changed!

In 1867 the Irish Republican Brotherhood, a revolutionary nationalist group supported by Irish-Americans, carried out a number of attacks in England. These were the first acts of "republican terrorism," which were to become a feature of British history for the next 120 years, and these Fenians were the forerunners of the Irish Republican Army. The ideology of the group was Irish nationalism and their continued aim was to drive the British out of Ireland.

About the same time, the intelligentsia in Russia, impatient with the slow pace of Tsarist reforms, worked to transform peasant discontent into open revolution. Anarchists like Mikhail Bakunin argued that progress was impossible without destruction. Their objective was the complete destruction of the state. Anything that contributed to this goal was regarded as moral. As explosives become more available, the anarchists had a powerful weapon to challenge those charged with protecting the state—the military and the police. Organized into secret societies like the People's Will, Russian revolutionaries launched a campaign of terror against the state that climaxed in 1881 when Tsar Alexander II was killed in a bomb attack. The authority's response was swift and ruthless and led to more than twenty years of state-sponsored terrorism. However, rather than stamping out dissent, the repression fueled the revolutionaries and in 1918 they overthrew the Tsarists and set up their own communist government.

In 1893 the Internal Macedonian Revolutionary Organization (IMRO) was founded in Thessaloniki, now in Greece but then part of the Ottoman Empire. Driven by Slavic nationalism, the organization acquired a reputation for ferocious attacks, including the 1934 assassination of Alexander I of Yugoslavia during a state visit to France.

The Fenians and the IMRO were the prototypes of today's terrorism. Both were fighting for what they believed in—national freedom. Ironically neither was totally successful. While an independent Irish Republic was created, the six northeastern counties (Ulster) continue to be part of the British Isles and Macedonia became absorbed into Yugoslavia and didn't achieve its independence until 1991.

One problem with the term "terrorist" is that it depends on which side you are in. It is ironic that many people once branded as terrorists are now called freedom fighters and considered heroes by their own people. These include Israeli statesman Yitzhak Shamir, Palestine Liberation Organization leader Yasser Arafat, and former South African President Nelson Mandela; all at one time were hunted men with prices on their head.

Three major events in the first half of the twentieth century—the Russian Revolution and the two world wars—helped fuel nationalist sentiments. The Russian Revolution showed that a popular armed movement could topple an authoritarian, repressive regime. And peace treaties signed after the two world wars inflamed many populations as they saw their countries disappearing or being annexed.

As a result, many groups emerged in the twentieth century. They had different goals and different methods for achieving them. For some the goal was politics, for others it was religion and some were simply groups of anarchists out to destroy the established order.

The most striking difference between terrorism in the first half and the second half of the twentieth century was tactics. In the first five decades, targets were singled out and by and large, care was taken to minimize innocent casualties. The last five decades of the century were marked by a crescendo in violence with the aim being to cause maximum damage and destruction. The perpetrators wanted to cause maximum economic damage and truly cause terror.

State sponsored terrorism became a feature of the Cold War. Different countries supported different terrorist groups, providing bases, funding, arms, and training, with the aim that these groups would destabilize rival governments.

The year 1966 is one of the dates that can be attributed to the internationalizing of terrorism. That year Cuba hosted the Tri-Continental Conference although the Soviet Union laid down the agenda. Terrorist groups from Europe, the Middle East, Asia, Africa, and Latin America attended and many started to build alliances. This collaboration included sharing finances, resources, and intelligence and led to a number of joint operations. It also launched a decade-long wave of terror in Europe and the Middle East.

In Germany, the Red Army Faction (German) allied itself with Black September (Palestinian); in France, Action Direct (French) allied with the Red Army Faction and the Red Army Brigade (Italian); in Japan, the Japanese Red Army allied with the Popular Front for the Liberation of Palestine.

The late 1960s also saw the rapid rise of religious-based terrorism and the expansion of radical Islam in the Middle East. After the 1967 war in which Israel defeated Jordan, Egypt, and Syria, and took control of the Golan Heights (from Syria), East Jerusalem and the West Bank (from Jordan), and the Gaza Strip and the Sinai Peninsula (from Egypt), the use of conventional war as a means to destroy Israel ended and the use of terror to focus attention on Israel and the Palestinians in the occupied territories began. This was accompanied by an increase in the use of suicide bombers.

Suicide bombing became a widely used target by the Japanese military in World War II. Kamikaze pilots caused massive damage by crashing their explosive-packed aircraft into targets, especially ships at sea. Following World War II, Viet Minh "death volunteers" were used against the French colonial army. In 1972 three Japanese used grenades and automatic rifles at Lod airport in Tel Aviv killing twenty-six people and wounding more than a hundred. The attackers were members of the Japanese Red Army (JRA) a terrorist organization created in 1969 and allied to the Popular Front for the Liberation of Palestine. It was the first suicide operation conducted in Israel.

Suicide bombing, however, was not a product of the twentieth century. It has been a tactic used by extremely committed groups for more than 1,000 years.

It emerged again in the 1980s as a highly effective, highly lethal weapon, particularly among extreme militant groups whose members were willing to die as martyrs for the cause. It was a tactic used extensively by the Tamil Tigers in Sri Lanka and the Kurdistan Workers Party and later adopted by Hamas, Hezbollah, al-Qaeda, and other extremist Islamic terrorist groups.

It is also a very cost-effective weapon if you have enough volunteers. A walking bomb can gain access to crowded places that a vehicle cannot. The bomber is usually wearing a specially constructed vest designed to carry up to 40 lbs. of explosives and shrapnel—usually nails, screws, and ball bearings—to inflict maximum damage. The vest is worn under everyday clothing and when detonated, the blast fans out in all directions. Shrapnel is responsible for about 90 percent of all casualties.

As terrorism increased globally, a number of states actively supported particular groups. Iran supported Hezbollah; Libya supported Abu Nidal; and Iraq, Cuba, Sudan, and Algeria provided training camps and economic and political support to other terrorist groups.

If the 1960s was the decade of the internationalization of terrorism, the 1970s was the decade of air terrorism with more than a score of American and European passenger aircraft being hijacked.

The rise to power of Ayatollah Khomeini in 1979 further inflamed the situation as militant Islamic groups increased their attacks in the Middle East.

The 1980s was the decade of hostage taking and this was also a significant period because for the first time the United States became a principal target. Americans overseas were taken hostage and US embassies and interests were bombed.

Another major change in terrorist tactics between the 1960s and 1980s, and the 1980s to the end of the century was communications. Terrorist acts are committed to draw attention to a cause as part of the struggle to achieve their stated objectives. From the 1960s to 1980s, terrorist acts were almost always followed by a statement from one group or another, claiming responsibility. The rationale (a) commit a terrorist attack to get publicity and (b) claim responsibility to get further publicity and then, hopefully, encourage people to think about the merits of your cause. Toward the end of the century, the aim switched to getting maximum publicity courtesy of satellite television, cable news, and round the clock coverage of events.

A major downside of terrorism, however, is that the more horrific the atrocity, the more coverage it generates, so terrorists have a vested interest in planning and staging ever bigger events.

When the first US soldier was killed in Iraq, it was a major front page news story, subsequent deaths as tragic as they were got pushed further and further down an inside page until they no longer even merited a mention. In order to get back on to the front pages, the terrorists have to carry out ever more horrific attacks. And in the process, the publicity also attracts more recruits and sympathizers to their cause.

TYPES OF TERRORISM

Security experts have identified four main types of terrorism, which are as follows:

- Nationalist-Separatist Terrorism (N): acts of violence undertaken by those seeking to establish a separate state for their own national or ethnic group, for example, the Irish Republican Army, Basque Homeland & Liberty, and the Kurdish Workers Party.

- Religious Terrorism (R): the use of violence by those seeking to further what they see as divinely commanded purposes, often targeting broad categories of "enemies" in an attempt to bring about sweeping changes, for example, al-Qaeda, Hezbollah, and Hamas.
- Left-Wing Terrorism (L): violence undertaken by those seeking to destroy capitalism and replace it with a communist or socialist regime, for example, Red Army Faction, German Red Brigades, Prima Linea, and Symbionese Liberation Army; and
- Right-Wing Terrorism (RW): the use of violence by those seeking to dispense with liberal democratic governments, for example, Combat 18, KKK, and Timothy McVeigh.

Some of the most prominent of these groups in the latter part of the twentieth century are as follows:

Action Directe (L) is a French left-wing urban terrorist group that committed a series of assassinations and violent attacks in France in the 1980s. It was founded in 1977 as a merger by two extreme left wing groups, GARI (Groupes d'Action Révolutionnaire Internationalistes) and NAPAP (Noyaux Armés pour l'Autonomie Populaire). It carried out more than fifty assassinations and attacks, including a machine gun assault on the headquarters of the French Employers Federation. The group was outlawed by the French government in 1982; and in 1984 it allied with the German Red Army Faction. The three leaders were captured in 1987 and sentenced to life imprisonment. One was released on health grounds in 2004 but the other two remain incarcerated.

Angry Brigade (L) is a British communist urban guerilla group founded in 1967 and responsible for around twenty-five bombings in the early 1970s, targeting banks, embassies, and the homes of politicians. While property damage was caused, only one person was slightly injured in all the attacks. The group's leaders were arrested in 1972, and following a lengthy, highly publicized trial they were each sentenced to ten years imprisonment.

Armed Islamic Group (N, R) is an Algerian Islamist extremist group committed to overthrowing.

Baader-Meinhof (L) is also known as the Red Army Faction (RAF). The German group had its roots in the student riots of the late 1960s. Calling themselves urban guerrillas, the RAF was responsible for killing scores of prominent Germans during their twenty-year terror campaign. In 1968 four activists, including Gudrun Ensslin and Andreas Baader, decided to firebomb a number of German department stores. They were arrested, and while on trial, journalist Ulrike Meinhof published a number of sympathetic articles. Baader went into hiding after the trail but was captured in April 1970. The next month, in a violent shootout, he was sprung from jail by Meinhof and his lawyer, Horst Mahler. Baader, Ensslin, Mahler, and Meinhoff fled to the Middle East but were expelled from a Palestinian guerrilla training camp because they refused to follow the rules. They returned to Germany and robbed banks to fund their operation. They also carried out a number of arson attacks before being captured in 1972. While in prison their RAF supporters carried out kidnappings and hostage taking to try to force the authorities to release their jailed colleagues, including a deadly attack on the German embassy in Sweden. Their trial started in 1975 and RAF attacks continued against government figures and prosecutors. In 1976 Meinhof committed suicide in her cell during the trial which continued until April 28 when all defendants were found guilty of murder and terrorist acts and sentenced to life imprisonment. RAF attacks, including kidnappings, continued in an attempt to get the leaders released. In October 1977 a Lufthansa plane was hijacked by four Arab sympathizers and eventually flown to Mogadishu, Somalia, where an elite German police unit stormed the plane, killing three of the four hijackers and releasing all the hostages. That night Baader was found dead in his cell from a gunshot wound and Ensslin was found hanged in her cell. A third jailed member died the next day in the prison hospital from wounds. The authorities claimed they had all committed suicide using weapons smuggled in by their lawyers. In the early 1980s, RAF survivors allied with Action Directe and the group continued to murder industrialists and bomb. They claimed responsibility for bombing the US Air Force Base at Ramstein which killed three people. After German unification, evidence was discovered that the group had received financial and logistical support from the Stasi, East Germany's secret police. The group's last major attack was in 1993 when it bombed a prison. In 1998 it published a memorandum, declaring the group dissolved.

Combat 18 (RW) is a neo Nazi organization founded in Britain in 1991, between far right groups and football hooligans. The "18" in the name comes from Adolf Hitler's initials, the first and eighth letter of the alphabet. The group claimed it had been set up to defend far right groups from attacks by anti-fascist organizations. It encouraged and participated in violence against immigrants and left wing groups. In a nationwide police operation between 1998 and 2000, scores of C18 members were arrested and charged and many sent to prison. In 1999 a one-time member of C18 conducted a brief nail bomb campaign against black, Asian, and gay communities in London. One bomb killed three people including a pregnant woman and 100 others. The group was widely discredited by the end of the 1990s because of suggestions that it had been infiltrated by undercover MI5 agents.

Contras (N, L), a short term for contrarrevolucionarios (counter-revolutionaries), were the armed opponents of Nicaragua's Sandinista regime which in 1979 overthrew Anastasio Somoza Debayle and ended the family's forty-three-year rule. The Contras received support from both the Argentinian government and the CIA, and carried out a series of indiscriminate attacks over an eight-year period. A ceasefire negotiated in 1988 led to the Contras disarming, and in 1990 elections an anti-Sandinista coalition was voted in. A small group of disaffected Contras briefly took up arms again in the 1990s before being persuaded to disarm permanently.

EOKA (N) (Ethniki Organosis Kyprion Agoniston/National Organization of Cypriot Fighters) was a Greek Cypriot organization set up in 1950 to end British sovereignty of the island and to form a union with Greece. EOKA mostly targeted British troops and military bases on the island, and during their four-year campaign, more than 100 British soldiers were killed. The British had more than 30,000 troops on the island, but the EOKA men had popular support; and in 1959 an agreement was reached to establish an independent republic of Cyprus although the settlement specifically rules out union (enosis) with Greece. EOKA-B was formed in 1971 as a Greek Cypriot fascist paramilitary organization with the aim of overthrowing the Greek Cypriot government and achieving enosis. This terrorist group had little public support and was further estranged when it murdered a government minister and tried to assassinate the island's President Makarios. In 1974, backed by the Greek Junta and the National Guard, EOKA-B overthrew the government and ousted Makarios. Rather than bring about union with Greece, the coup prompted Turkey to invade the island and Turkish troops still control one third of Cyprus.

FALN (N): From the late 1970s to the mid-1980s the Fuerzas Armadas Liberacion Nacional Puertoriquena (Armed Forces of Puerto Rican National Liberation) and the Ejercito Popular Boricua (Popular Boricua Army) known as the Macheteros claimed responsibility for a wave of bombings and robberies in both the United States and Puerto Rico. The FALN operated in the continental United States, and the Macheteros were active mostly in Puerto Rico. In October 1974, FALN claimed responsibility for five bombings in New York. Over the next decade, FALN were responsible for more than seventy bombings and forty incendiary attacks, resulting in five deaths, eighty-three injuries, and over $3 million in property damage. The Macheteros, who wanted complete autonomy for Puerto Rica, focused their attacks on the military and police. In 1981 they infiltrated a Puerto Rican Air National Guard base and blew up eleven planes. The number of attacks steadily decreased as more members were caught and imprisoned. In 1999 clemency was offered to sixteen terrorists still in federal prisons.

Japanese Red Army (L) was founded by Fusako Shigenobu in 1971 after she broke with the Japanese Communist League's Red Army Faction. Although it never numbered more than forty or so members, it carried out numerous terrorist attacks around the world. Its aim was to overthrown the Japanese government, but by the early 1980s it was no longer active in Japan and was strongly linked with the Popular Front for the Liberation of Palestine which it relied on for funding and weapons. During the 1970s and 1980s, it carried out a series of attacks including bombings and several hijackings. They were responsible for the machine gun and grenade attack at Tel Aviv's Lod airport which killed seventeen and injured eighty others. In several instances, the group took hostages who were freed only after jailed members of the JRA were released. In 1987 they bombed British and US embassies in Rome, and in 1988 they bombed a USO club in Naples, killing five people. At the same time another member of the group was arrested with a large quantity of explosives on the New Jersey turnpike, and he is imprisoned in the United States.

Khmer Rouge (L) is a Communist organization that ruled Cambodia from 1975 to 1979, and one of the most ruthless and violent regimes of the twentieth century. The Cambodian Communist Party was formed in the 1950s and in the mid-1960s started to attack US troops across the border in Vietnam. When the military took control of Cambodia in a coup in 1970, the Khmer Rouge turned its attention on them and by 1973 controlled much of the country. In April 1975, the government collapsed, and the Khmer Rouge took over. Over the next three years, it was responsible for the deaths of between 1.7 million and 3 million people as it closed schools, hospitals, factories, and churches, tried to wipe out intellectuals and the middle classes, and forced people out of cities and towns and back into force labor camps in the countryside. Although it changed its name to the Communist Party of Kampuchea in the 1970s and then the Party of Democratic Kampuchea in the 1980s, it was universally referred to as the Khmer Rouge. Pol Pot led the group and its leadership changed little between the 1960s and 1990s. The group was funded by Vietnam and then by China.

At the end of 1978, Vietnam invaded Cambodia and in January 1979 ousted the Khmer Rouge regime. For the next ten years, the remnants of the Khmer Rouge waged a guerrilla war in an area close to the Thai border which they controlled. They were funded and armed by sections within the Thai army and China. A peace treaty signed in 1991 was shattered when the Khmer Rouge resumed hostilities, but by the mid-1990s they were fighting among themselves. In 1997, Pol Pot was tried and imprisoned by the Khmer Rouge itself and he died in 1998. Later that year, the remaining leaders of the Khmer Rouge apologized for the deaths in the 1970s and the Khmer Rouge effectively ceased to exist.

Red Brigade (L) was a Marxist group formed in 1969 to establish an independent revolutionary state of Italy. Claiming to be defending workers and trade unions, they conducted a campaign of kidnappings and attacks against factories, with occasional bombings in Greece and elsewhere. In 1974 with most of the original membership dead or in prison, a new leadership took over with a more violet agenda of attacks against the police, army, and prominent businessmen and politicians. In 1978 they kidnapped and murdered former Prime Minister Aldo Moro. In the early 1980s the group split into two groups, the Communist Combatant Party and the Union of Combatant Communists. Both continued the campaign of high-profile political murders until 1988 when a massive crack down by the Italian police led to the arrest of most of the leaders of the two factions. There have been sporadic incidents of violence since and the police continue to track down and arrest surviving members.

Symbionese Liberation Army (L) was an American revolutionary organization that in the early 1970s carried out a series of assassinations, murders, bank robberies, and acts of violence. It was formed in 1973 by African American Donald DeFreeze after he escaped from prison. It is best known for the kidnapping of Patty Hearst, the newspaper heiress, who was initially snatched in a bid to get two of the SLA's leaders released from prison. The swap never took place, and Hearst, who became known as Tania, took part in a bank robbery after her capture. She was later found guilty of bank robbery but pardoned after serving twenty-one months. She was brutally treated during her captivity and was thought to be suffering from Stockholm syndrome when she took part in the robbery. The group never had more than thirteen members. Many were killed in a shootout with the police in Los Angeles in May 1974, and the remaining members hid out in the Bay area, carrying out minor bomb attacks and bank robberies. All have subsequently been captured and convicted for their crimes.

Tonton Macoutes (RT) was founded in Haiti in 1959 under the control of Papa Doc Duvalier who came to power in 1957. He modeled it on the Fascist Blackshirts of Italy and used it to brutally suppress all political opposition. He even rewrote the constitution to give an automatic amnesty to any Tonton Macoutes member who committed any crime while on "official business." Members were noted for wearing sunglasses and wielding machetes; they had a virtual license to torture and kill and they would frequently hang their victims and leave the bodies in public places as a warning to others. Jean Claude Duvalier (Baby Doc) took over on his father's death in 1971 and changed their name to the National Security Volunteers, although they were still as brutal as ever. The Tonton

Macoutes were responsible for hundreds if not thousands of deaths before they were officially disbanded when Baby Doc was overthrown in 1986. Bands of former members of the organization, however, continued to terrorize the population.

Weathermen (L) was a group of militant left wing revolutionaries who believed violence could overthrow the government. Founded in the late 1960s, it collapsed following the US withdrawal from Vietnam. During that time their tactics were mainly bombings, jail breaks, and causing riots. In 1970 they "declared war" on the United States but their first mission, to plant a bomb at a dance at Fort Dix, failed when three of the group were killed while they were building the bomb in their Greenwich Village safe house. Subsequent bombings targeted the US Capital, the Pentagon, and police and prison buildings, but warnings were given to prevent casualties. The group also broke LSD guru Timothy Leary out of prison and escorted him to Algeria. The group started to dissolve in the late 1970s. Some members joined other militant groups while many surrendered to the police. Few served prison time because of their activities in the organization.

MULTILATERAL EFFORTS TO COUNTER INTERNATIONAL TERRORISM

The United States continues to work through multilateral organizations to strengthen regional and international efforts to counter terrorism and violent extremism, including by developing and promoting global norms and building the capacities of states to implement them.

The Global Counterterrorism Forum (GCTF)

Since its launch in September 2011, the GCTF has mobilized more than $300 million to support national and regional efforts to strengthen civilian institutions to counter terrorism and violent extremism. This includes support for the development and implementation of GCTF framework documents at both the regional and country levels.

The GCTF works with partners around the globe to change how states—particularly those emerging from authoritarian rule—respond to the challenges of terrorism and the violent extremist ideologies that underpin them. The GCTF, with its thirty founding members (twenty-nine countries and the EU), regularly convenes counterterrorism policymakers and practitioners, as well as experts from the United Nations and other multilateral and regional bodies, to identify urgent counterterrorism needs, devise solutions, and mobilize expertise and resources to address such needs and enhance global cooperation.

With its primary focus on countering violent extremism (CVE) and strengthening civilian criminal justice and other rule of law institutions that deal with terrorism, the GCTF aims to diminish terrorist recruitment and increase countries' capacity for dealing with terrorist threats within their borders and regions. The United Nations is a close partner of, and participant in, the GCTF and its activities. The GCTF serves as a mechanism for furthering the implementation of the universally agreed UN Global Counter-Terrorism Strategy and, more broadly, to complement and reinforce existing multilateral counterterrorism efforts, starting with those of the United Nations. The GCTF also partners with a wide range of regional multilateral organizations, including the Council of Europe, the Organization for Security and Cooperation in Europe, the African Union, and the Inter-Governmental Authority on Development (IGAD).

GCTF launched two new initiatives in 2016:

- Protection of Soft Targets in a Counterterrorism Context Initiative: Co-led by the United States and Turkey, this initiative has two objectives: (1) to raise awareness, identify needs, and leverage the expertise and experience of governments and industry to better protect potential soft targets—like restaurants, sports arenas, and hotels from extremist attack—whether by organized groups or by individuals intent on doing harm both to their neighbors and to themselves; and (2) to develop a set of internationally recognized, non-binding good practices, which can serve as the basis for international engagement, assistance, and training to enhance the security and resilience of sites that are potential soft targets.
- Dialogue on Countering Foreign Terrorist Fighters and Radicalization to Violence in Central Asia: This initiative aims at assisting central Asian states in addressing the challenge of countering foreign terrorist fighters and in implementing "The Hague-Marrakech Memorandum on Good Practices for a More Effective Response to the FTF Phenomenon."

The GCTF also continues to work on two earlier initiatives, the *Lifecycle Initiative* and *The International CT/CVE Clearinghouse Mechanism*:

- "Initiative to Address the Lifecycle of Radicalization to Violence" Toolkit: In September 2016, GCTF ministers adopted six new documents developed as part of the "Initiative to Address the Life Cycle of Radicalization to Violence" ("Lifecycle Initiative") first launched in September 2015 as a response to the then emerging issue of returning foreign terrorist fighters and ISIS-inspired home-grown violent extremists. Co-led by Turkey and the United States, the crosscutting Lifecycle Initiative equips policymakers and practitioners with conceptual tools they can apply at various points of the life cycle of radicalization to violence: from prevention, to intervention, to rehabilitation, and reintegration. The six new documents include the role of families in identifying and intervening with members at risk of radicalization; identifying and countering terrorist recruiters and facilitators; comprehensive reintegration programs for returning foreign terrorist fighters; juvenile justice in a counterterrorism context; appropriate alternative measures for terrorism-related offenses; and guidelines for legal frameworks for rehabilitation and reintegration. The new documents complement some twenty other existing GCTF and related documents that together make up the Lifecycle Initiative conceptual "toolkit." To make the tools as accessible and useful as possible to stakeholders, the United States has funded the creation of a web-based Toolkit. The Toolkit gives policymakers and practitioners access to consolidated information on the Lifecycle Initiative and related good practices, program models, research, and other counterterrorism and countering violent extremism resources. The Toolkit will eventually include a web-based forum that will allow users to connect with counterparts and experts in other countries, as well as a mobile "app" version that will enhance its accessibility.

- The International CT-CVE Clearinghouse Mechanism (ICCM): In September 2015, GCTF Ministers approved the launch of the International Counterterrorism and Countering Violent Extremism Capacity-Building Clearinghouse Mechanism (ICCM). During 2016, the ICCM developed an up-to-date database of recent/ongoing counterterrorism and countering violent extremism capacity-building assistance programs in three pilot countries: Kenya, Nigeria, and Tunisia. A key objective of the ICCM is to identify gaps in programming and to de-conflict overlapping programs. Officially launched at the September 2016 GCTF Ministers' meeting, the password-protected ICCM database is available to GCTF members, pilot countries, and key donors. The United States is providing $2.25 million of funding for the initial two-year pilot effort.

The International Criminal Police Organization (INTERPOL)

Through its I-24/7 secure global police communications system, INTERPOL connects its member countries' law enforcement officials to its array of investigative and analytical databases, as well as its system of messages, diffusions, and notices. As a result of technical expertise provided by INTERPOL Washington, a number of member countries are now establishing or extending connections between I-24/7 in their national central bureaus to their respective national border security and law enforcement systems to increase their ability to help screen and interdict the international transit of foreign terrorist fighters and other transnational criminals. With continued financial and staffing support from the United States, the INTERPOL Counter-Terrorism Fusion Centre's Foreign Terrorist Fighter project manages an analytical database containing identity particulars that supports law enforcement and border control authorities' abilities to determine the terrorist threat posed by subjects located in, or attempting to enter, their respective jurisdictions. More than sixty countries contribute to INTERPOL's foreign terrorist fighter (FTF) database, which now contains more than 8,000 foreign terrorist fighter identities. From these records, dedicated analysis has been delivered to INTERPOL's membership to combine, evaluate, and share intelligence on the capabilities, means, and emerging trends of FTFs to ensure that the right piece of data reaches the right officer on the frontlines.

European Union (EU)

In December 2016, the US-EU Justice and Home Affairs (JHA) Ministerial included a discussion on initiatives to improve counterterrorism efforts, including border security, screening of travelers, and information sharing, as well as cooperation, to better identify terrorist and foreign terrorist fighter (FTF) travel. The EU and the United States also agreed to reinforce their dialogue on chemical, biological, radioactive, and nuclear material and its possible use by terrorist networks.

In April 2016, the EU Commission published the "Communication on Stronger and Smarter Information Systems for Borders and Security," outlining a multiyear approach to greater information sharing in Europe. In June, the Council of the European Union endorsed a "Roadmap to Enhance Information Exchange and Information Management" that detailed a lengthy list of actions the EU and its member states must implement to improve information sharing. In line with this roadmap, the Commission proposed an array of counterterrorism-relevant regulations and initiatives in 2016, including the creation of an Entry-Exit System and European Travel Information and Authorization System to strengthen Schengen borders and track movements of persons of interest; requiring inspections of all travelers crossing Schengen borders; harmonizing counterterrorism legislation across EU member states and criminalizing various types of support to terrorists; and incorporating biometrics and new alerts into the Schengen Information System. The Council also passed an EU Passenger Name Record (PNR) Directive in April 2016, mandating member states' sharing of PNR data for intra-EU flights by April 2018.

The 2010 US-EU Agreement on the Terrorist Finance Tracking Program continued to enable the EU and the United States to share information related to financial messaging data for the purpose of identifying, tracking, and pursuing terrorists and their networks. The agreement's fourth joint review, conducted by the United States and EU during 2016, concluded that implementation of the agreement remained successful and vitally important to both sides' counterterrorism efforts.

The EU has also created two new institutions, the European Counter Terrorism Centre (ECTC) and the European Border and Coast Guard (EBCG). The ECTC will coordinate European information sharing on counterterrorism matters and will include representatives from DHS and the FBI. The EBCG, in cooperation with member states, will be responsible for security the EU's borders and sharing CT-related information with Europol.

Finally, the EU has invested hundreds of millions of euros in humanitarian and economic programs within and outside the EU as part of its effort to counter violent extremism. The EU also continues six military and law enforcement capacity-building missions in the Horn of Africa and the Sahel, working closely with US elements in counterterrorism, border security, and stabilization efforts.

Organization for Security and Cooperation in Europe (OSCE)

The OSCE is focused on preventing and countering terrorism and violent extremism and radicalization that lead to terrorism, and this culminated in the adoption of declarations at the OSCE Hamburg Ministerial Council meeting in December 2016. As part of the, "OSCE United in Countering Violent Extremism (#UnitedCVE)" campaign, OSCE missions organized discussions, workshops, social media outreach, and a regional competition under the "Peer-2-Peer (P2P): Challenging Extremism" initiative. The OSCE Counterterrorism Conference (May 31–June 1, 2016) brought together more than 300 deputy foreign ministers, secretaries of state, and counterterrorism coordinators from OSCE participating states with Asian and Mediterranean cooperating partners. The conference highlighted the role of youth, and, in particular, young women, with a focus on cooperation with youth-led and youth-focused initiatives, especially when designing strategies to counter violent extremism and radicalization that lead to terrorism.

The OSCE, enabled by US support, also promotes practical scenario-based multi-stakeholder seminars to discuss foreign terrorist fighters and countering violent extremism to enhance interagency and whole-of-society collaboration and develop good practices to share at regional, national, and community levels. These events help build interagency coordination, promote whole-of-society collaboration, implement national CVE strategies, and develop action plans.

North Atlantic Treaty Organization (NATO)

NATO's counterterrorism efforts focus on improving awareness of the threat, developing response capabilities, and enhancing engagement with partner countries and organizations. All NATO allies participate in the Global Coalition to Defeat ISIS. At the 2016 NATO Summit in Warsaw, allied leaders agreed that NATO as an institution would provide direct Airborne Warning and Control System support to the Coalition's efforts; provide training and capacity-building support inside Iraq, while continuing to train hundreds of Iraqi officers in Jordan; and provide logistical and intelligence support to EU naval forces conducting anti-trafficking operations in the Mediterranean; and Allies reaffirmed their commitment to provide training and capacity-building support for Libya upon request and when conditions permit.

NATO has also announced agreement to create a NATO Istanbul Cooperation Initiative Regional Centre in Kuwait City. This Center will serve as a hub for NATO's practical cooperation with Kuwait and other Istanbul Cooperation Initiative partners (Bahrain, Qatar, and United Arab Emirates), as well as with Saudi Arabia and Oman.

Building partner capacity and developing innovative technologies are part of NATO's core mission, and methods that address asymmetric threats like terrorism are of particular relevance. Much of this work is conducted through the Defence against Terrorism Programme of Work (DAT POW), which aims to protect troops, civilians, and critical infrastructure against terrorist attacks, including suicide bombers, improvised explosive devices, rockets against aircraft, and chemical, biological, and radiological materials. The DAT POW also supports the implementation of NATO's spearhead force—the Very High Readiness Joint Task Force—by developing projects to improve troop readiness and preparedness.

Group of Seven (G-7)

Within the context of the G-7 Roma-Lyon Group meetings on counterterrorism and counter-crime, the United States worked with RLG counterparts to draft the G-7 Action Plan on Countering Terrorism and Violent Extremism, which was included as annex to the G-7 Leader's Communiqué. Based on language in the action plan, the United States spearheaded an initiative to generate greater G-7 political and financial support to connect priority countries to INTERPOL's I-24/7 secure global communications system in an effort to help interdict foreign terrorist fighter travelers more effectively. The United States also sought to advance projects through the Roma-Lyon Group's expert groups on counterterrorism, transportation security, high-tech crime, migration, criminal legal affairs, and law enforcement. In particular, the United States made substantial contributions to the counterterrorism expert group's terrorist threat assessment, which surveys a variety of emerging threats to G-7 countries.

Organization of American States' Inter-American Committee against Terrorism (OAS/CICTE)

The OAS Inter-American Committee against Terrorism (CICTE) bolsters counterterrorism partnerships, cooperation, and information sharing in the Western Hemisphere through promoting counterterrorism policies, training, and capacity building. It focuses programs and activities on cyber and supply chain security as part of the OAS's long-standing work on critical infrastructure security and resilience. Its meetings bring together high-level experts and authorities from the region to discuss the use of the internet for terrorist and criminal purposes, critical cyber-security considerations for the development of the economy and a digital society, and confidence building measures in cyberspace.

The Caribbean Community (CARICOM)

CARICOM's Implementation Agency for Crime and Security (CARICOM/IMPACS) has developed a coordinated regional counterterrorism strategy for the Caribbean. CARICOM consulted with international partners including the United States, Canada, and the United Nations to identify best practices for inclusion in its counterterrorism strategy. This effort to improve cooperation and coordination builds on the CARICOM Crime and Security Strategy, approved in 2013.

Association of Southeast Asian Nations (ASEAN) and the ASEAN Regional Forum (ARF)

Counterterrorism activities of the ten-member ASEAN and twenty-seven-member ARF countries include annual meetings on counterterrorism and transnational crime and capacity building through ARF institutions. In 2016, the United States funded a grant for technical assistance, equipment, and training to improve and automate ASEAN member state reporting to INTERPOL's I-24/7 secure global communications system in an effort to help interdict foreign terrorist fighter travelers more effectively.

Asia-Pacific Economic Cooperation (APEC)

APEC continues to implement its comprehensive Consolidated Counterterrorism and Secure Trade Strategy. The Strategy, adopted in 2011, endorses the principles of security, efficiency, and resilience, and advocates for risk-based approaches to security challenges across its four crosscutting areas of supply chains, travel, finance, and infrastructure. The United States sponsored a follow-on workshop that built on the lessons learned and highlighted the threat that foreign terrorist fighter travel poses to the Asia-Pacific region and that explained why advance passenger information and passenger name record (API/PNR) systems are effective at helping mitigate that threat. In particular, the follow-on workshop illustrated how API/PNR data can improve the effectiveness of domestic terrorist watch lists.

The African Union (AU)

Funded by the United States, the AU has been working to help member states adopt its Model Law on Counterterrorism. The AU, in partnership with the UN Office on Drugs and Crime Terrorism Prevention Branch, currently is providing a team of legal experts to assist three pilot member states—Burkina Faso, Ghana, and Mauritius—in undertaking the process to update, strengthen, or create counterterrorism laws to ensure that member states have in place an effective rule of law-based criminal justice response to terrorism.

The Gulf Cooperation Council (GCC)

The US-GCC Counterterrorism Working Group continues to collaborate on key areas of mutual interest including countering threat from foreign terrorist fighters, countering violent extremism, and countering Hizballah.

The United Nations (UN)

Sustained and strategic engagement at the United Nations on counterterrorism facilitates the United States' ability to create strategic partnerships and coalitions, set global standards and norms, and build the political will and capacities abroad that are needed to deter and respond to terrorism. In this regard, an important priority has been to support the UN Secretary-General's Plan of Action to Prevent Violent Extremism (PVE), and encourage member states to implement its seventy-one recommendations in seven key areas:

- Dialogue and conflict prevention;
- Strengthening good governance, human rights, and the rule of law;
- Empowering youth;
- Engaging communities;
- Gender equality and empowering women;

- Education, skill development, and employment facilitation; and
- Strategic communications, the internet, and social media.

One core recommendation of the PVE Plan is that member states consider developing national and regional PVE Action Plans, which would provide a strong foundation and entry point for the United States, UN entities, and other donors to initiate capacity-building assistance projects where they are needed.

The UN General Assembly continues to be engaged in building norms and facilitating international consensus to counter terrorism. In addition to other annual resolutions on counterterrorism, member states reaffirm the UN Global Counter-Terrorism Strategy (GCTS) and discuss key trends in terrorism during its biennial review. The UN GCTS articulates principles and activities to address conditions conducive to the spread of terrorism, prevent and counter terrorism, build states' capacity, and strengthen the role of the United Nations, while respecting human rights and the rule of law. In that vein, the Human Rights Council convened a panel at its 31st session in March to discuss the human rights dimensions of preventing and countering violent extremism and adopted a US-supported resolution on the "Protection of Human Rights and Fundamental Freedoms While Countering Terrorism" in September.

The UN Security Council (UNSC) continues to promote international cooperation to stem the flow of foreign terrorist fighters by promoting implementation of UNSC resolution (UNSCR) 2178 (2014), a Chapter VII binding resolution that requires all states to "prevent and suppress the recruiting, organization, transporting, or equipping" of foreign terrorist fighters, as well as the financing of foreign terrorist fighter travel and activities. In 2016, the UNSC adopted several other counterterrorism-related resolutions, including UNSCR 2309 on aviation security, which called upon States to require that airlines operating in their territories provide advanced passenger information; UNSCR 2322 on strengthening international judicial cooperation; and UNSCRs 2325 and 2331, which focused on weapons of mass destruction and trafficking in persons, respectively, with a nexus to terrorism and terrorist financing.

In addition, the United States engages with a wide range of UN actors on counterterrorism, which include

- The Counter-Terrorism Committee (CTC) Counter-Terrorism Executive Directorate (CTED). CTED continues its efforts to analyze capacity gaps of member states to implement UNSCRs 1373 (2001), 1624 (2005), 2178 (2014), and other counterterrorism-related resolutions, and facilitate training and other technical assistance to UN member states. This includes the UN Security Council's Counter-Terrorism Committee thematic debates on a range of issues including stemming the flow of foreign terrorist fighters; depriving terrorists of accessing, moving, and raising funds; successful adjudication of terrorism-related cases; and preventing terrorists from exploiting the internet and social media to recruit and incite terrorist acts, while respecting human rights and fundamental freedoms.

- The Counter-Terrorism Implementation Task Force (CTITF). The United States supports CTITF efforts to create a capacity-building plan to assist member states' implementation of UNSCR 2178 and improve implementation of the UN Global Counter-Terrorism Strategy, including by serving on the Advisory Board of the UN Counter-Terrorism Centre (UNCCT), which delivers training and technical assistance. In 2016, the United States continued to fund a range of UNCCT and CTITF activities, including promoting effective use of advance passenger information to stem the flow of foreign terrorist fighters; capacity building for Mali's security and justice sectors; a training initiative to secure open borders; implementing good practices on addressing and preventing terrorist kidnapping for ransom; supporting the working group on protecting human rights while countering terrorism; supporting media workshops for victims of terrorism; and supporting community engagement through human rights-led policing.
- The UN ISIL (Da'esh) and Al-Qaeda Sanctions Committee. The United States works closely with the UN 1267/1989/2253 ISIL (Da'esh) and al-Qaeda Sanctions Committee and its Monitoring Team by proposing listings and de-listings, providing amendments, engaging the Committee's Ombudsperson in de-listings, and providing input to the Committee to enhance its procedures and implementation of sanctions measures. The United States also assists the Monitoring Team with information for its research and reports. There are 256 individuals and 75 entities on the Committee's sanctions list. The Committee also works to ensure the integrity of the list by conducting regular reviews and by endeavoring to remove those individuals and entities that no longer met the criteria for listing.
- The UN Office on Drugs and Crime (UNODC): The UNODC develops criminal justice responses to prevent and counter terrorism in all its forms and manifestations, including the financing of terrorism, and technical assistance in support of the implementation of relevant international conventions and protocols. The UNODC's Terrorism Prevention Branch (TPB) continues to provide assistance to countries seeking to ratify and implement the universal legal instruments against terrorism as well as providing assistance for countering the financing of terrorism in conjunction with the UNODC's Global Program against Money Laundering. The United States has engaged UNODC/TPB as a counterterrorism assistance implementer, supporting programming focused on strengthening the criminal justice system's response to terrorism by member states. Specifically,

the US-supported UNODC/TPB programs aimed at strengthening the legal regime against terrorism within a rule of law framework in the Sahel region.

- The UN Inter-Regional Crime Research Institute (UNICRI): The focus of UNICRI's work has been on rehabilitation efforts in prison. The United States has provided assistance support to a UNICRI-led global effort to strengthen state capacity to implement the good practices contained in the Global Counterterrorism Forum's *Rome Memorandum on Good Practices for the Rehabilitation and Reintegration of Violent Extremist Offenders* and to conduct a pilot program on diversion and use of alternatives for juvenile offenders charged with foreign terrorist fighter-related offenses. In addition, the United States provides funding to UNICRI to bring it into the existing initiative co-led by the International Institute for Justice and Rule of Law and Hedayah to rehabilitate and reintegrate returning foreign terrorist fighters.
- The UN Development Programme (UNDP): The UNDP engages countries to mitigate and prevent conflicts, including in East Africa, by developing national and regional strategies to counter and prevent violent extremism (PVE) and deepening research on PVE through its center in Oslo, Norway. In 2016, the United States provided assistance funding to UNDP to help strengthen community-police partnerships in high-risk communities.
- The UN Security Council (UNSC) 1540 Committee: The UN Security Committee monitors and fosters implementation of the obligations and recommendations of UN Security Council resolution (UNSCR) 1540, which establishes legally binding obligations on all UN member states related to the establishment of and enforcement of appropriate and effective measures against the proliferation of chemical, biological, and nuclear weapons, their means of delivery, and related materials to non-state actors, including terrorists. The 1540 Committee's program of work focuses on four main areas:
    - Monitoring and national implementation;
    - Assistance;
    - Cooperation with international organizations, including the UNSC committees established pursuant to UNSCRs 1267 and 1373; and
    - Transparency and media outreach.

  The 1540 Committee has completed its second comprehensive review of the implementation of UNSCR 1540. It found that, while many states have taken important steps to strengthen prohibitions and controls on weapons of mass destruction and their means of delivery in fulfillment of their 1540 obligations, key gaps remain in areas that include biological and chemical security, and proliferation finance. To address these and other issues, the UNSC unanimously adopted UNSCR 2325. The new resolution calls upon States to intensify their efforts to fulfill their 1540 obligations and to submit timely reports on their efforts. It also gives the 1540 Committee a mandate to focus its work on major gaps where it might have the most value added, particularly noting a need for more attention on enforcement measures; measures relating to biological, chemical, and nuclear weapons; proliferation finance measures; accounting for and securing related materials; and national export and transshipment controls. In addition, it calls for greater assistance for building State capacity, including through voluntary contributions, and for greater cooperation among all stakeholders, including civil society, academia, and industry. The Committee's Group of Experts also participates as part of the Counter-Terrorism Implementation Task Force, and cooperates with INTERPOL, the UN Office of Drugs and Crime, the Financial Action Task Force, and other bodies involved in counterterrorism efforts.

The International Civil Aviation Organization (ICAO)

UN Security Council resolution (UNSCR) 2309 calls on all States to ensure that effective, risk-based measures are in place at the airports within their jurisdiction and that such measures reflect the ever-evolving threat picture and are in accordance with international standards and recommended practices. The UNSCR calls on all States to strengthen information sharing among States and to require that airlines operating in their territories provide advance passenger information to appropriate national authorities to track the movement of individuals identified by the counterterrorism committees. It also urges all States to ensure cooperation among their domestic departments, agencies, and other entities, and encouraged continued close cooperation between ICAO and the Counter-Terrorism Executive Directorate on identifying gaps and vulnerabilities in aviation security. ICAO's Universal Security Audit Program (USAP) continued to support US security in its efforts to evaluate the capabilities of ICAO's member states in establishing, implementing, and performing oversight of National Civil Aviation Security Programs in compliance with Annex 17 security standards.

ICAO operates fully under its new USAP-Continuous Monitoring Approach, which aims to enable greater focus of resources on states requiring more assistance in meeting the standards, allowing additional flexibility in where audits are focused and how follow-on activities are planned. Additionally, ICAO's Implementation Support and Development-Security Section continues to conduct training and assistance missions to help states correct security problems revealed by surveys and audits. ICAO is also working with member states to encourage incorporation of advance passenger information and passenger name record in the travel decision process and with priority countries on implementation of ICAO's public key directory program, as a means to validate e-passports at key ports of entry. ICAO and a task force of member states have begun work on drafting the Global Aviation Security Plan (GASeP) to reflect the reality and needs of the current threat and risk environment. Also, ICAO, with the World Customs Organization, is working to establish standard practices for enhanced screening of cargo.

LONG-TERM PROGRAMS AND INITIATIVES DESIGNED TO COUNTER TERRORIST SAFE HAVENS AND RECRUITMENT COUNTERING VIOLENT EXTREMISM (CVE)

CVE refers to proactive actions to counter efforts by violent extremists to radicalize, recruit, and mobilize followers to violence and to address specific factors that facilitate violent extremist recruitment and radicalization to violence.

State and the US Agency for International Development (USAID) leverage a range of available diplomatic, development, and foreign assistance tools to have a demonstrable impact on preventing and countering radicalization and recruitment to violence, both online and offline.

The following five objectives guide our CVE assistance and engagement:

1. Expand international political will, partnerships, and expertise to better understand the drivers of terrorist radicalization and recruitment and to mobilize effective interventions.
2. Encourage and assist partner governments to adopt more effective policies and approaches to prevent and counter the spread of violent extremism, including changing unhelpful practices where necessary.
3. Employ foreign assistance tools and approaches, including development, to reduce specific factors that contribute to community support for violent extremism in identifiable areas or put particular segments of a population at high risk.
4. Empower and amplify locally credible voices that can change the perception of violent extremist groups and their ideology among key demographic segments.
5. Strengthen the capabilities of government and nongovernmental actors to isolate, intervene with, and promote the rehabilitation and reintegration of individuals caught in the cycle of radicalization to violence.

State also works to counter the use of the internet for terrorist purposes, including countering violent extremism online. This includes

1. Developing and promoting with foreign partners US policy on countering the use of the internet for terrorist purposes;

2. Partnering with private sector, civil society, academia, and other governments to find creative and effective solutions;
3. Providing law enforcement training for foreign partners to more effectively counter terrorist activities online, including through Anti-Terrorism Assistance training;
4. Promoting and providing training for counter-messaging and alternative messages;
5. Providing foreign policy guidance and diplomatic coordination on US counterterrorism-related cyber operations, as appropriate; and
6. Developing and providing training for Department and US federal employees on CVE online and offline, including through the Foreign Service Institute.

The Department of State and USAID are pursuing a range of programs to assist partners around the world. Key areas of programming include the following:

- Supporting the Development and Implementation of National CVE Action Plans: The United States provides technical assistance to governments as they design and implement national CVE action plans, in partnership with civil society and the private sector. To reinforce these national action plan efforts, the United States supports Hedayah, the international CVE center in Abu Dhabi, by providing capacity building and technical expertise to interested governments.
- Researching Drivers of Violent Extremism and Effective Interventions: The United States supports innovative regional, country-based, and thematic research on the drivers of radicalization and recruitment to violence and on programming approaches. The United States supports the Researching Solutions to Violent Extremism Network, which connects academics and researchers to study the dynamics of CVE in specific, local contexts and identify effective CVE interventions. At the same time, the United States works with the Global Counterterrorism Forum (GCTF) to implement an expanded toolkit for addressing the life cycle of radicalization to violence.
- Building the Capacity of Criminal Justice Actors and Institutions: The United States is supporting programs, especially in the Horn, Sahel, and Maghreb regions of Africa to strengthen the CVE capacity of law enforcement, including police deployed to peace and stabilization operations, prison management and justice sector actors, and to help address drivers of terrorist radicalization such as corruption and human rights abuses. The United States also supports programs to train and assist corrections officials to counter radicalization to violence in prison settings and promote rehabilitation, including addressing returning foreign terrorist fighters.

- Strengthening CVE Efforts by Sub-National, City, and Local Partners: The United States supports the Strong Cities Network, a global network of municipal and other sub-national leaders and local government practitioners involved in building community resilience. The United States also contributes to the Global Community Engagement and Resilience Fund, the first multilateral fund supporting community–based projects that counter local drivers of recruitment and radicalization to violence.
- Enhancing Civil Society's Role in Countering Violent Extremism: The United States supports programs that empower youth as positive change agents in preventing violent extremism among their peers. The United States also supports programs that elevate the role of women in recognizing and preventing the spread of violent extremism in their families and communities.

Counter-Messaging and Promoting Alternative Narratives

With the leadership of the Department of State's Global Engagement Center (GEC), the United States supports efforts to help government and nongovernmental partners to counter ISIS and other terrorist messaging and to promote alternative narratives. The United States supports a network of messaging centers that expose, refute, and counter online terrorist propaganda. These centers harness the creativity and expertise of local actors to generate positive content that challenges the rhetoric of ISIS and its supporters. The United States is also degrading ISIS's ability to operate in cyberspace due to an improved knowledge of its propaganda and recruiting networks, and to voluntary private sector efforts. Counter-ISIS content is more prevalent online and pro-ISIS content has declined. The GEC works with Facebook to send highly targeted messages to young Muslims showing signs of radicalization to violence. Videos in a recent campaign targeting specific people in Morocco, Saudi Arabia, and Tunisia were watched more than 14 million times. The effort has since expanded to other nations, including France, Jordan, and Libya.

The GEC also counters terrorist messaging through a network of civil society partners. The GEC launched its first partner-driven network, Families against Terrorism and Extremism (FATE) in March 2016. The network, which supports skills building and intervention training for families at the forefront of violent radicalization, operates in key source countries for foreign terrorist fighters. FATE runs social media and programming in Arabic, English, French, German, and Russian, and has grown to include more than eighty organizations united to curb radicalization to violence and ISIS recruitment. One other such partnership is the GEC's *Amplifying Credible Voices* (ACV) program, which establishes and activates networks of CVE influencers in priority countries. ACV was launched in Albania and Kosovo with a counter-messaging workshop that convened and trained forty-six CVE influencers from across sectors to design and execute locally resonant messaging campaigns.

CIVILIAN COUNTERTERRORISM CAPACITY-BUILDING PROGRAMS

As the terrorist threat has evolved and grown more geographically diverse in recent years, it has become clear that our success depends in large part on the political will and capabilities of our partners to counter terrorism. To succeed over the long term, we must have partners who can not only militarily disrupt threats and degrade networks in a way that comports with international laws and norms, but who have strong civilian capabilities, as well. We need law enforcement, justice sector, and corrections partners, who can disrupt attacks and investigate, arrest, prosecute, and incarcerate terrorists and their facilitation networks.

The United States uses various funding authorities and programs to build the capacity of law enforcement, justice, and corrections officials to counter terrorism. The Department of State's Bureau of Counterterrorism and Countering Violent Extremism oversees the following capacity-building programs: Antiterrorism Assistance, Countering the Financing of Terrorism, Counterterrorism Engagement with Allies, the Regional Strategic Initiative, and the Terrorist Interdiction Program. For further information on these programs see Annual Report on Assistance Related to International Terrorism: https://www.state.gov/j/ct/rls/crt/267763.htm.

Congress has appropriated additional funding for the Department of State's Counterterrorism Partnerships Fund. This funding will allow the United States to significantly expand civilian counterterrorism capacity-building activities with key partner nations in the Middle East, North Africa and the Sahel, the Horn of Africa, South and Central Asia, Southeast Asia, and other regions as required to mitigate the threat posed by foreign terrorist fighters, prevent and counter terrorist safe havens and recruitment, and counter Iranian-sponsored terrorism.

Regional Strategic Initiative (RSI)

The Department of State's Bureau of Counterterrorism and Countering Violent Extremism uses RSI to promote a more regional and trans-regional approach to counterterrorism, to rapidly respond to and address emerging terrorism threats and related counterterrorism capacity gaps, and to complement and expand upon our targeted capacity-building programs funded through the Counterterrorism Partnerships Fund, Antiterrorism Assistance program, and other accounts. Current RSI efforts focus on stemming the flow of foreign terrorist fighters to Syria and Iraq, countering terrorist safe havens, and countering Hizballah's activities.

RSI funding has supported a wide variety of projects focused on regional law enforcement cooperation and effectiveness against transnational threats:

- Department of Justice advisors assisted Kosovo in processing foreign terrorist fighter cases and taking them to trial; this support contributed to more than fifty prosecutions.
- DOJ provided targeted technical assistance to Cote D'Ivoire's investigations into the al-Qaeda in the Islamic Maghreb (AQIM)-linked attack at Grand Bassam, producing valuable leads and resulting in improved regional cooperation on the investigation.
- RSI funding strengthened Brazilian counterterrorism preparedness through focused counterterrorism training prior to the Olympics.
- RSI funding was used to counter Hizballah. These funds assisted Gulf Cooperation Council (GCC) partners to implement the GCC designation of Hizballah, enhanced the Cypriot government's ability to investigate and prosecute Hizballah cases (resulting in two successful prosecutions), and provided a forum for investigators and prosecutors to share information about Hizballah with their counterparts in other countries.

Programs to Counter Foreign Terrorist Fighters

A core component of the United States' Counter-ISIS strategy has been to prevent the travel of foreign terrorist fighters (FTFs) to and from the conflicts in Iraq and Syria. Since 2011, an estimated 40,000 FTFs have traveled to Iraq or Syria from more than 120 countries to fight alongside a number of violent extremist groups, including ISIS and al-Qaeda's affiliate in Syria, al-Nusrah Front. Over the past year, the flow of FTFs travelling to Iraq or Syria declined significantly. As a result of declining FTFs flows, ISIS can no longer replenish its ranks lost on the battlefield in Iraq and Syria with new foreign recruits. The number of ISIS fighters has been reduced to its lowest level in more than three years. While the sustained military campaign and ISIS's loss of territory and resources are key factors, governments enacted a number of reforms and measures to make it demonstrably more difficult for FTFs to transit to and from Iraq or Syria.

The United States undertook numerous bilateral and multilateral efforts to encourage partners to strengthen measures to counter foreign terrorist fighters. These efforts, both multilateral and bilateral, are guided by UN Security

Council resolution (UNSCR) 2178, passed in September 2014. UNSCR 2178 addresses fundamental measures necessary to stem the flow of FTFs, including increasing international information sharing, having in place and enforcing appropriate counterterrorism legislation, strengthening border security, and increasing efforts on counter-messaging. As a result, a number of countries took steps to implement UNSCR 2178 and as of the end of 2016:

- More than sixty countries have laws in place to prosecute and penalize FTFs activities.
- At least sixty-five countries have prosecuted or arrested FTFs or FTF facilitators.
- The United States has information-sharing arrangements with fifty-nine international partners to help identify, track, and deter known and suspected terrorists.
- At least twenty-six partners share financial information within their country or with partners that could provide actionable leads to prosecute or target FTFs.
- At least thirty-one countries use enhanced traveler screening measures.

With the appeal of fighting in Iraq or Syria diminished, thousands of FTFs returned home, and many in the international community are concerned that these returning fighters will commit terrorist attacks at home. Since 2012, many of these returning fighters sought to reintegrate into society. A small number of these returning FTFs, however, relied on operational training, connections, and experience gained in Iraq or Syria to plot terrorist attacks or carry out an attack once returning home. In the last three years, returning FTFs have been involved in numerous incidents including a bombing in Istanbul, Turkey, and attacks in Brussels, London, and Paris.

The Department of State plays a pivotal role in the creation of an international framework for addressing the threat from FTFs and will continue to expand and deepen bilateral and multilateral engagement to counter the FTF threat and related radicalization and recruitment to violence. Over the course of the last four years, the Department of State has led interagency delegations to countries in Europe, North Africa, the Middle East, and Southeast Asia, and continues to work with our interagency partners to achieve the following objectives:

- Encourage and assist partners to employ more robust border and aviation security procedures to identify and interdict potential FTFs and those returning.
- Work with partners to identify, monitor, and address the travel of FTFs to and from Iraq and Syria and new areas affected by ISIS expansion, such as Libya.

- Enable partners to track and interdict travel by FTFs through more robust information sharing, watchlisting, and traveler screening.
- Encourage and assist partners to establish and employ the necessary legislative, administrative, policy, and criminal justice frameworks, capabilities, and tools to investigate, interdict, divert, prosecute, adjudicate, and incarcerate aspirant or returning FTFs.
- Encourage and assist partners to counter violent extremism and prevent radicalization to violence, including through community resilience programs and counter-messaging.
- Encourage and assist partners to develop comprehensive diversion and rehabilitation and reintegration programs for aspirant and returning FTFs, both inside and outside the prison setting.

Chapter 8

The Spread of Global Terrorism

Every nation in every region now has a decision to make. Either you are with us, or you are with the terrorists.

—President Bush

We share the common goal of denying al-Qaeda a safe haven in Pakistan or Afghanistan. The world has come too far to let this region backslide, and to let al-Qaeda terrorists plot further attacks. That's why we are committed to a more focused effort to disrupt, dismantle, and defeat al-Qaeda. That is why we are increasing our efforts to train Afghans to sustain their own security, and to reconcile former adversaries. That's why we are increasing our support for the people of Afghanistan and Pakistan, so that we stand on the side not only of security, but also of opportunity and the promise of a better life.

—President Obama

The fact remains—America is awash in desirable targets—those that are symbolic like the U.S. Capitol and the White House—as well as the many infrastructure targets, like nuclear power plants, mass transit systems, bridges and tunnels, shipping and port facilities, financial centers, and airports—that if successfully hit, would cause both mass casualties and a crippling effect on our economy.

It remains the FBI's overriding priority to predict and prevent terrorist attacks. The threat posed by international terrorism, and in particular from al Qaida and related groups, continues to be the gravest we face.

—Robert S. Mueller III, director, Federal Bureau of Investigation, before the Senate Committee on Intelligence of the United States Senate, February 16, 2005

Today we begin a conversation about how to Make America Safe Again. In the 20th Century, the United States defeated Fascism, Nazism, and Communism. Now, a different threat challenges our world: Radical Islamic Terrorism.

Overseas, ISIS has carried out one unthinkable atrocity after another. Children slaughtered, girls sold into slavery, men and women burned alive. Crucifixions, beheadings, and drownings. Ethnic minorities targeted for mass execution. Holy sites desecrated. Christians driven from their homes and hunted for extermination. ISIS rounding-up what it calls the "nation of the cross" in a campaign of genocide. We cannot let this evil continue.

Nor can we let the hateful ideology of Radical Islam—its oppression of women, gays, children, and nonbelievers—be allowed to reside or spread within our own countries.

We will defeat Radical Islamic Terrorism, just as we have defeated every threat we have faced in every age before.

—President Donald Trump, speaking in Ohio during his presidential campaign, August 15, 2016

Although there were indications during the last two decades of the twentieth century that terrorism was crossing borders and continents, it wasn't until the start of the twenty-first century that the spread of terrorism had become truly global.

Leading this terror assault were extremist Islamic groups in the Middle East, Far East, and North Africa with break away and copy cat cells in Europe and possibly North America. How much cooperation there is between these groups is subject to speculation but there is no doubt that alliances have been forged between them, that their numbers have grown, and that the United States is their main target.

The London bombings proved for the first time that there were al-Qaeda cells in Western Europe whose members were prepared to give their lives for their cause. More troubling still was that these bombers were British citizens, born and bred in the United Kingdom and, as far as everyone else was concerned, were leading ordinary, law abiding lives. Intelligence sources estimate that there could be up to 700 British born extremist Muslims with thousands more in all the major western European countries. There could be many thousands in the United States and Canada.

Further evidence of the global spread of the Islamic insurgency comes from attacks in the Russia town of Nalchik by Chechen rebels. Shamil Basayev, the most ruthless Chechen warlord, acknowledged that he ordered the storming of a southern Russian town in October 2005 which left 137 militants and officials dead. Basayev admitted that a local Islamic cell, not ethnic Chechens, carried out the assault. This was the first concrete evidence that there had been a fundamental shift in the fighting in the Caucasus. The issue was no longer simply independence for Chechnya but the formation of an Islamic republic that would take in several Russian provinces, including Chechnya, Dagestan, Ingushetia, and Kabardino-Balkaria. Russian officials confirmed that the dead terrorists were locals not Chechen fighters.

Today, terrorism remains a global threat from which no nation is immune. Despite ongoing improvements in US homeland security, military campaigns against insurgents and terrorists in Iraq, Syria, and Afghanistan, and deepening counterterrorism cooperation among the nations of the world, international terrorism continues to pose a significant threat to the United States and its partners.

GLOBAL OVERVIEW

Africa

Africa has experienced significant levels of terrorist activity. In East Africa, the Somalia-based terrorist group al-Shabaab remains the most potent threat to regional stability, having regained territory in parts of southern and central Somalia. As the African Union Mission in Somalia (AMISOM) limited offensive operation to proactively counter al-Shabaab in southern Somalia, the terrorist group gained the time and space needed to regroup and recruit new fighters. The organization maintains its allegiance to al-Qaeda and continues its campaign to isolate and defeat the growing faction of ISIS supporters in Somalia. Although ISIS has claimed responsibility for small-scale attacks in Somalia and Kenya, the group failed to launch a major attack in East Africa. Al-Shabaab uses safe havens and towns it reclaimed in 2016 to refine its asymmetric tactics. The group continued to focus on targeting AMISOM forces by launching large-scale raids against its forward-operating bases. In January 2016, al-Shabaab claimed responsibility for a raid against Kenyan forces in El Adde, Somalia, that was reported to have killed more than 140 soldiers. In its attempt to disrupt the Federal Government of Somalia's national electoral process, al-Shabaab increased the use of suicide bombings and ambush attacks against Somali government facilities and select hotels popular with government officials and business people. Al-Shabaab also launched vehicle-borne improvised explosive devices (VBIEDs) against the Mogadishu International Airport that killed UN, AMISOM, and Somali officials. Al-Shabaab claimed responsibility for the targeted assassination of several Somali parliamentarians, security officials, and other government personnel in Mogadishu and elsewhere in Somalia, including Puntland. The United States continues to support counterterrorism capacity building for its East African partners, including advisory assistance for AMISOM, training and mentoring the staff of Somalia's security sector, and improving regional law enforcement agencies' crisis response capabilities.

In response to al-Shabaab's mass-casualty attacks in Kenya's Northeast that claimed hundreds of lives over the last few years, Kenya increased its law enforcement and military presence along the border to detect, deter, and disrupt terrorist travel and cross-border activity. East African governments leveraged US assistance to improve counterterrorism-focused investigations and prosecutions, particularly in Djibouti, Kenya, Somalia, and Uganda. In the Lake Chad Basin, the Multi-National Joint Task Force (MNJTF), comprising Benin, Cameroon, Chad, Niger, and Nigeria, successfully coordinated defeat-ISIS-West Africa and counter-Boko Haram efforts that increased pressure on the groups. Although degraded, Boko Haram continues an asymmetric campaign of kidnappings, killings, bombings, suicide bombers, and attacks on civilian and military targets throughout northeast Nigeria, resulting in a significant number of deaths, injuries, and destruction of property. ISIS-West Africa consolidated its presence in the Lake Chad area following its August split from Boko Haram and targeted primarily vulnerable regional military and government targets. France's Operation Barkhane, a counterterrorism operation focused on countering terrorists operating in the Sahel, continues and is supported by the important contributions of the UN Multinational Integrated Stabilization Mission in Mali.

Trans-Sahara Counterterrorism Partnership: Established in 2005, the Trans-Sahara Counterterrorism Partnership (TSCTP) is a US-funded and -implemented, multi-faceted, multi-year effort designed to build the capacity and cooperation of military, law enforcement, and civilian actors across North and West Africa to counter terrorism. Areas of support include (1) enabling and enhancing the capacity of North and West African militaries and law enforcement to conduct counterterrorism operations; (2) integrating the ability of North and West African militaries and law enforcement, and other supporting partners, to operate regionally and collaboratively on counterterrorism efforts; (3) enhancing border security capacity to monitor, restrain, and interdict terrorist movements; (4) strengthening the rule of law, including access to justice, and law enforcement's ability to detect, disrupt, respond to, investigate, and prosecute terrorist activity; (5) monitoring and countering the financing of terrorism (such as that related to kidnapping for ransom); and

(6) countering local support for violent extremism. TSCTP partners include Algeria, Burkina Faso, Cameroon, Chad, Mali, Mauritania, Morocco, Niger, Nigeria, Senegal, and Tunisia.

TSCTP has built capacity and cooperation despite setbacks caused by a restive political climate, violent extremism, ethnic rebellions, and extra-constitutional actions that interrupted work and progress with select partner countries. Regional cooperation, a strategic objective of US assistance programming globally, has increased substantially in West and Central Africa among most of the partners of the TSCTP. Nigeria and its neighbors agreed to form a Multinational Joint Task Force to combat Boko Haram, and remained actively engaged in countering the group throughout the region. The TSCTP partners were joined in this effort by the African Union and by the country of Benin, which is not a member of the TSCTP. In North Africa, the United States, and Moroccan, training for Senegalese law enforcement has improved crisis response management, while Algeria has offered counter-improvised explosive device (IED) training to neighboring states.

Partnership for Regional East Africa Counterterrorism: First established in 2009, the Partnership for Regional East Africa Counterterrorism (PREACT) is a US-funded and -implemented multi-year, multi-faceted program designed to build counterterrorism capacity and cooperation of military, law enforcement, and civilian actors across East Africa to counter terrorism. Areas of support include (1) reducing the operational capacity of terrorist networks; (2) developing a rule of law framework for countering terrorism in partner nations; (3) enhancing border security; (4) countering the financing of terrorism; and (5) reducing the appeal of radicalization and recruitment to violent extremism. Active PREACT partners include Djibouti, Ethiopia, Kenya, Somalia, Tanzania, and Uganda. Burundi, Comoros, Rwanda, Seychelles, South Sudan, and Sudan are inactive members of PREACT and did not receive PREACT assistance in 2016. In 2016, the US government, through PREACT, continued to build the capacity and resilience of East African governments to contain the spread of and ultimately counter the threat posed by al-Qaeda, al-Shabaab, and other terrorist organizations. PREACT complements the US government's dedicated efforts, including support for AMISOM and promoting stability and governance in Somalia and the greater East Africa region with an emphasis on respect for human rights and the rule of law.

PREACT additionally serves as a broader US government interagency mechanism to coordinate counterterrorism and countering violent extremism programming to address gaps in criminal justice, defense, and financial sector reform. Joint training exercises for Kenyan, Tanzanian, and Ugandan first responders and law enforcement professionals support efforts to enhance regional coordination and cooperation, protect shared borders, and respond to terrorist incidents responsibly and effectively.

TERRORIST SAFE HAVENS

Somalia

Terrorists use under-governed areas in northern, central, and southern Somalia as safe havens from where they conducted, planned, and facilitated operations with little resistance. Despite having made significant progress toward formally federating its member states, Somalia continues to struggle with the provision of security, justice, and governance capacity at all levels needs to limit terrorists' freedom of movement, access to resources, and capacity to operate. Despite these limitations, all levels of Somali government continue to cooperate with US counterterrorism efforts.

Al-Shabaab's ability to retain its safe haven in southern Somalia and regain new territory in central Somalia was due largely to lapses in offensive counterterrorism operations during 2016. The group carried out a series of complex assaults and raids against the AMISOM forward-operating bases that resulted in the deaths of hundreds of AMISOM and Somali soldiers, the deadliest of these occurred in January against a Kenyan forward-operating base in the Gedo region town of Ceel Adde.

Although select US-trained Somali forces have made limited progress against al-Shabaab, the Somali National Army, as a whole, remains incapable of securing and retaking towns from al-Shabaab independently. Ethiopian forces have largely withdrawn from some areas it controlled in central Somalia, leaving previously liberated towns in Bakool, Hiiraan, and other regions vulnerable to al-Shabaab.

In northern Somalia, a small group of former al-Shabaab members who had aligned with ISIS in October 2015 established a presence in remote areas that provided access to rudimentary ports along the coast of the Gulf of Aden and proximity to the Arabian Peninsula. Less than a year after pledging allegiance to ISIS, the group established a safe haven in the Golis Mountains of Puntland and temporarily took the port town of Qandala, which was retaken by Puntland's security forces in December 2016. The ISIS-aligned faction retreated to the mountains where a group of roughly 200 fighters remains out of the reach of regional security forces.

Al-Shabaab continues to keep much of its safe haven in the Jubba River Valley as a primary base of operations for plotting and launching attacks throughout Somalia and northern Kenya. The group's operatives control several villages and towns throughout Jubaland region, including Janaale, Jilib, and Kunyo Barow, and exploit the porous border regions further south between Kenya and Somalia to launch cross-border attacks.

Following one of the deadliest terrorist attacks in its history in April 2015 in Garissa, Kenya, increased its security presence along its northeastern border with Somalia to detect and counter al-Shabaab's cross-border operations, especially in and around the Boni Forest area. Al-Shabaab has capitalized on Kenya's limited border security capacities by conducting smaller-scale attacks against soft targets and law enforcement facilities in Mandera and other northeastern counties in Kenya. The group continues to extort money from locals via illegal checkpoints and local businesses, while also taxing livestock and commodities, such as sugar, to finance its operations and pay recruits.

Despite chronically low capacity and human capital, the Federal Government of Somalia remains committed to regional counterterrorism efforts to eliminate al-Shabaab's access to safe haven in Somalia. Still heavily dependent on regional and international partners, Somalia has progressed politically and has taken steps toward advancing its national security objectives through law enforcement and criminal justice reform in its urban areas.

According to independent sources and nongovernmental organizations engaged in demining activities on the ground, there is little cause for concern for the presence of weapons of mass destruction (WMD) in Somalia.

The Trans-Sahara

Terrorist organizations—including al-Qaeda in the Islamic Maghreb, al-Mulathamun Battalion (AMB), Movement for Justice and Oneness in West Africa, Ansar al-Dine, and Macina Liberation Front—can no longer claim safe haven in the Trans-Sahara region.

Following their degrading and scattering in 2013 by combined African and French operations, these terrorist groups took a year to reorganize and began a campaign of asymmetric warfare that included light raids, soft target attacks, and use of improvised explosive devices, land mines, and suicide bombers.

The groups were no longer able to conduct major military-style campaigns. Nor were they able to claim safe haven in northern Mali as they are under constant pressure by French forces and denied free movement by the French and the UN Multidimensional Integrated Stabilization Mission in Mali. Partner forces challenge their movements and operations when found throughout the Maghreb and Sahel. While these terrorist groups have not grown and they are unable to expand their zones of operation, terrorists' asymmetric warfare in this region continues, making progress restoring governance, services, and security in parts of the Sahel difficult.

The Malian government does not support or facilitate the proliferation or trafficking of WMD in and through its territory.

EAST ASIA AND PACIFIC

Terrorist attacks have occurred in Australia, Indonesia, Malaysia, the Philippines, and Thailand. ISIS-affiliated operatives attempted but failed to conduct additional attacks in Indonesia and the Philippines, and a number of other terrorist attacks were foiled by arresting or deporting individuals who were in various stages of attack planning in these countries, as well as Australia and Malaysia. The flow of foreign terrorist fighters to Iraq or Syria from the region has declined, consistent with global trends. As a result, regional concerns have shifted to foreign terrorist fighters potentially returning to Southeast Asia and using the operational training, skills, connections, and experience gained in Iraq or Syria to launch domestic attacks.

Governments in East Asia and the Pacific continue to work to strengthen legal frameworks, investigate and prosecute terrorism cases, increase regional cooperation and information sharing, and address critical border and aviation security gaps throughout the year. Cooperation between domestic law enforcement and judicial authorities throughout Southeast Asia, especially in Indonesia, Malaysia, and Singapore, has resulted in high numbers of terrorism-related arrests and, in many cases, successful prosecutions. Despite these efforts, Southeast Asia remains a target for terrorist group recruitment.

East Asian countries actively participated in regional and international efforts to counter terrorism. Australia, Japan, Malaysia, New Zealand, the Republic of Korea, Singapore, and Taiwan are partners in the Global Coalition to Defeat ISIS. Australia and Indonesia continue their co-chairmanship of the Global Counterterrorism Forum's Detention and Reintegration Working Group. The Japanese government continues to participate in international counterterrorism efforts at global, regional, and bilateral levels. Japan identified counterterrorism as one of its priorities for its G-7 presidency and joined the UN Security Council as an elected member. China's counterterrorism efforts focuses primarily on the East Turkistan Islamic Movement (ETIM); it continues to claim ETIM has influence in the Xinjiang Uighur Autonomous Region (XUAR) and was responsible for several domestic attacks, as well as an attack on the Chinese Embassy in the Kyrgyz Republic. In July, the XUAR became the only provincial level government to pass specific implementing measures of a new counterterrorism law. China continues to express concerns that Chinese citizens have traveled to Syria and Iraq to associate with ISIS.

TERRORIST HAVENS

The Sulu/Sulawesi Seas Littoral

The number of islands and maritime traffic in the Sulawesi Sea and the Sulu Archipelago make it a difficult region to secure. Traditional smuggling and piracy groups supported terrorist networks, including the movement of personnel, equipment, and funds. Kidnapping for ransom remains an ongoing threat and a source of funding for terrorist networks in the region.

Indonesia, Malaysia, and the Philippines have made concerted new efforts to control their shared maritime boundaries. In 2016, the three countries signed a trilateral agreement that envisions joint air and maritime patrols, information sharing, and standard operating procedures for "hot pursuit" of criminal and terrorist elements actively conducting attacks.

Southeast Asia serves as a global trade hub, with some of the highest volume transit and transshipment ports in the world. Lack of political will, incomplete legal and regulatory frameworks, weak strategic trade controls, inadequate law enforcement and security capabilities, and emerging and reemerging infectious disease and burgeoning bioscience capacity make Southeast Asia an area of concern for WMD proliferation and transit.

Malaysia, the Philippines, and Singapore are the only countries in the region with strategic trade control laws, and countries across Southeast Asia struggle with controls over dual-use items, as well as end-use or "catch-all" provisions. Assisting countries in the region to develop strong laws that meet international standards and helping to build effective targeting and risk management systems are major goals of the Department of State's Export Control and Related Border Security program over the next few years.

The Southern Philippines

The Philippines' geographical composition, with more than 7,500 islands, makes it difficult for the central government to maintain a presence in all areas. Counterterrorism operations have succeeded in isolating the geographic influence and largely constraining the activities of transnational terrorist groups to the south. Members of numerous groups present in Mindanao have pledged allegiance to ISIS, including parts of the Abu Sayyaf Group (ASG), the Dawlah Islamiyah Lanao (DIL)—commonly referred to as the Maute Group—and Ansar-al Khalifah Philippines. The Bangsamoro Islamic Freedom Fighters (BIFF) and other splinter groups were also present in large areas of Mindanao.

EUROPE

Europe has endured terrorist threats from a variety of sources, including foreign terrorist organizations operating out of Iraq and Syria, such as ISIS and al-Nusrah Front (al-Qaeda's affiliate in Syria), and from European foreign terrorist fighters who had returned to Europe to conduct attacks. In the face of increased military pressure, ISIS pursued mass-casualty terrorist attacks against European symbolic targets and public spaces. Various European Union (EU) member states have reported a marked increase in the rate of returning foreign terrorist fighters from Syria and Iraq. Indeed, approximately 30 percent of foreign terrorist fighters from EU member states are believed to have returned, although the overall number of departed foreign terrorist fighters originating from Western European and Balkan countries has declined significantly from the previous year. Concurrently, violent extremist groups espousing left-wing and nationalist ideologies, such as the Turkey-based Revolutionary People's Liberation Party/Front (DHKP/C) and the Kurdistan Workers' Party (PKK), respectively, continue to operate within Europe. By virtue of its location, the presence of international transport hubs on its territory, and its long border with Syria and Iraq, Turkey remains an important transit route for foreign terrorist fighters entering and departing the conflict areas. Turkey continues to deepen its cooperation with foreign terrorist fighter source countries to counter the threat, and expanded its "Banned from Entry List" to prevent travel into Turkey by nefarious actors. Risk Analysis Units continue to operate at major ports of entry and internal transit points to interdict potential foreign terrorist fighters.

To maintain tighter physical control of its border with Syria, Turkey has constructed 270 km of concrete wall along the 911-km Turkey-Syria border. European countries contribute immensely to worldwide counterterrorism efforts. Thirty-nine European countries, the EU, and Lyon-based INTERPOL collaborated within the force-multiplying framework of the Global Coalition to Defeat ISIS. Correspondingly, European officials, along with civil society members, participate in counterterrorism and CVE policies and programs within multilateral and regional fora, such as the United Nations, the Global Counterterrorism Forum, the Organization for Security and Cooperation in Europe, and the Council of Europe.

The 2015 attacks in Paris and elsewhere galvanized the EU to increase traveler information sharing and agree to an EU Passenger Name Record (PNR) system. The EU Parliament and Council presidents on April 27, 2016, signed the Directive, regulating the use of EU Passenger Name Record for the prevention, detection, investigation, and prosecution of terrorist offenses and serious crimes. EU member states had until May 2018 to fully apply the Directive package. Also, the European Commission issued a Communication on "Stronger and Smarter Information Systems for Borders and Security," and the EU Justice and Home Affairs Council has endorsed a "Roadmap to Enhance Information Exchange and Information Management Including Interoperability Solutions in the Justice and Home Affairs Area." The Commission's Communication provides a thorough review of existing European information systems and recommendations to improve them, while the Roadmap provides a comprehensive plan that identifies short, medium, and long-term actions required of member states to improve the cross-border exchange of information. In July 2016, the European Commission presented a comprehensive set of regulatory amendments, known as the Fourth Anti-Money Laundering Directive to help counter money laundering and the financing of terrorism, which were ratified in 2017.

THE MIDDLE EAST AND NORTH AFRICA

Significant terrorist activity and safe havens persist in the Middle East and North Africa. ISIS has lost most of its territory in Iraq and Syria and all of the territory it controlled in Libya. ISIS branches in the Sinai Peninsula, Saudi Arabia, and Yemen heighten sectarian tensions and they continue to conduct numerous attacks in the region. Al-Qaeda and its affiliates also maintain safe havens amid the fragile political and security climate across the region, particularly in Yemen and Syria, and conducted attacks in multiple countries. ISIS's core continues to operate in Iraq and Syria, from which it projected its "caliphate."

At one stage ISIS maintained a formidable force in Iraq and Syria, including thousands of foreign terrorist fighters from more than 100 countries, although that force has been significantly depleted as it has lost territory and equipment and its leadership hunted down and eliminated.

The United States has successfully been leading the seventy-three-member coalition to liberate territory in Iraq and Syria from ISIS, cut off ISIS's financing, disrupt its plots, counter its narrative, and stop the flow of foreign terrorist fighters. Thousands of air strikes have targeted ISIS's key leaders, heavy weapons, oil tankers, training camps, and its economic infrastructure.

In North Africa, the conflict in Libya remains the most pressing regional problem. Libya's Government of National Accord (GNA) demonstrated its willingness to partner with the United States in the fight against ISIS. Prime Minister Fayez al Sarraj invited US air support in the fight against ISIS and cooperated with international counterterrorism efforts. Most significantly, local Libyan forces—with the support of US air strikes—removed ISIS from its stronghold of Sirte. Although more than 1,700 ISIS terrorists were killed during the Sirte counterterrorism operations, many members of the terrorist organization fled to Libya's western and southern deserts, abroad, or into neighboring urban centers. Other terrorist organizations, including Ansar al-Shari'a Darnah, Ansar al-Shari'a Benghazi, and al-Qaeda in the Islamic Maghreb (AQIM) retain a presence in Libya. AQIM's focus is to fight against General Khalifa Haftar and his "Libyan National Army" forces even as Haftar and his forces publicly opposed the GNA. In Algeria, the government pursues an aggressive counterterrorism campaign against all terrorist activity within its borders. AQIM, AQIM-allied groups, and ISIS elements including the Algerian branch known as Jund al-Khilafah-Algeria (JAK-A, Soldiers of the Caliphate) are active terrorist organizations within Algeria and along its borders but struggle to accomplish their goals due to Algeria's largely effective counterterrorism pressure. In Tunisia, government counterterrorism efforts have intensified, although terrorism remains a serious challenge. Terrorist attacks persist in the Tunisian mountains and along the border with Libya. The government has developed a national counterterrorism strategy and reached out to the international community, particularly to the United States, to bolster its counterterrorism capability.

In Egypt, terrorists conduct numerous deadly attacks on government, military, and civilian targets throughout the country. Several high-profile attacks indicate that the threat level remains high despite a focus on counterterrorism by the government. ISIS continues its terrorist campaign in the Sinai through the local ISIS-affiliate, ISIL-Sinai Province (ISIL-SP, formerly Ansar Bayt al-Maqdis). The Egyptian Armed Forces conducted a counterterrorism campaign against ISIL-SP in North Sinai, known as Operation "Right of the Martyr" starting in September 2015 through 2016.

In Yemen, the ongoing conflict between the Government of Yemen, supported by the Saudi-led coalition, and the Houthi-led opposition continued the security vacuum that has enabled al-Qaeda in the Arabian Peninsula (AQAP) and ISIS's Yemen branch to deepen their inroads across much of the country. Despite losing its control of the port city of Mukalla to local Yemeni and Saudi-led coalition forces, AQAP used its tribal connections to continue recruiting, conduct attacks, and operate in areas of southern and central Yemen with relative impunity. It released several videos reiterating its intent to attack the West. Although significantly smaller than AQAP, ISIS's Yemen affiliate is able to conduct large-scale attacks targeting Yemeni soldiers in Aden and Mukalla.

Israel faces constant threats from Palestinian terrorists from Gaza and the West Bank. Since 2015, a series of lone-offender attacks by Palestinians in Israel, Gaza, and the West Bank has increased tensions between Israel and the Palestinians. Israeli and Palestinian security forces continue their coordination in an effort to mitigate the ongoing violence. Gaza-based Palestinian terrorist organizations continue rocket and mortar attacks into

Israeli territory, and multiple terrorist attacks have been launched along Israel's security barrier with Gaza. Israeli counterterrorism officials report that Hamas and other Gaza terrorists have made significant advances in their military capabilities. Terrorists continue their arms and dual-use smuggling efforts through the Sinai into Gaza via tunnels, although the Government of Egypt has undertaken efforts to prevent such smuggling from its side, which Israeli officials welcome. Other threats to Israel included Hizballah in Lebanon and Syria; al-Qaeda (AQ) and its affiliates, and ISIS and its affiliates along its borders—such as ISIL-SP and the Jaysh Khalid ibn al-Waleed group (JKW, formerly the al-Yarmouk Martyrs Brigade) in the Syrian Golan Heights.

Iran's state sponsorship of terrorism worldwide remains undiminished through the Islamic Revolutionary Guard Corps-Qods Force, its Ministry of Intelligence and Security, and Tehran's ally Hizballah, which remains a significant threat to the stability of Lebanon and the broader region.

TERRORIST HAVENS

Egypt

Portions of Egypt's Sinai region remain a safe haven for terrorist organizations. The situation in Sinai continues as a low-grade armed conflict involving forces linked to external terrorist movements and those with longstanding local grievances. The December 2016 suicide bombing of a Coptic church in Cairo, for which ISIS claimed responsibility, created increased concern that ISIS is making greater efforts to carry out terrorist attacks on mainland Egypt.

The Egyptian army has increased combat operations in northeastern Sinai but has not been able to defeat fewer than 1,000 ISIL-Sinai Province (ISIL-SP) fighters. Successful airstrikes in the Sinai, however, have removed key terrorist leadership, including ISIL-SP Sinai leader Abu Duaa Al-Ansari, although ISIL-SP quickly reconstituted its leadership after these airstrikes.

The United States supports Egypt's efforts to combat ISIL-SP by providing Apache helicopters, Mine-Resistant Ambush Protected vehicles, counter-improvised explosive devices training, mobile sensor towers, and border security training programs. The United States has also intensified military-military discussions on how it can help Egypt defeat the group. The United States remains concerned about the deteriorating security situation and potential impacts on the Multinational Force and Observers peacekeeping mission in the Sinai.

Through the Department of State's Export Control and Related Border Security Program, the United States works with the Government of Egypt to enhance its border security capabilities through the provision of land, air, and maritime border enforcement and targeting and risk management training for Egyptian Customs, Ministry of Defense, Ministry of Interior, Ministry of Transportation, and Ministry of Foreign Affairs officials. In addition, since 2009, the Department of State's Nonproliferation and Disarmament Fund has assisted Egypt with the provision of passenger and cargo vehicle x-ray detection equipment with the capability to inspect vehicular and truck traffic at fixed transportation checkpoints for WMD-related materials, conventional weapons, and other illicit items.

Iraq

Supported by the sixty-eight-member Global Coalition to Defeat ISIS, the Government of Iraq has retaken almost all of Iraqi territory once controlled by ISIS. ISIS showcased cruel savagery as it retreated, brutally executing hundreds of Iraqi civilians, publishing macabre videos of the murders to terrorize Iraqis, and forcing Mosul residents to remain as human shields to discourage airstrikes. However, ISIS has lost most of its ability to generate new revenue or resupply itself militarily.

ISIS used the territory under its control in 2016 to produce sulfur mustard and improvised explosive devices filled with chlorine. The United States has been proactively working with our allies to dismantle this chemical weapons capability, as well as deny ISIS access to chemical, biological, radiological, and nuclear (CBRN) materials and expertise through interdictions and strengthening the ability of regional governments to detect, disrupt, and respond effectively to suspected CBRN activity.

The United States and Iraq have strengthened their bilateral partnership to counter nuclear terrorism by concluding the "Joint Action Plan between the Government of the Republic of Iraq and the Government of the United States of America on Combating Nuclear and Radioactive Materials Smuggling," which expresses the intention of the two governments to work together to enhance Iraq's capabilities to prevent, detect, and respond to nuclear smuggling incidents, and ultimately prevent terrorist groups from acquiring nuclear or other radioactive materials. The Joint Action Plan provides a useful framework for US capacity-building efforts. For example, in line with this effort, the Department of Energy and the National Nuclear Security Administration's Office of Nuclear Smuggling Detection and Deterrence has provided Iraq with several mobile detection vans and training that can be used to alert officials to the presence of nuclear and radiological materials.

Lebanon

Lebanon remains a safe haven for certain terrorist groups. The Lebanese government has not taken significant action to disarm Hizballah or eliminate its safe havens on Lebanese territory, nor does it seek to limit Hizballah's travel to and from Syria to fight in support of the Assad regime or to and from Iraq. The Lebanese government does not have complete control of all regions of the country, or fully control its borders with Syria and Israel. Hizballah controls access to parts of the country and has influence over some elements within Lebanon's security services, which allows it to operate with relative impunity.

Ungoverned areas along the un-demarcated Lebanese-Syrian border also serve as safe havens for al-Nusrah Front, ISIS, and other Sunni terrorist groups, which operate in mountainous, mostly uninhabited zones where the government has limited reach. Although the Lebanese government undertook sustained military operations to counter these safe havens and ISIS, al-Nusrah Front, and other non-Hizballah terrorist activity, these groups maintain a presence in parts of Lebanon. Other terrorist groups, including Hamas, the Popular Front for the Liberation of Palestine, the Popular Front for the Liberation of Palestine General Command, Asbat al-Ansar, Fatah al-Islam, Fatah al-Intifada, Jund al-Sham, Palestinian Islamic Jihad, the Abdullah Azzam Brigades, and several other splinter groups continue to operate within Lebanon, although primarily out of Lebanon's twelve Palestinian refugee camps. These groups use the Palestinian camps as safe havens to house weapons, shelter wanted criminals, and plan terrorist attacks. Lebanese security services also note the ability of terrorists to use Syrian refugee settlements as places of refuge and transit.

The United States works closely with the Lebanese Armed Forces (LAF) and Internal Security Forces to counter terrorist threats within Lebanon and along its border with Syria by providing counterterrorism training, military equipment, and weaponry.

Lebanon is not a source country for WMD components, but its porous borders make the country vulnerable for use as a transit and transshipment hub for proliferation-sensitive transfers, particularly with the conflict in Syria. The LAF Engineer Regiment partners with US government agencies to detect and prevent proliferation and trafficking of WMD along the Syrian border.

The Department of State's Export Control and Related Border Security program (EXBS) provides commodity identification training for items that could be used in chemical, biological, radiological, and nuclear weapons, in order to keep these items from transiting through Lebanon. EXBS also proceeded with a frontier border security interdiction training program, in partnership with the Department of Defense, to strengthen LAF and Internal

Security Forces border security and interdiction capabilities. In addition, the US Department of Energy's Office of Nuclear Smuggling Detection and Deterrence equipped the Port of Beirut with radiation detection equipment to scan cargo for the presence of radiation.

Libya

Due to the inability of the Libyan GNA to effectively administer its territory, many areas of Libya's vast ungoverned space constitute potential safe havens for terrorists and violent extremist organizations, including Benghazi, Darnah, and the deserts in the south and the west. The GNA devotes significant effort to wiping out ISIL-Libya, but due to the difficulties of controlling the southern and desert borders in particular, it is unable to track flows of foreign terrorist fighters in and out of its territory. Rival factions and political stakeholders outside of the GNA, including in the House of Representatives and the "Libyan National Army," has also not stemmed or tracked the flow of foreign terrorist fighters.

In 2016, Libya requested international assistance with the destruction of its remaining chemical weapons program components, approximately 500 metric tons of category 2 chemical weapons precursors, to prevent its acquisition by terrorists operating less than 60 km away. In total, nine countries and the Organization for the Prohibition of Chemical Weapons contributed to the removal and destruction effort, with Denmark removing the chemicals from Libya, and Germany completing the actual destruction. The United States made a sizable financial contribution to the overall effort. Libya retained a stockpile of natural uranium ore concentrate (yellowcake), however, stored in a former military facility near Sebha in Libya's south. This material represents a limited risk of trafficking and proliferation due to the bulk and weight of the storage containers and the need for extensive additional processing before the material would be suitable for weapons purposes.

The Department of State's Export Control and Related Border Security program efforts in Libya focus on border enforcement capacity building through the provision of land, air, and sea border security training for Libyan officials, with the aim of advancing the GNA's ability to mitigate the threat posed by ISIS, its affiliates, and related entities. Targeted technical training and equipment for land, air, and sea border enforcement authorities will improve the Libyan government's ability to detect, identify, and interdict illicitly trafficked WMD, related items, and conventional weapons along Libya's porous borders. These activities aim to encourage interagency cooperation and promote regional and international cooperation to counter illicit trafficking in strategic items.

Yemen

Al-Qaeda in the Arabian Peninsula (AQAP) and ISIS in Yemen (ISIS-Y) continue to exploit the political and security vacuum created by the ongoing conflict between the Yemeni government under President Abdu Rabbu Mansour Hadi and the Houthi-Saleh opposition forces. Due to the instability and violence in Yemen, the internationally recognized government under Hadi could not effectively enforce counterterrorism measures. A large security vacuum persists giving AQAP and ISIS-Y more room to operate. Both groups manipulate and emphasize the conflict as part of a broader Sunni-Shia sectarian divide to increase their support bases and strengthen footholds in the country.

AQAP, in particular, benefits from the conflict by significantly expanding its presence in the southern and eastern governorates. It successfully insinuates itself among multiple factions on the ground, making the group more difficult to counter. AQAP has managed to exacerbate the effects of the conflict, fighting against the Houthi-Saleh alliance, while at the same time attempting to prevent the Yemeni government from consolidating control over southern governorates.

While ISIS-Y demonstrates a violent operational pace, it does not occupy significant territory. The Government of Yemen continues to work toward a peaceful solution to the conflict while cooperating with US counterterrorism efforts. Yemen's political instability continues to hinder efforts to enact or enforce strategic trade controls, leaving the country vulnerable as a transit point for WMD-related materials.

SOUTH AND CENTRAL ASIA

Although al-Qaeda in Afghanistan and Pakistan has been seriously degraded, remnants of AQ's global leadership, as well as its regional affiliate al-Qaeda in the Indian Subcontinent (AQIS), continue to operate from remote locations in the region that the group has historically exploited for safe haven. International, Afghan, and Pakistani forces continue to contest AQ's presence in the region, and Pakistan's continued military offensive in North Waziristan further degraded the group's freedom to operate. Pressure on AQ's traditional safe havens has constrained the leadership's ability to communicate effectively with affiliate groups outside of South Asia. Afghanistan, in particular, continues to experience aggressive and coordinated attacks by the Afghan Taliban, including the affiliated Haqqani Network (HQN) and other insurgent and terrorist groups. A number of these attacks were planned and launched from safe havens in Pakistan. Afghan National Defense and

Security Forces (ANDSF) retain full responsibility for security in Afghanistan, and prevented the Taliban from capturing a provincial capital, although it suffered an unprecedented number of casualties in intense fighting. The ANDSF and Coalition Forces, in partnership, took aggressive action against terrorist elements across Afghanistan. A peace agreement between Hizb-e Islami Gulbuddin and the Afghan government in September 2016 was the first signed by an insurgent group since the 2001 fall of the Taliban. While terrorist-related violence in Pakistan has declined, the country continued to suffer significant terrorist attacks, particularly against vulnerable civilian and government targets. The Pakistani military and security forces undertake operations against groups that conducted attacks within Pakistan, such as Tehrik-e Taliban Pakistan. Pakistan does not take substantial action against the Afghan Taliban or HQN, or substantially limit their ability to threaten US interests in Afghanistan, although Pakistan supports efforts to bring both groups into an Afghan-led peace process. Pakistan does not take sufficient action against other externally focused groups, such as Lashkar e-Tayyiba and Jaish-e-Mohammad (JeM), which continue to operate, train, organize, and fundraise in Pakistan. ISIS's formal branch in Afghanistan and Pakistan, Islamic State's Khorasan Province, remains active, although counterterrorism pressure from Afghan and US Forces has removed hundreds of fighters from the battlefield and restricted the group's ability to control territory. Nevertheless, the group continues to conduct a number of high-profile, mass-casualty attacks in Kabul against sectarian and Afghan government targets. The group also claimed a number of mass-casualty attacks in Pakistan's settled areas, likely conducted in collaboration with anti-Shia terrorist groups like Lashkar i Jhangvi.

India continues to experience attacks, including attacks by Maoist insurgents and Pakistan-based terrorists. Indian authorities continue to blame Pakistan for cross-border attacks in Jammu and Kashmir. In January 2016, India experienced a terrorist attack against an Indian military facility in Pathankot, Punjab, which was blamed by authorities on JeM. Over the course of 2016, the Government of India sought to deepen counterterrorism cooperation and information sharing with the United States. The Indian government continues to closely monitor the domestic threat from transnational terrorist groups like ISIS and AQIS, which made threats against India in their terrorist propaganda. A number of individuals were arrested for ISIS-affiliated recruitment and attack plotting within India. Bangladesh has experienced a significant increase in terrorist activity. Transnational groups such as ISIS and AQIS claimed several attacks targeting foreigners, religious minorities, police, secular bloggers, and publishers. Most notably, ISIS claimed responsibility for a July 1, 2016, attack on a restaurant in Dhaka's diplomatic enclave, which resulted in twenty-two deaths. The Government of Bangladesh primarily attributes these attacks to domestic terrorists and political opposition. People from Central Asia have travelled to Iraq or Syria to fight with militant and terrorist groups, including ISIS. Central Asians, like western Europeans, have been drawn to the fighting in Iraq and Syria for myriad reasons and fight on several sides. Central Asian leaders remain concerned about their involvement, but there is little evidence of Central Asian fighters returning home in significant numbers intent on attacking. Foreign terrorist fighters from Central Asian nation are suspected of committing attacks in third countries, however, including an attack at the Istanbul, Turkey, airport.

TERRORIST HAVENS

Afghanistan

Terrorist and insurgent groups are active in the border region of Afghanistan and Pakistan. The Government of Afghanistan struggles to assert control over this remote terrain, where the population is largely detached from national institutions. Afghanistan generally cooperates with US counterterrorism efforts, although there are some disagreements on the role of US nationals during combined counterterrorism operations. President Ghani has actively pursued cross-border security cooperation with the Government of Pakistan, including the prospect of joint operations to reduce safe havens on both sides.

Since the transition from the International Security Assistance Force (ISAF) to Resolute Support in January 2015, the trilateral border agreement that governed the Afghanistan-Pakistan border (ISAF was also a signatory) expired and the two countries were unable to finalize a bilateral agreement to replace it. While there have been some positive tactical-level steps taken by each country's military to improve operational coordination, regular cross-border shelling, and terrorist attacks on both sides of the border make formal agreement politically untenable.

The potential for WMD trafficking and proliferation remain a concern in Afghanistan because of its porous borders and the presence of terrorist groups. The United States and Afghanistan continue to work to finalize a bilateral framework to facilitate closer cooperation to counter nuclear terrorism and enhance Afghanistan's capabilities to prevent, detect, and respond to nuclear smuggling incidents. The Afghanistan and US governments also continue to work to implement comprehensive strategic trade controls and strengthen Afghanistan's border security system.

The Department of State's Export Control and Related Border Security (EXBS) Program contributed to strengthening Afghanistan's border enforcement capacity by providing training to the Afghan Customs Department.

EXBS also sponsored regional cross-border collaboration through trainings with its South and Central Asian neighbors through the US Department of Energy as well as Organization for Security and Cooperation in Europe and the UN Office on Drugs and Crime-World Customs Organization's Container Control Program. To increase the Government of Afghanistan's strategic trade control awareness, EXBS sponsored Afghan Ministry of Commerce and Industry's participation in the University of Georgia, Center for International Trade and Security, Strategic Security Trade Management Academy. In addition, the Department of Energy's Nuclear Smuggling Detection and Deterrence program provided Afghanistan with mobile detection vans and associated training to enhance capabilities to detect nuclear and radiological materials.

The United States continues to assist the Afghan government in building capacity to secure potentially dangerous biological materials and infrastructure housed at Afghan facilities, promote surveillance capabilities to detect and identify possibly catastrophic biological events, and engage Afghan scientists and engineers that have WMD or WMD-applicable expertise.

Pakistan

Numerous terrorist groups, including the Haqqani Network (HQN), Lashkar e-Tayyiba (LeT), and Jaish-e-Mohammad (JeM), continue to operate from Pakistan. Although LeT is banned in Pakistan, LeT's wings Jamaat-ud-Dawa (JuD) and Falah-i-Insaniat Foundation (FiF) openly engage in fundraising, including in the capital. LeT's chief Hafiz Saeed (a UN-designated terrorist) continues to address large rallies, although in February 2017, Pakistan proscribed him under relevant provisions of Schedule Four of the Anti-Terrorism Act, thus severely restricting his freedom of movement. The 2015 ban on media coverage of Saeed, JuD, and FiF continues and was generally followed by broadcast and print media. The Pakistani government has not publicly reversed its December 2015 declaration that neither JuD nor FiF is banned in Pakistan, despite their listing under UN sanctions regimes, although in January 2017, Pakistan placed both organizations "under observation" pursuant to Schedule Two of the Anti-Terrorism Act. While not a ban, this allows the government to closely scrutinize the activities of both organizations. On November 11, Pakistan's National Counterterrorism Authority published its own list of banned organizations that placed JuD in a separate section for groups that are "Under Observation," but not banned. Pakistan continues military operations to eradicate terrorist safe havens in the Federally Administered Tribal Areas, although their impact on all terrorist groups was uneven.

The Government of Pakistan also administers an Exit Control List intended to prevent terrorists from traveling abroad.

To combat WMD trafficking, Pakistan harmonized its national control list with items controlled by the Nuclear Suppliers Group, and continues to harmonize its control lists with other multilateral regimes, such as the Missile Technology Control Regime and the Australia Group. Pakistan improved legal and regulatory cooperation, industry outreach, and nonproliferation awareness for the Strategic Export Control Division and Pakistani Customs. In addition to industry outreach, Pakistan also delivers technical trainings to licensing and enforcement officials for the proper detection, interdiction, and identification of dual-use commodities that could be used to create WMDs.

Pakistan is a constructive and active participant in the Nuclear Security Summit process and in the Global Initiative to Combat Nuclear Terrorism, and worked to strengthen its strategic trade controls, including updating its national export control list. The State Department's Export Control and Related Border Security (EXBS) program increased the Government of Pakistan's enforcement capacity by sponsoring training for Pakistani Customs and the Strategic Export Control Division officials on how to properly identify strategic commodities of concern. These commodity identification and advanced interdiction trainings were implemented by the US Department of Energy.

EXBS also sponsored regional collaboration through nonproliferation fellowships and cross-border coordination with Afghanistan through the UN Office of Drugs and Crime-World Customs Organization's Container Control Program (CCP). Under the CCP, training was provided to enhance the targeting of skills of port control unit officials at the Jalalabad border-crossing and encouraged sharing of customs data between countries.

WESTERN HEMISPHERE

Many countries in Latin America have porous borders, limited law enforcement capabilities, and established smuggling routes. These vulnerabilities offer opportunities to local and international terrorist groups and pose challenges to governments in the region. Canada and the Caribbean—particularly Trinidad and Tobago—were sources of foreign terrorist fighters in 2016; the return of these trained foreign terrorist fighters is of great concern. Transnational criminal organizations continue to pose a significant threat to the region, and most countries made efforts to investigate possible connections between terrorist organizations and criminal activity. For many countries in the region, the challenges related to transnational crime remain the preeminent security concern.

Corruption, weak government institutions, insufficient interagency cooperation, weak or non-existent legislation, and a lack of resources remain the primary causes for the lack of significant political will to counter terrorism in some Western Hemisphere countries. Nevertheless, some countries, like Brazil and Trinidad and Tobago, have made significant progress in their counterterrorism efforts. The Revolutionary Armed Forces of Colombia and the Government of Colombia reached an agreement to end their five-decade conflict. The Colombian Congress approved the peace accord on November 30, 2016, which put in motion a six-month disarmament, demobilization, and reintegration process. Peru's Shining Path terrorist group remains active, but with reduced strength, and focused mostly on criminal activities. The Tri-Border Areas of Argentina, Brazil, and Paraguay remain an important regional nexus of arms, narcotics, pirated goods, human smuggling, counterfeiting, and money laundering—all potential funding sources for terrorist organizations.

Terrorist groups based in the Middle East find some support in Latin America despite the geographic distance. The call to fight in Iraq or Syria drew limited numbers of recruits from Latin America and parts of the Caribbean, which offered areas of financial and ideological support for ISIS and other terrorist groups in the Middle East and South Asia. In addition, Hizballah maintains some financial supporters, facilitators, and sympathizers in the region that it could tap for support in building and expanding its activities there. The Organization of American States' Inter-American Committee against Terrorism (CICTE) focuses programs and activities on cyber and supply chain security as part of the OAS' long-standing work on critical infrastructure security and resilience. The sixteenth CICTE Regular Session took place in February in Washington, DC, and brought together high level experts and authorities from the region to discuss the use of the internet for terrorist and criminal purposes, among other topics. The Caribbean Community Implementation Agency for Crime and Security (CARICOM/IMPACS) has taken steps to develop a coordinated regional counterterrorism strategy for the Caribbean. CARICOM also consults with international partners including the United States, Canada, and the United Nations to identify best practices for inclusion in its counterterrorism strategy. This effort to improve cooperation and coordination builds on the CARICOM Crime and Security Strategy, approved in 2013. The United States collaborates with both Canada and Mexico to protect our shared borders through regular exchanges of intelligence and other information.

TERRORIST HAVENS

Colombia

Rough terrain and dense forest cover, coupled with low population densities and historically weak government presence have defined Colombia's borders with Brazil, Ecuador, Peru, and Venezuela, and historically have allowed for safe havens for domestic terrorist groups, particularly the Revolutionary Armed Forces of Colombia (FARC) and the National Liberation Army (ELN). The Government of Colombia maintains pressure on these groups to deny safe haven, disrupt terrorist financing efforts, and degrade terrorist groups' logistics infrastructure. In addition, Colombia conducts operations to counter the ability of the FARC and ELN to conduct terrorist attacks. Coupled with the peace accords with the FARC, Colombia has experienced an overall decline in the total number of terrorist incidents. Despite these successes, the ELN and illegal armed groups, primarily known as "Bandas Criminales," continue to use the porous border, remote mountain areas, and jungles to maneuver, train, conduct kidnappings for ransom, cultivate and transport narcotics, operate illegal mines, "tax" the local populace, and engage in other illegal activities.

Improved relations with neighboring Ecuador have led to some increased cooperation on law enforcement issues. Colombia also continues to cooperate and share information with the Panamanian National Border Service. Additionally, Brazil continues implementing its Integrated Border Monitoring System in an effort to monitor its entire border, and along with continued cooperation with the Government of Colombia, addresses potential safe haven areas along their shared borders.

Venezuela

Venezuela's porous border with Colombia has made its territory useful to the Revolutionary Armed Forces of Colombia and the National Liberation Army, who both use it to transit in and out of its territory. There are credible reports that Venezuela maintains a permissive environment that allows for support of activities that benefited known terrorist groups.

STATE SPONSORS OF TERRORISM

The terrorist groups of greatest concern—because of their global reach—share many of the characteristics of a global insurgency: propaganda campaigns, grass roots support, transnational ideology, and political and territorial ambitions. Responding requires a comprehensive response that focuses on recruiters and their networks, potential recruits, the local population, and the ideology. An holistic approach incorporates efforts aimed at protecting and securing the population; politically and physically marginalizing insurgents; winning the support and cooperation of at-risk populations by targeted political and development measures; and conducting precise intelligence-led special operations to eliminate critical enemy elements with minimal risk to innocent civilians.

Significant achievements in this area have been made against terrorist leadership targets, notably the capturing or killing of key terrorist leaders. These efforts buy time to carry out the non-lethal and longer term elements of a comprehensive counterterrorist strategy: disrupting terrorist operations, communications, propaganda, subversion efforts, planning and fundraising, and preventing radicalization before it takes root by addressing the grievances that terrorists exploit and discrediting the ideology that provides their legitimacy. Actions that advance these strategic objectives include building and strengthening networks among governments, multilateral cooperation, business organizations, and working within civil society. It is crucial to empower credible voices and provide alternatives to joining extremist organizations.

Working with allies and partners across the world, the United States has created a less permissive operating environment for terrorists, keeping terrorist leaders on the move or in hiding, and degrading their ability to plan and mount attacks. Canada, Australia, the United Kingdom, Germany, Spain, Jordan, the Philippines, Pakistan, Afghanistan, Iraq, and many other partners have played major roles in this success. Dozens of countries have continued to pass counterterrorism legislation or strengthen pre-existing laws that provide their law enforcement and judicial authorities with new tools to bring terrorists to justice. The United States has expanded the number of foreign partners with which it shares terrorist screening information. This information serves as an important tool for disrupting and tracking travel of known and suspected terrorists. Saudi Arabia has implemented one of the first rehabilitation programs for returning extremists to turn them against violent extremism and to reintegrate them as peaceful citizens.

Through the Regional Strategic Initiative, the State Department and other US agencies are working with US ambassadors overseas in key terrorist theaters of operation to assess threats and devise collaborative strategies, action plans, and policy recommendations. Considerable progress has been made in organizing regional responses to terrorists who operate in ungoverned spaces or across national borders. This initiative has produced better intragovernmental coordination among US government agencies, greater cooperation with and between regional partners, and improved strategic planning and prioritization, allowing us to use all tools of statecraft to establish long-term measures to marginalize terrorists.

Radicalization continues in immigrant populations, youth and alienated minorities in Europe, the Middle East, and Africa. A special focus on new approaches in Europe has been productive and has informed the way we understand government's role in combating radicalization. It is increasingly clear that radicalization does not occur by accident, or because such populations are innately prone to extremism. Rather, there is increasing evidence of terrorists and extremists manipulating the grievances of alienated youth or immigrant populations, and then cynically exploiting those grievances to subvert legitimate authority and create unrest.

Efforts to manipulate grievances represent a conveyor belt through which terrorists seek to convert alienated or aggrieved populations, by stages, to increasingly radicalized and extremist viewpoints, turning them into sympathizers, supporters, and ultimately, in some cases, members of terrorist networks. In some regions, this includes efforts by ISIS, AQ, and other terrorists to exploit insurgency and communal conflict as radicalization and recruitment tools, using the internet to convey their message. Counter-radicalization is a priority for the United States, particularly in Europe, given the potential for Europe-based violent extremism to threaten the United States and its key interests directly. The leaders of ISIS, AQ and its affiliates are interested in recruiting terrorists from and deploying terrorists to Europe. They are especially interested in people familiar with Western cultures who can travel freely in the region and to the United States. However, countering such efforts requires that we treat immigrant and youth populations not as a threat to be defended against, but as a target of enemy subversion to be protected and supported. It requires that community leaders take responsibility for the actions of members within their communities and act to counteract extremist propaganda and subversion. It also requires governments to serve as facilitators, conveners, and intellectual partners to credible organizations/people who can do what governments cannot. Finally, bilateral, regional, and multilateral cooperation is essential.

To designate a country as a State Sponsor of Terrorism, the Secretary of State must determine that the government of such country has repeatedly provided support for acts of international terrorism. Once a country is designated, it remains a State Sponsor of Terrorism until the designation is rescinded in accordance with statutory criteria. A wide range of sanctions are imposed as a result of a State Sponsor of Terrorism designation.

Countries determined by the Secretary of State to have repeatedly provided support for acts of international terrorism are designated pursuant to three laws: section 6(j) of the Export Administration Act, section 40 of the Arms Export Control Act, and section 620A of the Foreign Assistance Act. Taken together, the four main categories of sanctions resulting from designation under these authorities include restrictions on US foreign assistance; a ban on defense exports and sales; certain controls over exports of dual-use items; and miscellaneous financial and other restrictions.

Designation under the above-referenced authorities also implicates other sanctions laws that penalize persons and countries engaging in certain trade with state sponsors. Currently there are four countries designated under these authorities: the Democratic People's Republic of Korea (North Korea), Iran, Sudan, and Syria.

IRAN

Designated as a State Sponsor of Terrorism in 1984, Iran continues its terrorist-related activity, including support for Hizballah, Palestinian terrorist groups in Gaza, and various groups in Syria, Iraq, and throughout the Middle East. Iran uses the Islamic Revolutionary Guard Corps-Qods Force (IRGC-QF) to implement foreign policy goals, provide cover for intelligence operations, and create instability in the Middle East. Iran has acknowledged the involvement of the IRGC-QF in the conflicts in Iraq and Syria and the IRGC-QF is Iran's primary mechanism for cultivating and supporting terrorists abroad.

Iran supports various Iraqi Shia terrorist groups, including Kata'ib Hizballah, as part of an effort to fight ISIS in Iraq and bolster the Assad regime in Syria. Iran views the Assad regime in Syria as a crucial ally and Syria and Iraq as crucial routes to supply weapons to Hizballah, Iran's primary terrorist partner. Iran has facilitated and coerced, through financial or residency enticements, primarily Shia fighters from Afghanistan and Pakistan to participate in the Assad regime's brutal crackdown in Syria. Iranian-supported Shia militias in Iraq have committed serious human rights abuses against primarily Sunni civilians and Iranian forces have directly backed militia operations in Syria with armored vehicles, artillery, and drones.

Since the end of the 2006 Israeli-Hizballah conflict, Iran has supplied Hizballah with thousands of rockets, missiles, and small arms, in direct violation of UN Security Council resolution (UNSCR) 1701. Iran provides the majority of financial support for Hizballah in Lebanon and has trained thousands of its fighters at camps in Iran. Hizballah fighters have been used extensively in Syria to support the Assad regime and in support of operations against ISIS in Iraq. Hizballah has also carried out several attacks against Israeli Defense Forces in along the Lebanese border with Israel.

Iran has historically provided weapons, training, and funding to Hamas and other Palestinian terrorist groups, including Palestine Islamic Jihad and the Popular Front for the Liberation of Palestine General Command. These Palestinian terrorist groups have been behind a number of deadly attacks originating in Gaza and the West Bank, including attacks against Israeli civilians and Egyptian security forces in the Sinai Peninsula.

Iran has provided weapons, funding, and training to Bahraini militant Shia groups that have conducted attacks on the Bahraini security forces. In January 2016, Bahraini security officials dismantled a terrorist cell, linked to IRGC-QF, planning to carry out a series of bombings throughout the country.

The Iranian government maintains a robust cyberterrorism program and has sponsored cyberattacks against foreign government and private sector entities.

Iran remains unwilling to bring to justice senior al-Qaeda members; it continues to detain and has refused to publicly identify the members in its custody. Since at least 2009, Iran has allowed AQ facilitators to operate a core facilitation pipeline through the country, enabling AQ to move funds and fighters to South Asia and Syria.

NORTH KOREA

The Democratic People's Republic of Korea (DPRK) is not known to have sponsored any terrorist acts since the bombing of a Korean Airlines flight in 1987. At a summit with Japanese Prime Minister Koizumi in Pyongyang in September 2002, National Defense Commission Chairman Kim Jong Il acknowledged the involvement of DPRK "special institutions" in the kidnapping of Japanese citizens and said that those responsible had already been punished. Five surviving abductees were allowed to return to Japan in 2003.

Four Japanese Red Army members remain in the DPRK following their involvement in a jet hijacking in 1970. Although it is a party to six international conventions and protocols relating to terrorism, Pyongyang has not taken substantial steps to cooperate in efforts to combat international terrorism.

As part of the Six-Party Talks process, the United States reaffirmed its intent to fulfill its commitments regarding the removal of the designation of the DPRK as a state sponsor of terrorism in parallel with the DPRK's actions on denuclearization and in accordance with criteria set forth in US law. North Korea's development of a nuclear program and its willingness to sell this technology have been matters of grave international concern.

Following harsh sanctions and intense public pressure, the first meeting of the leaders of North and South Korea since the Korean War ended in 1953 took place in the Demilitarized Zone on April 27, 2018. North Korea's Kim Jong-un and South Korea's Moon Jae-in committed to achieving "a nuclear-free Korean Peninsular" and finally declaring an end to the Korean War. A meeting between Kim Jong Un and President Donald Trump took place in Singapore on June 12. A second meeting was being arranged at the time of publication.

SUDAN

Sudan was designated as a State Sponsor of Terrorism in 1993 due to concerns about support to international terrorist groups, including the Abu Nidal

Organization, Palestine Islamic Jihad, Hamas, and Hizballah. Notwithstanding this history, countering terrorism is today a national security priority for Sudan, and Sudan is a cooperative partner of the United States on counterterrorism, despite its continued presence on the State Sponsors of Terrorism List. The Government of Sudan continues to pursue counterterrorism operations alongside regional partners, including operations to directly counter threats to US interests and personnel in Sudan. Sudan takes seriously the threat posed by ISIS, and the US government has acknowledged the Government of Sudan's significant efforts to counter ISIS and other terrorist groups and to prevent their movement into and through Sudan. Reports indicate that the Government of Sudan has ceased providing Hamas any direct support as they did in years past.

As a result of a US engagement plan with the Government of Sudan in 2016 that offered potential economic sanctions relief if Sudan took positive actions in a number of areas—including enhanced counterterrorism cooperation—the Sudanese government has taken steps to improve its counterterrorism efforts through enhanced interagency and international cooperation to address the ISIS threat.

SYRIA

Designated in 1979 as a State Sponsor of Terrorism, the Assad regime continues its political and military support to a variety of terrorist groups affecting the stability of the region. The regime continues to provide political and weapons support to Hizballah and continues to allow Iran to rearm the terrorist organization. The Assad regime's relationship with Hizballah, Iran, and Russia has grown stronger as the regime became more reliant on external actors to militarily fight the Syrian opposition. These groups, especially the Russian military, played a critical role in the regime's recovery of territory from opposition rebels groups. President Bashar al-Assad remains a staunch defender of Iran's policies, while Iran exhibits equally energetic support for the Syrian regime. Statements supporting terrorist groups, particularly Hizballah, are often in Syrian government speeches and press statements.

Over the past decade, the Syrian government has played an important role in the growth of terrorist networks in Syria through the Assad regime's permissive attitude toward al-Qaeda and other terrorist groups' foreign terrorist fighter facilitation efforts during the Iraq conflict. Syria served for years as a hub for foreign terrorist fighters, and the Syrian government's awareness and encouragement for many years of violent extremists' transit through Syria to enter Iraq for the purpose of fighting Coalition troops, is well documented. Those very networks are among the violent extremist elements which have terrorized the Syrian and Iraqi populations. While there has been a general decline in the flow of Sunni extremist linked foreign terrorist fighters traveling to Syria, ISIS and other terrorist groups continue to attract some new recruits to Syria. Additionally, foreign terrorist fighters aligned with Iran continue to travel to Syria to fight for or with the regime.

As part of a broader strategy, the regime portrays Syria itself as a victim of terrorism, characterizing all of the internal armed opposition as "terrorists." From Syria, ISIS plotted or inspired external terrorist operations. Additionally, the Syrian regime has purchased oil from ISIS through various middlemen, adding to the terrorist group's revenue.

Syria is not in compliance with its obligations under the Chemical Weapons Convention (CWC). The United States assesses that Syria has used chemical weapons repeatedly against the Syrian people every year since acceding to the Convention, and is therefore in violation of its obligations under Article I of the CWC. There have been numerous reports of chemical weapons use by the regime during the current conflict. In 2015, the UN Security Council established the Organization for the Prohibition of Chemical Weapons-UN Joint Investigative Mechanism (JIM) to identify individuals, entities, groups, or governments involved in the confirmed uses of chemical weapons in Syria. The JIM has concluded that the Syrian Arab Armed Forces used toxic chemicals as weapons, likely chlorine, multiple times in Syrian opposition held areas.

Part III

BE PREPARED

Chapter 9

Preparing for and Coping with a Disaster

Every year, millions of people around the world lose their lives or have them seriously disrupted because of floods, tsunamis, wildfires, earthquakes, blizzards, hurricanes, and so on. In today's troubled world, we face even greater threats from terrorism, civil unrest, explosions and the possibility of chemical, biological, or even nuclear attack.

In the United States, the terrorist threat is not only from overseas groups; domestic terrorism has been with us for decades, and violence in the workplace and schools is a serious area of concern. The threat is real and growing.

Even after 9/11 and deadly, devastating like Hurricanes Katrina and Irma, many businesses in the United States do not have a crisis management plan—what to do in the event of an emergency—and many of those that do have one do not keep it up-to-date. Even fewer businesses and organizations have a physical security plan to protect their property and the people who work in it.

In the United States there are more than 1,800 government-owned buildings and more than 6,200 leased locations throughout the fifty continental states and Washington, DC. They employ almost a million federal workers and host tens of millions of visitors.

There are 327,000 educational buildings in the fifty states and DC There are more than 132,000 public and private schools with 55 million children enrolled and employing around 5 million teachers.

There are 5,800 hospitals nationwide employing 2.7 million registered nurses, 1.5 million nursing aides, 819,000 physicians and surgeons, and 350,000 therapists. On any given day there are about 540,000 hospital inpatients plus visitors. There are over 125,000 additional health care facilities nationwide offering inpatient/outpatient treatment.

There are 133,000 malls and strip malls, 534,000 large stores, 349,000 food-service facilities, and 153,000 hotels and motels nationwide.

Throughout the United States, there are more than 1 million office buildings, and 603,000 warehouses and storage facilities. There are more than 307,000 churches nationwide and more than 305,000 public assembly buildings. And, there are more than 115 million households.

Terrorism is not a new challenge, and it is not going to go away any time soon so we have a duty to our families, friends, and loved ones to ensure that the places where we live, work, learn, and play are secure and that the people using them are safe. The precautions you take to protect you and your family and fellow workers are the same whether you are guarding against a terrorist attack or a disaster, whether natural or man-made. Planning and preparation are critical for survival.

EMERGENCY PREPAREDNESS

Why Prepare?

Immediately after an emergency, essential services may be cutoff and local disaster relief and government responders may not be able to reach you right away. Even if they can reach you, knowing what to do to protect yourself and your household is essential. One of the most important steps you can take in preparing for emergencies is to develop a household disaster plan.

- Learn about the natural disasters that could occur in your community from your local emergency management office or American Red Cross chapter. Learn whether hazardous materials are produced, stored, or transported near your area. Learn about possible consequences of deliberate acts of terror. Ask how to prepare for each potential emergency and how to respond.
- Talk with employers and school officials about their emergency response plans.
- Talk with your household about potential emergencies and how to respond to each. Talk about what you would need to do in an evacuation.
- Plan how your household would stay in contact if you were separated. Identify two meeting places: the first should be near your home—in case of fire, perhaps a tree or a telephone pole; the second should be away from your neighborhood in case you cannot return home.
- Pick a friend or relative, who lives out of the area, for household members to call to say they are OK.
- Draw a floor plan of your home. Mark two escape routes from each room.
- Post emergency telephone numbers by telephones. Teach children how and when to call 911.

- Make sure everyone in your household knows how and when to shut off water, gas, and electricity at the main switches. Consult with your local utilities if you have questions.
- Take a first aid and CPR class. Local American Red Cross chapters can provide information. Official certification by the American Red Cross provides "good Samaritan" law protection for those giving first aid.
- Reduce the economic impact of disaster on your property and your household's health and financial well-being.
- Review property insurance policies before disaster strikes—make sure policies are current and be certain they meet your needs (type of coverage, amount of coverage, and hazard covered—flood, earthquake).
- Protect your household's financial well-being before a disaster strikes—review life insurance policies and consider saving money in an "emergency" savings account that could be used in any crisis. It is advisable to keep a small amount of cash or traveler's checks at home in a safe place where you can quickly gain access to it in case of an evacuation.
- Be certain that health insurance policies are current and meet the needs of your household.
- Consider ways to help neighbors who may need special assistance, such as the elderly or the disabled.
- Make arrangements for pets. Pets are not usually allowed in public shelters. Service animals, for those who depend on them, are allowed.

Emergency Planning for People with Special Needs

If you have a disability or special need, you may have to take additional steps to protect yourself and your household in an emergency. If you know of friends or neighbors with special needs, help them with these extra precautions. Examples include the following:

- Hearing impaired may need to make special arrangements to receive a warning.
- Mobility impaired may need assistance in getting to a shelter.
- Households with a single, working parent may need help from others both in planning for disasters and during an emergency.
- Non-English speaking people may need assistance planning for and responding to emergencies. Community and cultural groups may be able to help keep these populations informed.
- People without vehicles may need to make arrangements for transportation.
- People with special dietary needs should have an adequate emergency food supply.

1. Find out about special assistance that may be available in your community. Register with the office of emergency services or fire department for assistance, so needed help can be provided quickly in an emergency.
2. Create a network of neighbors, relatives, friends, and coworkers to aid you in an emergency. Discuss your needs and make sure they know how to operate necessary equipment.
3. Discuss your needs with your employer.
4. If you are mobility impaired and live or work in a high-rise building, have an escape chair.
5. If you live in an apartment building, ask the management to mark accessible exits clearly and to make arrangements to help you evacuate the building.
6. Keep extra wheelchair batteries, oxygen, catheters, medication, food for Seeing Eye or hearing ear dogs, or other items you might need. Also, keep a list of the type and serial numbers of medical devices you need.
7. Those who are not disabled should learn who in their neighborhood or building is disabled, so that they may assist them during emergencies.
8. If you are a caregiver for a person with special needs, make sure you have a plan to communicate if an emergency occurs.

What You Can Do to Prepare

Finding out what can happen is the first step. Once you have determined the events possible and their potential in your community, it is important that you discuss them with your family or household. Develop a disaster plan together.

1. Create an emergency communications plan.

Choose an out-of-town contact your family or household will call or e-mail to check on each other should a disaster occur. Your selected contact should live far enough away that they would be unlikely to be directly affected by the same event, and they should know they are the chosen contact. Make sure every household member has that contact's, and each other's, e-mail addresses and telephone numbers (home, work, pager, and cell). Leave these contact numbers at your children's schools, if you have children, and at your workplace. Your family should know that if telephones are not working, they need to be patient and try again later or try e-mail. Many people flood the telephone lines when emergencies happen but e-mail can sometimes get through when calls don't.

2. Establish a meeting place.

Having a predetermined meeting place away from your home will save time and minimize confusion should your home be affected or the area evacuated. You may even want to make arrangements to stay with a family member or friend in case of an emergency. Be sure to include any pets in these plans, since pets are not permitted in shelters and some hotels will not accept them.

3. *Assemble a disaster supplies kit.*

If you need to evacuate your home or are asked to "shelter in place," having some essential supplies on hand will make you and your family more comfortable. Prepare a disaster supplies kit in an easy-to-carry container such as a duffel bag or small plastic trash can. Include "special needs" items for any member of your household (infant formula or items for people with disabilities or older people), first aid supplies (including prescription medications), a change of clothing for each household member, a sleeping bag or bedroll for each, a battery-powered radio or television and extra batteries, food, bottled water, and tools. It is also a good idea to include some cash and copies of important family documents (birth certificates, passports, and licenses) in your kit.

Copies of essential documents—like powers of attorney, birth and marriage certificates, insurance policies, life insurance beneficiary designations, and a copy of your will—should also be kept in a safe location outside your home. A safe deposit box or the home of a friend or family member who lives out of town is a good choice.

You may need to survive on your own for three days or more. This means having your own water, food, and emergency supplies. Try using backpacks or duffel bags to keep the supplies together.

Assembling the supplies you might need following a disaster is an important part of your disaster plan. You should prepare the following emergency supplies:

- Keep a disaster supply kit with essential food, water, and supplies for at least three days—this kit should be kept in a designated place and be ready to "grab and go" in case you have to leave your home quickly because of a disaster, such as a flash flood or major chemical emergency. Make sure all household members know where the kit is kept.
- Consider having additional supplies for use in shelters or home confinement for up to two weeks.
- You should also have a disaster supply kit at work. This should be in one container, ready to "grab and go" in case you have to evacuate the building.
- Keep a car kit of emergency supplies, including food and water, stored in your vehicle at all times. This kit would also include flares, jumper cables, and seasonal supplies.

The following checklists will help you assemble disaster supply kits that meet the needs of your household. The basic items that should be in a disaster supply kit are water, food, first-aid supplies, tools and emergency supplies, clothing and bedding, and specialty items. You will need to change the stored water and food supplies every six months, so be sure to write the date you store it on all containers. You should also re-think your needs every year and update your kit as your household changes. Keep items in airtight plastic bags and put your entire disaster supply kit in one or two easy-to-carry containers such as an unused trashcan, camping backpack, or duffel bag.

Water: The Absolute Necessity

- Stocking water reserves should be a top priority. Drinking water in emergency situations should not be rationed. Therefore, it is critical to store adequate amounts of water for your household.
- Individual needs vary, depending on age, physical condition, activity, diet, and climate. A normally active person needs at least two quarts of water daily just for drinking. Children, nursing mothers, and ill people need more. Very hot temperatures can double the amount of water needed.
- Because you will also need water for sanitary purposes and, possibly, for cooking, you should store at least one gallon of water per person per day.
- Store water in thoroughly washed plastic, fiberglass or enamel-lined metal containers. Don't use containers that can break, such as glass bottles. Never use a container that has held toxic substances. Sound plastic containers, such as soft drinks bottles, are best. You can also purchase food-grade plastic buckets or drums.
- Containers for water should be rinsed with a diluted bleach solution (1 part bleach to 10 parts water) before use. Previously used bottles or other containers may be contaminated with microbes or chemicals. Do not rely on untested devices for decontaminating water.
- If your water is treated commercially by a water utility, you do not need to treat water before storing it. Additional treatments of treated public water will not increase storage life.
- If you have a well or public water that has not been treated, follow the treatment instructions provided by your public health service or water provider.
- If you suspect that your well may be contaminated, contact your local or state health department or agriculture extension agent for specific advice.
- Seal your water containers tightly, label them and store them in a cool, dark place.
- It is important to change stored water every six months.

Food: Preparing an Emergency Supply

- If activity is reduced, healthy people can survive on half their usual food intake for an extended period or without any food for many days. Food, unlike water, may be rationed safely, except for children and pregnant women.
- You don't need to go out and buy unfamiliar foods to prepare an emergency food supply. You can use the canned goods, dry mixes, and other staples on your cupboard shelves. Canned goods do not require cooking, water, or special preparation. Be sure to include a manual can opener.
- Keep canned goods in a dry place where the temperature is fairly cool. To protect boxed foods from pests and to extend their shelf life, store the food in tightly closed plastic or metal containers. Replace items in your food supply every six months. Throw out any canned good that becomes swollen, dented, or corroded. Use foods before they go bad, and replace them with fresh supplies. Date each food item with a marker. Place new items at the back of the storage area and older ones in front.
- Food items that you might consider including in your disaster supply kit: ready-to-eat meats, fruits, and vegetables; canned or boxed juices, milk, and soup; high-energy foods like peanut butter, jelly, low-sodium crackers, granola bars, and trail mix; vitamins; foods for infants or people on special diets; cookies and hard candy; and instant coffee, cereals, and powdered milk.

You may need to survive on your own after a disaster. Local officials and relief workers will be on the scene after a disaster, but they cannot reach everyone immediately. You could get help in hours, or it may take days. Basic services, such as electricity, gas, water, sewage treatment, and telephones, may be cut off for days, even a week or longer. Or you may have to evacuate at a moment's notice and take essentials with you. You probably won't have the opportunity to shop or search for the supplies you'll need. Your household will cope best by preparing for disaster before it strikes.

First-Aid Supplies

Assemble a first-aid kit for your home and for each vehicle. The basics for your first-aid kit should include the following:

- First-aid manual
- Sterile adhesive bandages in assorted sizes
- Assorted sizes of safety pins
- Cleansing agents (isopropyl alcohol, hydrogen peroxide), soap, germicide
- Antibiotic ointment
- Latex gloves (two pairs)

- Petroleum jelly
- 2-inch and 4-inch sterile gauze pads (4–6 each size)
- Triangular bandages (3)
- 2-inch and 3-inch sterile roller bandages (three rolls each)
- Cotton balls
- Scissors
- Tweezers
- Needle
- Moistened towelettes
- Antiseptic
- Thermometer
- Tongue depressor blades (2)
- Tube of petroleum jelly or other lubricant
- Sunscreen

It may be difficult to obtain prescription medications during a disaster because stores may be closed or supplies may be limited. Ask your physician or pharmacist about storing prescription medications. Be sure they are stored to meet instructions on the label and be mindful of expirations dates—be sure to keep your stored medication up-to-date. Have an extra pair or two of prescription glasses or contact lens.

Have the following nonprescription drugs in your disaster supply kit:

- Aspirin and non-aspirin pain reliever
- Anti-diarrhea medication
- Antacid (for stomach upset)
- Syrup of ipecac (to induce vomiting if advised by the poison control center)
- Laxatives
- Vitamins

Tools and Emergency Supplies

It will be important to assemble these items in a disaster supply kit in case you have to leave your home quickly. Even if you don't have to leave your home, if you lose power it will be easier to have these item already assembled and in one place.

Tools and other items:

- A portable, battery-powered radio or television and extra batteries (also have a NOAA weather radio, if appropriate for your area)
- Flashlight and extra batteries

- Signal flare
- Matches in a waterproof container (or waterproof matches)
- Shut-off wrench, pliers, shovel, and other tools
- Duct tape and scissors
- Plastic sheeting
- Whistle
- Small canister, A-B-C-type fire extinguisher
- Tube tent
- Compass
- Work gloves
- Paper, pens, and pencils
- Needles and thread
- Battery-operated travel alarm clock

Kitchen items:

- Manual can opener
- Mess kits or paper cups, plates, and plastic utensils
- All-purpose knife
- Household liquid bleach to treat drinking water
- Sugar, salt, and pepper
- Aluminum foil and plastic wrap
- Re-sealing plastic bags
- If food must be cooked, small cooking stove and a can of cooking fuel

Sanitation and hygiene items:

- Washcloth and towel
- Towelettes, soap, hand sanitizer, and liquid detergent
- Toothpaste, toothbrushes, shampoo, deodorants, comb, brush and razor, shaving cream, lip balm, sunscreen, insect repellent, contact lens solutions, mirror, and feminine supplies
- Heavy-duty plastic garbage bags and ties, for personal sanitation use
- Toilet paper
- Medium-sized plastic bucket with tight lid
- Disinfectant and household chlorine bleach
- Consider including a small shovel for digging a latrine

Household documents and contact numbers:

- Personal identification, cash (including change), or traveler's checks, and a credit card

- Copies of important documents: birth certificate, marriage certificate, driver's license, Social Security cards, passport, wills, deeds, inventory of household goods, insurance papers, immunizations records, bank and credit card account numbers, stocks and bonds (be sure to store these in a watertight container.)
- Emergency contact list and phone numbers
- Map of the area and phone numbers of places you could go
- An extra set of car keys and house keys

Clothes and bedding:

- One complete change of clothing and footwear for each household member
- Shoes should be sturdy work shoes or boots
- Rain gear, hat and gloves, extra socks, extra underwear, thermal underwear, and sunglasses
- Blankets or a sleeping bag for each household member and pillows

Specialty items:

Remember to consider the needs of infants, elderly people, disabled people and pets, and to include entertainment and comfort items for children.

- For baby
- For the elderly
- For pets
- Entertainment: books, games, quiet toys, and stuffed animals

It is important for you to be ready, wherever you may be, when disaster strikes. With the checklists provided you can now put together appropriate disaster supply kits for your household:

- A disaster supply kit kept in the home with supplies for at least three days.
- Although it is unlikely that food supplies would be cut off for as long as two weeks, consider storing additional water, food, clothing, bedding and other supplies to expand your supply kit to last up to two weeks.
- A workplace disaster supply kit. (It is important to store a personal supply of water and food at work; you will not be able to rely on water fountains or coolers. Women who wear high heels should be sure to have comfortable flat shoes at their workplace in case an evacuation requires walking long distances.)
- A car disaster supply kit. (Keep a smaller disaster supply kit in the trunk of your car. If you become stranded or are not able to return home, having

these items will help you be more comfortable until help arrives. Add items for severe winter weather during months when heavy snow or icy roads are possible—salt, sand, shovels, and extra winter clothing, including hats and gloves.)

4. Check on the school emergency plan of any school-age children you may have.

You need to know if they will they keep children at school until a parent or designated adult can pick them up or send them home on their own. Be sure that the school has updated information about how to reach parents and responsible caregivers to arrange for pickup. And, ask what type of authorization the school may require to release a child to someone you designate, if you are not able to pick up your child. During times of emergency the school telephones may be overwhelmed with calls.

For more information on putting together a disaster plan, request a copy of the brochure titled *Your Family Disaster Plan* (A4466) from your local American Red Cross chapter. You may also want to request a copy of *Before Disaster Strikes . . . How to Make Sure You're Financially Prepared* (A5075) for specific information on what you can do now to protect your assets.

If Disaster Strikes:

- Remain calm and be patient.
- Follow the advice of local emergency officials.
- Listen to your radio or television for news and instructions.
- If the disaster occurs near you, check for injuries. Give first aid and get help for seriously injured people.
- If the disaster occurs near your home while you are there, check for damage using a flashlight. Do not light matches or candles or turn on electrical switches. Check for fires, fire hazards, and other household hazards. Sniff for gas leaks, starting at the water heater. If you smell gas or suspect a leak, turn off the main gas valve, open windows, and get everyone outside quickly.
- Shut off any other damaged utilities.
- Confine or secure your pets.
- Call your family contact—do not use the telephone again unless it is a life-threatening emergency.
- Check on your neighbors, especially those who are elderly or disabled.

A Word on What Could Happen

As we learned from the events of September 11, 2001, the following things can happen after a terrorist attack or major disaster:

- There can be significant numbers of casualties and/or damage to buildings and the infrastructure. So employers need up-to-date information about any medical needs you may have and on how to contact your designated beneficiaries.
- Heavy law enforcement involvement at local, state, and federal levels follows a terrorist attack due to the event's criminal nature.
- Health and mental health resources in the affected communities can be strained to their limits, maybe even overwhelmed.
- Extensive media coverage, strong public fear, and international implications and consequences can continue for a prolonged period.
- Workplaces and schools may be closed, and there may be restrictions on domestic and international travel.
- You and your family or household may have to evacuate an area, avoiding roads blocked for your safety.
- Cleanup may take many months.

Evacuation

Evacuations are more common than many people realize. Hundreds of times each year, transportation and industrial accidents release harmful substances, forcing thousands of people to leave their homes. Fires and floods cause evacuations even more frequently. And almost every year, people along the Gulf and Atlantic coasts evacuate in the face of approaching hurricanes.

When community evacuations become necessary, local officials provide information to the public through the media. In some circumstances other warning methods, such as sirens or telephone calls, are also used.

Government agencies, the American Red Cross, Salvation Army, and other disaster relief organizations provide emergency shelter and supplies. To be prepared for an emergency, you should have enough water, food, clothing, and emergency supplies to last at least three days. In a catastrophic emergency, you might need to be self-sufficient for even longer.

The amount of time you have to evacuate will depend on the disaster. If the event can be monitored, such as a hurricane, you might have a day or two to get ready. However, many disasters allow no time for people to gather even the most basic necessities. This is why you should prepare now.

Planning for Evacuation

1. Ask your local emergency management office about community evacuation plans. Learn evacuation routes. If you do not own a car, make transportation arrangements with friends or your local government.
2. Talk with your household about the possibility of evacuation. Plan where you would go if you had to leave the community. Determine how you would get there. In your planning, consider different scales of evacuations. In a hurricane, for example, entire counties would evacuate, while a much smaller area would be affected by a chemical release.
3. Plan a place to meet your household in case you are separated from one another in a disaster. Ask a friend outside your town to be the "checkpoint" so that everyone in the household can call that person to say they are safe.
4. Find out where children will be sent if schools are evacuated.
5. Assemble a disaster supplies kit. Include a battery-powered radio, flashlight, extra batteries, food, water, and clothing. See the "Emergency Planning and Disaster Supplies" chapter for a complete list.
6. Keep fuel in your car if an evacuation seems likely. Gas stations may be closed during emergencies and unable to pump gas during power outages.
7. Know how to shut off your home's electricity, gas, and water supplies at main switches and valves. Have the tools you would need to do this (usually adjustable pipe and crescent wrenches).

If local authorities ask you to leave your home, they have a good reason to make this request, and you should heed the advice immediately.

What to Do When You Are Told to Evacuate

Listen to a battery-powered radio and follow local instructions. If the danger is a chemical release and you are instructed to evacuate immediately, gather your household members and go. Take one car per household when evacuating. This will keep your household together and reduce traffic congestion and delay. In other cases, you may have time to follow these steps:

1. Gather water, food, clothing, emergency supplies, and insurance and financial records.
2. Wear sturdy shoes and clothing that provides some protection, such as long pants, long-sleeved shirts, and a cap.
3. Secure your home. Close and lock doors and windows. Unplug appliances. If a hard freeze is likely during your absence, take actions needed to prevent damage to water pipes by freezing weather, such as the following:
   - Turn off the water main

- Drain faucets
- Turn off inside valves for external faucets and opening the outside faucets to drain

4. Turn off the main water valve and electricity, if instructed to do so.
5. Let others know where you are going.
6. Leave early enough to avoid being trapped by severe weather.
7. Follow recommended evacuation routes. Do not take shortcuts. They may be blocked. Be alert for washed-out roads and bridges. Do not drive into flooded areas. Stay away from downed power lines.

Disaster situations can be intense, stressful, and confusing. Should an evacuation be necessary, local authorities will do their best to notify the public, but do not depend entirely on this. Often, a disaster can strike with little or no warning, providing local authorities scant time to issue an evacuation order.

Also, it is possible that you may not hear of an evacuation order due to communications or power failure or not listening to your battery-powered radio. Local authorities and meteorologists could also make mistakes, including underestimating an emergency or disaster situation. In the absence of evacuation instructions from local authorities, you should evacuate if you feel you and your household are threatened or endangered. Use pre-designated evacuation routes and let others know what you are doing and your destination.

Listen to Local Authorities

Your local authorities will provide you with the most accurate information specific to an event in your area. Staying tuned to local radio and television if possible, and following their instructions is your safest choice. Which is why you need a radio and spare batteries.

If you're sure you have time:

- Call your family contact to tell them where you are going and when you expect to arrive.
- Shut off water and electricity before leaving, if instructed to do so. Leave natural gas service ON unless local officials advise you otherwise. You may need gas for heating and cooking, and only a professional can restore gas service in your home once it's been turned off. In a disaster situation it could take weeks for a professional to respond.

Shelter

Taking shelter is often a critical element in protecting yourself and your household in times of disaster. Sheltering can take several forms. In-place sheltering is appropriate when conditions require that you seek protection in your home, place of employment, or other location where you are located when disaster strikes. In-place sheltering may either be short-term, such as going to a safe room for a fairly short period while a tornado warning is in effect or while a chemical cloud passes. It may also be longer-term, such as when you stay in your home for several days without electricity or water services following a winter storm. We also use the term "shelter" for Mass Care facilities that provide a place to stay along with food and water for people who evacuate following a disaster.

If you are advised by local officials to "shelter in place," what they mean is for you to remain inside your home or office and protect yourself there. Close and lock all windows and exterior doors. Turn off all fans, heating, and air-conditioning systems. Close the fireplace damper. Get your disaster supplies kit, and make sure the radio is working. Go to an interior room without windows that's above ground level. In the case of a chemical threat, an above-ground location is preferable because some chemicals are heavier than air, and may seep into basements even if the windows are closed. Using duct tape, seal all cracks around the door and any vents into the room. Keep listening to your radio or television until you are told all is safe or you are told to evacuate. Local officials may call for evacuation in specific areas at greatest risk in your community.

Additional Positive Steps You Can Take

Raw, unedited footage of terrorism events and people's reaction to those events can be very upsetting, especially to children. We do not recommend that children watch television news reports about such events, especially if the news reports show images over and over again about the same incident. Young children do not realize that it is repeated video footage, and think the event is happening again and again. Adults may also need to give themselves a break from watching disturbing footage. However, listening to local radio and television reports will provide you with the most accurate information from responsible governmental authorities on what's happening and what actions you will need to take. So you may want to make some arrangements to take turns listening to the news with other adult members of your household.

Another useful preparation includes learning some basic first aid. To enroll in a first aid and AED/CPR course, contact your local American Red Cross chapter. In an emergency situation, you need to tend to your own well-being first and then consider first-aid for others immediately around you, including possibly assisting injured people to evacuate a building if necessary.

People who may have come into contact with a biological or chemical agent may need to go through a decontamination procedure and receive medical attention. Listen to the advice of local officials on the radio or television to determine what steps you will need to take to protect yourself and your family. As emergency services will likely be overwhelmed, only call 9-1-1 about life-threatening emergencies.

Long-term In-place Sheltering

Sometimes disasters make it unsafe for people to leave their residence for extended periods. Winter storms, floods, and landslides may isolate individual households and make it necessary for each household to take care of its own needs until the disaster abates, such as when snows melt and temperatures rise, or until rescue workers arrive. Your household should be prepared to be self-sufficient for three days when cut off from utilities and from outside supplies of food and water.

1. Stay in your shelter until local authorities say it's OK to leave. The length of your stay can range from a few hours to two weeks.
2. Maintain a 24-hour communications and safety watch. Take turns listening for radio broadcasts. Watch for fires.
3. Assemble an emergency toilet, if necessary.
   - Use a garbage container, pail or bucket with a snug-fitting cover. If the container is small, use a larger container with a cover for waste disposal. Line both containers with plastic bags.
   - After each use, pour or sprinkle a small amount of regular household disinfectant, such as chlorine bleach, into the container to reduce odors and germs.

Managing Water Supplies

Water is critical for survival. Plan to have about one gallon of water per person per day for drinking, cooking, and personal hygiene. You may need more for medical emergencies.

1. Allow people to drink according to their need. The average person should drink between two and two-and-one-half quarts of water or other liquids per day, but many people need more. This will depend on age, physical activity, physical condition, and time of year.
2. Never ration water unless ordered to do so by authorities. Drink the amount you need today and try to find more for tomorrow. Under no circumstances should a person drink less than one quart of water each day. You can minimize the amount of water your body needs by reducing activity and staying cool.

3. Drink water that you know is not contaminated first. If necessary, suspicious water, such as cloudy water from regular faucets or muddy water from streams or ponds, can be used after it has been treated. If water treatment is not possible, put off drinking suspicious water as long as possible, but do not become dehydrated.
4. In addition to stored water, other sources include the following:
   - Melted ice cubes
   - Water drained from the water heater faucet, if the water heater has not been damaged
   - Water dipped from the flush tanks (not the bowls) of home toilets (Bowl water can be used for pets.)
   - Liquids from canned goods, such as fruit and vegetable juices
5. Carbonated beverages do not meet drinking water requirements. Drinks containing caffeine and alcohol dehydrate the body, which increases the need for drinking water.
6. If water pipes are damaged or if local authorities advise you, turn off the main water valves to prevent water from draining away in case the water main breaks.
   - The pipes will be full of water when the main valve is closed.
   - To use this water, turn on the faucet at the highest point in your house (which lets air into the system).
   - Then draw water, as needed, from the lowest point in your house, either a faucet or the hot water tank.
7. Unsafe water sources include the following:
   - Radiators
   - Hot water boilers (home-heating system)
   - Water beds (fungicides added to the water or chemicals in the vinyl may make water unsafe to use)
   - Swimming pools and spas (chemicals used in them to kill germs are too concentrated for safe drinking, but can be used for personal hygiene, cleaning, and related uses)

Water Treatment

Treat all water of uncertain purity before using it for drinking, food washing or preparation, washing dishes, brushing teeth, or making ice. In addition to having a bad odor and taste, contaminated water can contain microorganisms that cause diseases, such as dysentery, cholera, typhoid, and hepatitis.

There are many ways to treat water. None is perfect. Often the best solution is a combination of methods. Before treating, let any suspended particles settle to the bottom, or strain them through layers of clean cloth. Following are four treatment methods. The first three methods—boiling, chlorination, and water treatment tablets—will kill microbes but will not remove other contaminants, such as heavy metals, salts, most other chemicals and radioactive fallout. The final method—distillation—will remove microbes, as well as most other contaminants, including radioactive fallout. Boiling is the safest method of treating water.

- Boiling water kills harmful bacteria and parasites. Bringing water to a rolling boil for one minute will kill most organisms. Let the water cool before drinking.
- Boiled water will taste better if you put oxygen back into it by pouring it back and forth between two containers. This will also improve the taste of stored water. Chlorination uses liquid chlorine bleach to kill microorganisms, such as bacteria.
- Use regular household liquid bleach that contains no soap or scent. Some containers warn, "Not For Personal Use." You can disregard these warnings if the label states sodium hypochlorite as the only active ingredient and if you use only the small quantities mentioned in these instructions.
- Add 16 drops of unscented bleach per gallon of water stir and let stand for 30 minutes. If the water does not taste and smell of chlorine at that point, add another dose and let stand for another 15 minutes. This treatment will not kill parasitic organisms.
- If you do not have a dropper, use a spoon and a square-ended strip of paper or thin cloth about a 1/4 inch by 2 inches. Put the strip in the spoon with an end hanging down about a 1/2 inch below the scoop of the spoon. Place bleach in the spoon and carefully tip it. Drops the size of those from a medicine dropper will drip off the end of the strip.

Water treatment "purification" tablets release chlorine or iodine. They are inexpensive and available at most sporting goods stores and some drugstores. Follow the package directions carefully. NOTE: People with hidden or chronic liver or kidney disease may be adversely affected by iodized tablets and may experience worsened health problems as a result of ingestion. Iodized tablets are safe for healthy, physically fit adults and should be used only if you lack the supplies for boiling, chlorination, and distillation.

Distillation involves boiling water and collecting the vapor that condenses back to water. The condensed vapor may include salt or other impurities.

- Fill a pot halfway with water.
- Tie a cup to the handle on the pot's lid, so that the cup hangs right-side up when the lid is upside-down (make sure the cup is not dangling into the water).
- Boil for 20 minutes. The water that drips from the lid into the cup is distilled.

Managing Food Supplies

1. It is important to be sanitary when storing, handling, and eating food:
   - Keep food in covered containers.
   - Keep cooking and eating utensils clean.
   - Keep garbage in closed containers and dispose outside. Bury garbage, if necessary. Avoid letting garbage accumulate inside, both for fire and sanitation reasons.
   - Keep hands clean. Wash frequently with soap and water that has been boiled or disinfected. Be sure to wash:
     - Before preparing or eating food
     - After toilet use
     - After participating in flood cleanup activities
     - After handling articles contaminated with floodwater or sewage
2. Carefully ration food for everyone except children and pregnant women. Most people can remain relatively healthy with about half as much food as usual and can survive without any food for several days.
3. Try to avoid foods high in fat and protein, since they will make you thirsty. Try to eat salt-free crackers, whole-grain cereals, and canned foods with high liquid content.
4. For emergency cooking, heat food with candle warmers, chafing dishes, and fondue pots, or use a fireplace. Charcoal grills and camp stoves are for outdoor use only.
5. Commercially canned food can be eaten out of the can without warming. Before heating food in a can, remove the label, thoroughly wash the can, and then disinfect it with a solution consisting of one cup of bleach in five gallons of water, and open before heating. Re-label your cans, including expiration date, with a marker:
   - Do not eat foods from cans that are swollen, dented, or corroded even though the product may look okay to eat.
   - Do not eat any food that looks or smells abnormal, even if the can looks normal.
   - Discard any food not in a waterproof container if there is any chance that it has come into contact with contaminated floodwater.
   - Food containers with screw-caps, snap-lids, crimped caps (soda pop bottles), twist caps, flip tops, snap-open, and home canned foods should be discarded if they have come into contact with floodwater because they cannot be disinfected. For infants, use only pre-prepared canned baby formula. Do not use powdered formulas with treated water.
6. Your refrigerator will keep foods cool for about four hours without power if it is left unopened. Add block or dry ice to your refrigerator if the

electricity will be off longer than four hours. Thawed food usually can be eaten if it is still "refrigerator cold," or re-frozen if it still contains ice crystals. To be safe, remember, "When in doubt, throw it out." Discard any food that has been at room temperature for two hours or more, and any food that has an unusual odor, color, or texture. If you are without power for a long period:

- If friends have electricity, ask them to store your frozen foods in their freezers.
- Inquire if freezer space is available in a store, church, school, or commercial freezer that has electrical service.
- Use dry ice, if available. Twenty-five pounds of dry ice will keep a 10-cubic-foot freezer below freezing for 3–4 days. Use care when handling dry ice, and wear dry, heavy gloves to avoid injury.

Staying in a Mass Care Shelter

The American Red Cross and Salvation Army, assisted by the community and other disaster relief groups, work with local authorities to set up public shelters in schools, municipal buildings, and churches. While they often provide water, food, medicine, and basic sanitary facilities, you should plan to have your own supplies as well—especially water. See the "Emergency Planning and Disaster Supplies" chapter for more details.

1. Living with many people in a confined space can be difficult and unpleasant, so try to cooperate with shelter managers and others staying in the shelter.
2. Restrict smoking to designated areas that are well ventilated. Ensure that smoking materials are disposed of safely.
3. If you go to an emergency shelter, remember that alcoholic beverages and weapons are prohibited in shelters. Pets, except for service animals, are also not usually allowed in public shelters.

Mitigation

One of the most effective means of protection is to take steps to make your home and your household safe from the potential effects of disasters, such as floods, tornadoes, hurricanes, and earthquakes. This is called mitigation. Ideally, mitigation measures are implemented before disaster strikes since they can help protect your household, as well as your property. However, even after a disaster strikes, actions can be taken to avoid or reduce the impact of the next disaster.

1. If your home was damaged during the disaster, consider implementing mitigation measures while you repair your home.
2. Be sure that all upgrade construction projects comply with local building codes that pertain to seismic, flood, fire and wind hazards. Make sure your contractors follow the codes, including periodic building inspections of the construction.
3. If you live in a flood-prone area, consider purchasing flood insurance to reduce your risk of flooding. Buying flood insurance to cover the value of a building and its contents will not only provide greater peace of mind, but will also speed recovery if a flood occurs. You can call #1-888-FLOOD29 to learn more about flood insurance.

Also consider options for reducing your future flood losses (see *Homeowner's Guide to Retrofitting: Six Ways to Protect Your House from Flooding*, FEMA Publication # 312). The appropriate flood mitigation measure will depend upon the degree of flood risk to which your home is subject. For moderate degrees of flooding, incorporating flood-proofing techniques to meet National Flood Insurance Program criteria may be the most practical approach to flood damage reduction. These techniques include taking the following steps to protect your utilities from flood damages:

- Relocating electric, telephone, and cable lines to the upper level of your home
- Putting heating, ventilation, and air-conditioning units in the upper story or the attic
- Anchoring or bolting oil tanks to prevent flotation

If the homes within your community have a history of severe, repetitive, flooding, it may be necessary to consider more substantial measures. Consider the following measures:

- Elevate the structure to or above the Base Flood Elevation
- Relocate the structure to a new site located outside of the 100-year floodplain, outside of any regulatory erosion zones, and in conformance with any other applicable state or local land-use regulations

In areas prone to severe flooding, it may be appropriate to work directly with your local emergency management official to develop a community-based approach. Additionally, your local representative will be able to identify potential federal, state, and/or local funding sources for the implementation of elevation, acquisition or relocation activities. For example, FEMA offers three state-administered grant programs to help states and local governments significantly reduce or permanently eliminate future flood losses: the Hazard

Mitigation Grant Program, Flood Mitigation Assistance Program and Pre-Disaster Mitigation Program. Individuals may not apply directly to the state or FEMA, but local governments or private non-profit organizations may apply on behalf of local citizens.

4. If you live in an area prone to high winds, make sure your roof is firmly secured to the main frame of the residence. Consider building a wind "Safe Room or Shelter" in your home to protect your household. There are several additional steps you can take to reduce wind damages and losses, including the following:
   - Secure light fixtures and other items that could fall or shake loose in such events.
   - Move heavy or breakable objects to low shelves.
   - Anchor water heaters and bolt them to wall studs.
   - Purchase storm shutters for exterior windows and doors to protect your home against high winds.
5. If you live in an area likely to have an earthquake, consider using straps or other restraints to secure cabinets, bookshelves, large appliances (especially water heater and furnace), and light fixtures to prevent damage and injury.
6. Determine ways to prevent other types of hazards in your home, such as installing a fire sprinkler system.
7. Obtain information specific to your area and home. Ask local emergency management, fire and police departments, zoning and building offices, the American Red Cross, hardware dealers, home inspectors, structural engineers, and architects.
8. Ask your local government, a hardware dealer or a private home inspector for technical advice on these and other mitigation measures.
9. Check the list of available publications from FEMA mentioned in this section and at the end of this guide.

ANIMALS IN A DISASTER

Disaster disrupts and affects everything in its path, including pets, livestock, and wildlife. The following section provides general guidelines for handling animals in emergency and disaster situations.

Pets in a Disaster

Pets need to be included in your household disaster plan since they depend on you for their safety and well-being. It is important to consider and prepare for your pets before disaster strikes. Consider the following preparedness measures:

1. If you must evacuate, do not leave pets behind—there is a chance they may not survive, or they may get lost before you return.
2. With the exception of service animals, pets are not typically permitted in emergency shelters for health reasons.
3. Find out before a disaster which local hotels and motels allow pets and where pet boarding facilities are located. Be sure to include some outside your local area in case local facilities have closed.
4. Know that most boarding facilities require veterinarian records to prove vaccinations are current.
5. Only some animal shelters will provide care for pets during emergency and disaster situations. They should be used as a last resort. Use friends and family or keep them with you.
6. Be sure your pet has proper identification tags securely fastened to the collar. A current photo of your pet will assist identification should it become necessary.
7. Make sure you have a secure pet carrier or leash for your pets—they may need to be restrained during tense emergency situations.
8. Assemble a disaster kit for your pet. Include pet food, water, medications, veterinary records, litter box, can opener, food dishes, first-aid kit, other supplies that may not be available at a later time, and an information sheet that gives your pet's name and such things as behavior problems. Provide the kit to whoever assumes responsibility for your pet during a disaster.
9. Call your local emergency management office or animal shelter for further information.

Large Animals in a Disaster

If you have large animals, such as horses or cattle on your property, be sure to prepare before a disaster.

1. Evacuate animals whenever possible. Map out primary and secondary routes in advance.
2. Evacuation destinations should be prepared with, or ready to obtain, food, water, veterinary care, and handling equipment.
3. Vehicles and trailers needed for transporting and supporting each type of animal should be available along with experienced handlers and drivers. It is best to allow animals a chance to become accustomed to vehicular travel so they are less frightened and easier to move.

4. In case evacuation is not possible, animal owners must decide whether to move large animals to shelter or turn them outside. This decision should be based on the disaster type, quality and location of shelter, and the risks of turning them outside.
5. All animals should have some form of identification.

Wildlife in a Disaster

Disaster and life-threatening situations will exacerbate the unpredictable nature of wild animals. To protect yourself and your household, learn how to deal with wildlife.

1. Be cautious approaching wild animals during emergency situations. Do not corner them. Wild animals will likely feel threatened and may endanger themselves by dashing off into floodwaters, fire, etc.
2. If wild animals are trapped or no natural food source is available, you can leave food appropriate to individual animals (i.e., animals could become trapped on an "island" after seeking high ground as floodwaters rise).
3. Wild animals, such as snakes, opossums, and raccoons often seek refuge from floodwaters on upper levels of homes and have been known to remain after water recedes. If you encounter animals in this situation—open a window or other escape route and the animal will likely leave on its own. Do not attempt to capture or handle the animal. Should the animal stay, call your local animal control office or animal shelter.
4. If you see an injured or stranded animal, do not approach or attempt to help. Call your local animal control office or animal shelter.
5. Animal carcasses can present serious health risks. Contact your local emergency management office or health department for specific help and instructions.

Animals after a Disaster

Wild or stray domestic animals can pose a danger during or after many types of disaster. Remember, most animals are disoriented and displaced, too. Do not corner an animal. If an animal must be removed, contact your local animal control authorities.

If any animal bites you, seek immediate medical attention. If a snake bites you, try to accurately identify the type of snake so that, if poisonous, the correct anti-venom can be administered. Do not cut the wound or attempt to suck the venom out.

Certain animals may carry rabies. Although the virus is rare, care should be taken to avoid contact with stray animals and rodents. Health departments can provide information on the types of animals that carry rabies in your area.

Rats may also be a problem during and after many types of disaster. Be sure to secure all food supplies and contact your local animal control authorities to remove any animal carcasses in the vicinity.

Contact your local emergency manager for more information on animals in disasters. The Humane Society of the United States can be reached at: 2100 L Street, NW, Washington, DC, 20037, Attn: Disaster Services Program, by phone at 202-452-1100 or online at www.hsus.org/disaster.

First Aid Primer

If you encounter someone who is injured, apply the emergency action steps: Check-Call-Care.

Check the scene to make sure it is safe for you to approach. Then check the victim for unconsciousness and life-threatening conditions. Someone who has a life-threatening condition, such as not breathing or severe bleeding, requires immediate care by trained responders and may require treatment by medical professionals. **Call** out for help. There are some steps that you can take, however, to **care** for someone who is hurt, but whose injuries are not life threatening.

Control Bleeding

- Cover the wound with a dressing, and press firmly against the wound (direct pressure).
- Elevate the injured area above the level of the heart if you do not suspect that the victim has a broken bone.
- Cover the dressing with a roller bandage.
- If the bleeding does not stop:
  - Apply additional dressings and bandages.
  - Use a pressure point to squeeze the artery against the bone.
- Provide care for shock.

Care for Shock

- Keep the victim from getting chilled or overheated.
- Elevate the legs about 12 inches (if broken bones are not suspected).
- Do not give food or drink to the victim.

Tend Burns

- Stop the burning by cooling the burn with large amounts of water.
- Cover the burn with dry, clean dressings or cloth.

Care for Injuries to Muscles, Bones, and Joints

- Rest the injured part.
- Apply ice or a cold pack to control swelling and reduce pain.
- Avoid any movement or activity that causes pain.
- If you must move the victim because the scene is becoming unsafe, try to immobilize the injured part to keep it from moving.

Be Aware of Biological/Radiological Exposure

- Listen to local radio and television reports for the most accurate information from responsible governmental and medical authorities on what's happening and what actions you will need to take. The Web sites referenced at the end of this brochure can give you more information on how to protect yourself from exposure to biological or radiological hazards.

Reduce Any Care Risks

The risk of getting a disease while giving first aid is extremely rare. However, to reduce the risk even further:

- Avoid direct contact with blood and other body fluids.
- Use protective equipment, such as disposable gloves and breathing barriers.
- Thoroughly wash your hands with soap and water immediately after giving care.

It is important to be prepared for an emergency and to know how to give emergency care.

More Information

All of these recommendations make good sense, regardless of the potential problem. For more information on how to get ready for disaster and be safe when disaster strikes, or to register for a first aid and AED/CPR course, please contact your local American Red Cross chapter.

For information about your community's specific plans for response to disasters and other emergencies, contact your local office of emergency management.

For information on what a business can do to protect its employees and customers as well as develop business continuity plans, you may want to get a copy of the *Emergency Management Guide for Business and Industry* and/or *Preparing Your Business for the Unthinkable* from your local American Red Cross chapter or see http://www.redcross.org/services/disaster/beprepared.

For more information about the specific effects of chemical or biological agents, the following websites may be helpful:

Centers for Disease Control and Prevention: www.cdc.gov
U.S. Department of Energy: www.energy.gov
U.S. Department of Health and Human Services: www.hhs.gov
Federal Emergency Management Agency: www.fema.gov
Environmental Protection Agency: www.epa.gov

TERRORIST ATTACKS

Terrorism is the use of force or violence against people or property in violation of the criminal laws of the United States for purposes of intimidation, coercion or ransom. Terrorists often use threats to create fear among the public, to try to convince citizens that their government is powerless to prevent terrorism and to get immediate publicity for their causes.

Acts of terrorism can range from threats of terrorism to assassinations, kidnappings, hijackings, cyberattacks (computer-based), bomb scares and bombings, and the use of chemical, biological, and nuclear weapons.

High-risk targets include military and civilian government facilities, international airports, large cities and high-profile landmarks. Terrorists might also target large public gatherings, water and food supplies, utilities and corporate centers. Further, they are capable of spreading fear by sending explosives or chemical and biological agents through the mail.

Planning in the Event of a Terrorist Attack

Here are some other things you can do right now to be safer!

- Check and change the batteries in your smoke alarms and replace all alarms that are more than 10 years old.
- Make sure you know where your local fire department, police station, and hospital are and post a list of emergency phone numbers posted near all the telephones in your home.
- Organize and practice a family fire drill—make sure your children know what your smoke detector sounds like and what to do if it goes off when they are sleeping.
- Locate the utility mains for your home and be sure you know how to turn them off manually: gas, electricity, and water.
- Create an emergency plan for your household, including your pets. Decide where your family will meet if a disaster does happen: (1) right outside

your home in case of a sudden emergency, like a fire and (2) outside your neighborhood in case you can't return home—ask an out-of-town friend to be your "family contact" to relay messages.
- Prepare a 3-day disaster supply kit, complete with flashlights, batteries, blankets, and an emergency supply of water and food (and pet food!).
- Plan to hold a Neighborhood Watch meeting—your local Sheriffs' office or police station can help you get started.
- Check the expiration dates of all over-the-counter medications—discard all that are expired and replace any that are routinely needed.
- Make sure all cleaning products and dangerous objects are out of children's reach.
- Plan to sign up for a first-aid training course. Call your local American Red Cross chapter, the National Safety Council or American Safety & Health Institute to ask about courses in your area (www.redcross.org, www.nsc.org, or www.hsi.com).
- Visit with your neighbors and discuss how you would handle a disaster in your area. Talk to neighbors with special needs and help them become safer too!

Acts of terrorism include threats of terrorism; assassinations; kidnappings; hijackings; bomb scares and bombings; cyberattacks (computer-based); and the use of chemical, biological, nuclear, and radiological weapons.

The following are general guidelines:

- Be aware of your surroundings.
- Move or leave if you feel uncomfortable or if something does not seem right.
- Take precautions when traveling. Be aware of conspicuous or unusual behavior. Do not accept packages from strangers. Do not leave luggage unattended. You should promptly report unusual behavior, suspicious or unattended packages, and strange devices to the police or security personnel.
- Learn where emergency exits are located in buildings you frequent. Plan how to get out in the event of an emergency.
- Be prepared to do without services you normally depend on—electricity, telephone, natural gas, gasoline pumps, cash registers, ATMs, and Internet transactions.
- Work with building owners to ensure the following items are located on each floor of the building:
  - Portable, battery-operated radio and extra batteries.
  - Several flashlights and extra batteries.
  - First-aid kit and manual.
  - Hard hats and dust masks.
  - Fluorescent tape to rope off dangerous areas.

WHAT TO DO IN THE EVENT OF A TERRORIST ATTACK

Protection against Cyberattacks

Cyberattacks target computer or telecommunication networks of critical infrastructures, such as power systems, traffic-control systems or financial systems. Cyberattacks target information technologies (IT) in three different ways. First, is a direct attack against an information system "through the wires" alone (hacking). Second, the attack can be a physical assault against a critical IT element. Third, the attack can be from the inside as a result of compromising a trusted party with access to the system.

1. Be prepared to do without services you normally depend on that could be disrupted—electricity, telephone, natural gas, gasoline pumps, cash registers, ATM machines, and Internet transactions.
2. Be prepared to respond to official instructions if a cyberattack triggers other hazards, for example, general evacuation, evacuation to shelter, or shelter in place, because of hazardous-materials releases, nuclear power plant incident, dam or flood-control system failures.

Preparing for a Building Explosion

Explosions can collapse buildings and cause fires. People who live or work in a multi-level building can do the following:

1. Review emergency evacuation procedures. Know where emergency exits are located.
2. Keep fire extinguishers in working order. Know where they are located, and learn how to use them.
3. Learn first aid. Contact the local chapter of the American Red Cross for information and training.
4. Building owners should keep the following items in a designated place on each floor of the building:
   - Portable, battery-operated radio and extra batteries
   - Several flashlights and extra batteries
   - First-aid kit and manual
   - Several hard hats
   - Fluorescent tape to rope off dangerous areas

Bomb Threats

If you receive a bomb threat, get as much information from the caller as possible. Keep the caller on the line and record everything that is said. Then notify the police and the building management. If you are notified of a bomb threat, do not touch any suspicious packages. Clear the area around suspicious packages, and notify the police immediately. In evacuating a building, don't stand in front of windows, glass doors or other potentially hazardous areas. Do not block sidewalk or streets to be used by emergency officials or others still exiting the building.

Suspicious Parcels and Letters

Be wary of suspicious packages and letters. They can contain explosives, chemical or biological agents. Be particularly cautious at your place of employment. Some typical characteristics postal inspectors have detected over the years, which ought to trigger suspicion, include parcels that:

- Are unexpected or from someone unfamiliar to you
- Have no return address, or have one that can't be verified as legitimate
- Are marked with restrictive endorsements, such as "Personal," "Confidential," or "Do not X-ray"
- Have protruding wires or aluminum foil, strange odors or stains
- Show a city or state in the postmark that doesn't match the return address
- Are of unusual weight, given their size, or are lopsided or oddly shaped
- Are marked with any threatening language
- Have inappropriate or unusual labeling
- Have excessive postage or excessive packaging material, such as masking tape and string
- Have misspellings of common words
- Are addressed to someone no longer with your organization or are otherwise outdated
- Have incorrect titles or title without a name
- Are not addressed to a specific person
- Have handwritten or poorly typed addresses

With suspicious envelopes and packages other than those that might contain explosives, take these additional steps against possible biological and chemical agents:

- Refrain from eating or drinking in a designated mail-handling area.
- Place suspicious envelopes or packages in a plastic bag or some other type of container to prevent leakage of contents. Never sniff or smell suspect mail.

- If you do not have a container, then cover the envelope or package with anything available (e.g., clothing, paper, trashcan, etc.) and do not remove the cover.
- Leave the room and close the door, or section off the area to prevent others from entering.
- Wash your hands with soap and water to prevent spreading any powder to your face.
- If you are at work, report the incident to your building security official or an available supervisor, who should notify police and other authorities without delay.
- List all people who were in the room or area when this suspicious letter or package was recognized. Give a copy of this list to both the local public health authorities and law enforcement officials for follow-up investigations and advice.
- If you are at home, report the incident to local police.

Explosions

Terrorists have frequently used explosive devices as one of their most common weapons. Terrorists do not have to look far to find out how to make explosive devices; the information is readily available in books and other information sources. The materials needed for an explosive device can be found in many places including variety, hardware, and auto supply stores. Explosive devices are highly portable using vehicles and humans as a means of transport. They are easily detonated from remote locations or by suicide bombers.

If there is an explosion, you should:

- Get under a sturdy table or desk if things are falling around you. When they stop falling, leave quickly, watching for obviously weakened floors and stairways. As you exit from the building, be especially watchful of falling debris.
- Leave the building as quickly as possible. Do not stop to retrieve personal possessions or make phone calls.
- Do not use elevators.

Once you are out:

- Do not stand in front of windows, glass doors, or other potentially hazardous areas.
- Move away from sidewalks or streets to be used by emergency officials or others still exiting the building.

If you are trapped in debris:

- If possible, use a flashlight to signal your location to rescuers.
- Avoid unnecessary movement so you don't kick up dust.
- Cover your nose and mouth with anything you have on hand. (Dense-weave cotton material can act as a good filter. Try to breathe through the material.)
- Tap on a pipe or wall so rescuers can hear where you are.
- If possible, use a whistle to signal rescuers.
- Shout only as a last resort. Shouting can cause a person to inhale dangerous amounts of dust.

Biological Threats

Biological agents are organisms or toxins that can kill or incapacitate people, livestock, and crops. The three basic groups of biological agents that would likely be used as weapons are bacteria, viruses, and toxins. Most biological agents are difficult to grow and maintain. Many break down quickly when exposed to sunlight and other environmental factors, while others, such as anthrax spores, are very long lived. Biological agents can be dispersed by spraying them into the air, by infecting animals that carry the disease to humans, and by contaminating food and water.

Bioterrorism poses an enormous threat and is difficult to prevent. A small vial of anthrax or smallpox could be released and cause tens of thousands of casualties. Terrorists are most likely to use organisms that cause infectious diseases because they are easily spread among people.

Biological agents include bacteria, viruses, and rickettsia.

Toxins, which are poisons of biological origin and not living organisms, are sometimes grouped with biological agents and sometimes with chemical agents. Although there are hundreds of microorganisms that could be used as biological agents, the likely number is much smaller when the agents' effectiveness, reliability, availability, ease of manufacture, and stability in storage and dissemination are considered. When disseminated as aerosols, biological agents are most effective in the size range of 1 to 5 microns, because they can remain suspended for long periods. Smaller particles are less likely to survive as aerosols, and larger particles settle rapidly, making them less likely to enter the lungs.

Anthrax

The death rate for anthrax from inhaled sources is very high, approaching 100 percent. Death rates are lower for anthrax that enters your body through food or a wound. Anthrax is easy to produce, and it's readily available around the world. The spores don't require special handling procedures, so terrorists could take anthrax to many points for distribution. Unlike many disease-causing bacteria, spores can survive severe heat and cold.

Anthrax is an acute infectious disease caused by the spore-forming bacterium Bacillus anthracis. Anthrax most commonly occurs in hoofed mammals and can also infect humans.

Symptoms of disease vary depending on how the disease was contracted, but usually occur within seven days after exposure. The serious forms of human anthrax are inhalation anthrax, cutaneous anthrax, and intestinal anthrax.

Initial symptoms of inhalation anthrax infection may resemble a common cold. After several days, the symptoms may progress to severe breathing problems and shock. Inhalation anthrax is often fatal.

The intestinal disease form of anthrax may follow the consumption of contaminated food and is characterized by an acute inflammation of the intestinal tract. Initial signs of nausea, loss of appetite, vomiting, and fever are followed by abdominal pain, vomiting of blood, and severe diarrhea.

Direct person-to-person spread of anthrax is extremely unlikely, if it occurs at all. Therefore, there is no need to immunize or treat contacts of people ill with anthrax, such as household contacts, friends, or coworkers, unless they also were also exposed to the same source of infection.

In persons exposed to anthrax, infection can be prevented with antibiotic treatment. Early antibiotic treatment of anthrax is essential; delay lessens chances for survival. Anthrax usually is susceptible to penicillin, doxycycline, and fluoroquinolones. An anthrax vaccine also can prevent infection. Vaccination against anthrax is not recommended for the general public to prevent disease and is not available.

Smallpox

About 30 percent of those infected with smallpox die of it. Many people have never been vaccinated for smallpox, and no one knows whether those who received vaccinations 25 or more years ago are still protected. Smallpox is harder to propagate than anthrax and less tolerant of severe conditions. However, it can spread very rapidly from person to person.

Smallpox is a serious, contagious, and sometimes fatal infectious disease. There is no specific treatment for smallpox disease, and the only prevention is vaccination. The name smallpox is derived from the Latin word for "spotted" and refers to the raised bumps that appear on the face and body of an infected person.

There are two clinical forms of smallpox. Variola major is the severe and most common form of smallpox, with a more extensive rash and higher fever.

There are four types of variola major—ordinary; modified (mild and occurring in previously vaccinated persons); flat; and hemorrhagic (the last two are rare and very severe). Historically, variola major has an overall fatality rate of about 30 percent; however, flat and hemorrhagic smallpox usually are fatal. Variola minor is a less common presentation of smallpox, and a much less severe disease, with death rates historically of 1 percent or less.

Smallpox outbreaks have occurred from time to time for thousands of years, but the disease is now eradicated after a successful worldwide vaccination program. The last case of smallpox in the United States was in 1949. The last naturally occurring case in the world was in Somalia in 1977. After the disease was eliminated from the world, routine vaccination against smallpox among the general public was stopped because it was no longer necessary for prevention.

Where Smallpox Comes From

Smallpox is caused by the variola virus that emerged in human populations thousands of years ago. Except for laboratory stockpiles, the variola virus has been eliminated. However, in the aftermath of the events of September and October 2001, there is heightened concern that the variola virus might be used as an agent of bioterrorism. For this reason, the U.S. government is taking precautions for dealing with a smallpox outbreak.

Transmission

Generally, direct and fairly prolonged face-to-face contact is required to spread smallpox from one person to another. Smallpox also can be spread through direct contact with infected bodily fluids or contaminated objects such as bedding or clothing. Rarely, smallpox has been spread by virus carried in the air in enclosed settings such as buildings, buses, and trains. Humans are the only natural hosts of variola. Smallpox is not known to be transmitted by insects or animals. A person with smallpox is sometimes contagious with onset of fever (prodrome phase), but the person becomes most contagious with the onset of rash. At this stage the infected person is usually very sick and not able to move around in the community. The infected person is contagious until the last smallpox scab falls off.

Other threats

Other infectious diseases that pose a threat include plague, tularemia, botulism, viral hemorrhagic fever and tuberculosis.

Botulism

Botulism is a muscle-paralyzing disease caused by a toxin made by a bacterium called Clostridium botulinum. There are three main kinds of botulism:

Foodborne botulism occurs when a person ingests pre-formed toxin that leads to illness within a few hours to days. Foodborne botulism is a public health emergency because the contaminated food may still be available to other persons besides the patient.
Infant botulism occurs in a small number of susceptible infants each year who harbor C. botulinum in their intestinal tract.
Wound botulism occurs when wounds are infected with C. botulinum that secretes the toxin.

With foodborne botulism, symptoms begin within 6 hours to 2 weeks (most commonly between 12 and 36 hours) after eating toxin-containing food. Symptoms of botulism include double vision, blurred vision, drooping eyelids, slurred speech, difficulty swallowing, dry mouth, muscle weakness that always descends through the body: first shoulders are affected, then upper arms, lower arms, thighs, calves, etc. Paralysis of breathing muscles can cause a person to stop breathing and die, unless assistance with breathing (mechanical ventilation) is provided.

Botulism is not spread from one person to another. Foodborne botulism can occur in all age groups. A supply of antitoxin against botulism is maintained by CDC. The antitoxin is effective in reducing the severity of symptoms if administered early in the course of the disease. Most patients eventually recover after weeks to months of supportive care.

However, experts believe these organisms and diseases are unlikely to cause widespread illness because they're difficult to manufacture and distribute. These organisms are also less hardy than anthrax is. Alternatively, biological toxins derived from living organisms, such as the botulinum toxin derived from the bacterium *Clostridium botulinum*, or toxins derived from plants, such as ricin, from castor beans, could be used in terrorist attacks.

The bad news is that there is little citizens can do to prepare for a bioterror attack. The best defense against widespread disease is early detection which is handled by public health authorities, and that is where resources need to be focused.

Delivery methods include the following:

• Aerosols—biological agents are dispersed into the air, forming a fine mist that may drift for miles. Inhaling the agent may cause disease in people or animals.

- Animals—some diseases are spread by insects and animals, such as fleas, mice, flies, mosquitoes, and livestock.
- Food and water contamination—some pathogenic organisms and toxins may persist in food and water supplies. Most microbes can be killed, and toxins deactivated, by cooking food and boiling water. Most microbes are killed by boiling water for one minute, but some require longer. Follow official instructions.
- Person-to-person—spread of a few infectious agents is also possible. Humans have been the source of infection for smallpox, plague, and the Lassa viruses.

Specific information on biological agents is available at the Centers for Disease Control and Prevention's Web site, www.cdc.gov.

TAKE PROTECTIVE MEASURES

Before a Biological Attack

The following are guidelines for what you should do to prepare for a biological threat:

If you receive a telephoned bomb threat, you should do the following:

- Check with your doctor to ensure all required or suggested immunizations are up-to-date. Children and older adults are particularly vulnerable to biological agents.
- Consider installing a High Efficiency Particulate Air (HEPA) filter in your furnace return duct. These filters remove particles in the 0.3 to 10 micron range and will filter out most biological agents that may enter your house. If you do not have a central heating or cooling system, a stand-alone portable HEPA filter can be used.

Filtration in Buildings

Building owners and managers should determine the type and level of filtration in their structures and the level of protection it provides against biological agents. The National Institute of Occupational Safety and Health (NIOSH) provides technical guidance on this topic in their publication Guidance for Filtration and Air-Cleaning Systems to Protect Building Environments from Airborne Chemical, Biological, or Radiological Attacks. To obtain a copy, call 1 (800) 35NIOSH or visit www.cdc.gov/NIOSH/pubs/default.html and request or download NIOSH Publication 2003-136.

During a Biological Attack

In the event of a biological attack, public health officials may not immediately be able to provide information on what you should do. It will take time to determine what the illness is, how it should be treated, and who is in danger. Watch television, listen to radio, or check the Internet for official news and information including signs and symptoms of the disease, areas in danger, if medications or vaccinations are being distributed, and where you should seek medical attention if you become ill.

The first evidence of an attack may be when you notice symptoms of the disease caused by exposure to an agent. Be suspicious of any symptoms you notice, but do not assume that any illness is a result of the attack. Use common sense and practice good hygiene.

If you become aware of an unusual and suspicious substance nearby:

- Move away quickly.
- Wash with soap and water.
- Contact authorities.
- Listen to the media for official instructions.
- Seek medical attention if you become sick.

If You Are Exposed to a Biological Agent:

- Remove and bag your clothes and personal items. Follow official instructions for disposal of contaminated items.
- Wash yourself with soap and water and put on clean clothes.
- Seek medical assistance. You may be advised to stay away from others or even quarantined.

Using HEPA Filters

HEPA filters are useful in biological attacks. If you have a central heating and cooling system in your home with a HEPA filter, leave it on if it is running or turn the fan on if it is not running. Moving the air in the house through the filter will help remove the agents from the air. If you have a portable HEPA filter, take it with you to the internal room where you are seeking shelter and turn it on.

If you are in an apartment or office building that has a modern, central heating and cooling system, the system's filtration should provide a relatively safe level of protection from outside biological contaminants.

HEPA filters will not filter chemical agents.

After a Biological Attack

In some situations, such as the case of the anthrax letters sent in 2001, people may be alerted to potential exposure. If this is the case, pay close attention to all official warnings and instructions on how to proceed. The delivery of medical services for a biological event may be handled differently to respond to increased demand. The basic public health procedures and medical protocols for handling exposure to biological agents are the same as for any infectious disease. It is important for you to pay attention to official instructions via radio, television, and emergency alert systems.

Chemical Threats

Chemical agents are poisonous vapors, aerosols, liquids, and solids that have toxic effects on people, animals, or plants. They can be released by bombs or sprayed from aircraft, boats, and vehicles. They can be used as a liquid to create a hazard to people and the environment. Some chemical agents may be odorless and tasteless. They can have an immediate effect (a few seconds to a few minutes) or a delayed effect (2–48 hours). While potentially lethal, chemical agents are difficult to deliver in lethal concentrations. Outdoors, the agents often dissipate rapidly. Chemical agents also are difficult to produce.

A chemical attack could come without warning. Signs of a chemical release include people having difficulty breathing; experiencing eye irritation; losing coordination; becoming nauseated; or having a burning sensation in the nose, throat, and lungs. Also, the presence of many dead insects or birds may indicate a chemical agent release.

Most Likely Chemical Weapons Would Include:

Mustard Gas

Sulfur mustard is a type of chemical warfare agent. These kinds of agents are called vesicants or blistering agents, because they cause blistering of the skin and mucous membranes on contact. Sulfur mustard is also known as mustard gas or mustard agent, or by the military designations H, HD, and HT. Sulfur mustard sometimes smells like garlic, onions, or mustard and sometimes has no odor. It can be a vapor (the gaseous form of a liquid), an oily-textured liquid, or a solid. Sulfur mustard can be clear to yellow or brown when it is in liquid or solid form.

Where Sulfur Mustard Is Found and How It Is Used

Sulfur mustard is not found naturally in the environment. Sulfur mustard was introduced in World War I as a chemical warfare agent. Until recently, it was available for use in the treatment of a skin condition called psoriasis. Currently, it has no medical use.

How People Can Be Exposed To Sulfur Mustard

If sulfur mustard is released into the air as a vapor, people can be exposed through skin contact, eye contact, or breathing. Sulfur mustard vapor can be carried long distances by wind. If sulfur mustard is released into water, people can be exposed by drinking the contaminated water or getting it on their skin. People can be exposed by coming in contact with liquid sulfur mustard. Sulfur mustard can last from 1 to 2 days in the environment under average weather conditions and from weeks to months under very cold conditions. Sulfur mustard breaks down slowly in the body, so repeated exposure may have a cumulative effect (that is, it can build up in the body).

Sarin

Sarin is a human-made chemical warfare agent classified as a nerve agent. Nerve agents are the most toxic and rapidly acting of the known chemical warfare agents. They are similar to certain kinds of pesticides (insect killers) called organophosphates in terms of how they work and what kind of harmful effects they cause. However, nerve agents are much more potent than organophosphate pesticides. Sarin originally was developed in 1938 in Germany as a pesticide. Sarin is a clear, colorless, and tasteless liquid that has no odor in its pure form. However, sarin can evaporate into a vapor (gas) and spread into the environment. Sarin is also known as GB.

Where Sarin Is Found and How It Is Used

Sarin and other nerve agents may have been used in chemical warfare during the Iran-Iraq War in the 1980s. Sarin was used in two terrorist attacks in Japan in 1994 and 1995. Sarin is not found naturally in the environment.

How People Can Be Exposed To Sarin

Following release of sarin into the air, people can be exposed through skin contact or eye contact. They can also be exposed by breathing air that contains sarin. Sarin mixes easily with water, so it could be used to poison water. Following release of sarin into water, people can be exposed by touching or drinking water that contains sarin. Following contamination of food with sarin, people can be exposed by eating the contaminated food. A person's clothing can release sarin for about 30 minutes after it has come in contact with sarin vapor, which can lead to exposure of other people. Because sarin breaks down slowly in the body, people who are repeatedly exposed to sarin may suffer more harmful health effects. Because sarin vapor is heavier than air, it will sink to low-lying areas and create a greater exposure hazard there.

Tabun

Tabun is a man-made chemical warfare agent classified as a nerve agent. Nerve agents are the most toxic and rapidly acting of the known chemical warfare agents. They are similar to pesticides (insect killers) called organophosphates in terms of how they work and what kinds of harmful effects they cause. However, nerve agents are much more potent than organophosphate pesticides. Tabun was originally developed as a pesticide in Germany in 1936. Tabun is also known as "GA." Tabun is a clear, colorless, tasteless liquid with a faint fruity odor. Tabun can become a vapor if heated.

Where Tabun Is Found and How It Is Used

It is possible that tabun or other nerve agents were used in chemical warfare during the Iran-Iraq War in the 1980s. Tabun is not found naturally in the environment.

How People Can Be Exposed to Tabun

Following release of tabun into the air, people can be exposed through skin contact, eye contact, or inhalation (breathing in the tabun). Tabun mixes easily with water, so it could be used to poison water. Following release of tabun into water, people can be exposed by drinking contaminated water or getting contaminated water on their skin. Following contamination of food with tabun, people can be exposed by eating the contaminated food. A person's clothing can release tabun for about 30 minutes after contact with tabun vapor, which can lead to exposure of other people. Tabun breaks down slowly in the body, meaning that repeated exposures to tabun and/or other nerve agents can have a cumulative effect (build up in the body). Because tabun vapor is heavier than air, it will sink to low-lying areas and create a greater exposure hazard there.

VX

VX is a human-made chemical warfare agent classified as a nerve agent. Nerve agents are the most toxic and rapidly acting of the known chemical warfare agents. They are similar to pesticides (insect killers) called organophosphates in terms of how they work and what kinds of harmful effects they cause. However, nerve agents are much more potent than organophosphate pesticides. VX was originally developed in the United Kingdom in the early 1950s. VX is odorless and tasteless. VX is an oily liquid that is amber in color and very slow to evaporate. It evaporates about as slowly as motor oil.

Where VX Is Found and How It Is Used

It is possible that VX or other nerve agents were used in chemical warfare during the Iran-Iraq War in the 1980s. VX is not found naturally in the environment.

How People Can Be Exposed to VX

Following release of VX into the air, people can be exposed through skin contact, eye contact, or inhalation (breathing in the VX mist). Though VX does not mix with water as easily as other nerve agents do, it could be released into water. Following release of VX into water, people can be exposed by drinking contaminated water or getting contaminated water on their skin. Following contamination of food with VX, people can be exposed by eating the contaminated food. VX is primarily a liquid exposure hazard, but if it is heated to very high temperatures, it can turn into small amounts of vapor (gas). A person's clothing can release VX for about 30 minutes after contact with VX vapor, which can lead to exposure of other people. VX breaks down slowly in the body, meaning that repeated exposures to VX and/or other nerve agents can have a cumulative effect (build up in the body). Because VX vapor is heavier than air, it will sink to low-lying areas and create a greater exposure hazard there.

Other Chemical Agents Include the Following:

Abrin
Adamsite (DM)
Agent 15
Ammonia
Arsine (SA)
Bromobenzylcyanide (CA)
BZ
Cannabinoids
Chlorine (CL)
Chloroacetophenone (CN)
Chloropicrin (PS)

CNB (CN in Benzene and Carbon Tetrachloride)
CNC (CN in Chloroform)
Red Phosphorous (RP)
CNS (CN and Chloropicrin in Chloroform)
CR
CS
Cyanide
Cyanogen Chloride (CK)
CyclohexylSarin (GF)
Diphenylchloroarsine (DA)
Diphenylcyanoarsine (DC)
Diphosgene (DP)
Distilled Mustard (HD)
Ethyldichloroarsine (ED)
Fentanyls and Other Opioids
Hydrogen Chloride
Hydrogen Cyanide (AC)
Lewisite (L, L-1, L-2, L-3)
LSD
Methyldichloroarsine (MD)
Mustard Gas (H) (Sulfur Mustard)
Mustard/Lewisite (HL)
Mustard/T
Nitrogen Mustard (HN-1, HN-2, HN-3)
Nitrogen Oxide (NO)
Perflurorisobutylene (PHIB)
Phenodichloroarsine (PD)
Phenothiazines
Phosgene (CG)
Phosgene Oxime (CX)
Potassium Cyanide (KCN)
Red Phosphorous (RP)
Ricin
Sarin (GB)
Sesqui Mustard
Sodium Cyanide (NaCN)
Soman (GD)
Sulfur Mustard (H) (Mustard Gas)
Sulfur Trioxide-Chlorosulfonic Acid (FS)
Tabun (GA)
Teflon and Perflurorisobutylene (PHIB)
Titanium Tetrachloride (FM)

VX
Zinc Oxide (HC)

TAKE PROTECTIVE MEASURES

Before a Chemical Attack

The following are guidelines for what you should do to prepare for a chemical threat:

- Check your disaster supplies kit to make sure it includes the following:
  - A roll of duct tape and scissors.
  - Plastic for doors, windows, and vents for the room in which you will shelter in place. To save critical time during an emergency, pre-measure and cut the plastic sheeting for each opening.
- Choose an internal room to shelter, preferably one without windows and on the highest level.

During a Chemical Attack

The following are guidelines for what you should do in a chemical attack:
If you are instructed to remain in your home or office building, you should:

- Close doors and windows and turn off all ventilation, including furnaces, air conditioners, vents, and fans.
- Seek shelter in an internal room and take your disaster supplies kit.
- Seal the room with duct tape and plastic sheeting.
- Listen to your radio for instructions from authorities.

If you are caught in or near a contaminated area, you should:

- Move away immediately in a direction upwind of the source.
- Find shelter as quickly as possible.

After a Chemical Attack

Decontamination is needed within minutes of exposure to minimize health consequences. Do not leave the safety of a shelter to go outdoors to help others until authorities announce it is safe to do so.

A person affected by a chemical agent requires immediate medical attention from a professional. If medical help is not immediately available, decontaminate yourself and assist in decontaminating others.

Decontamination Guidelines Are as Follows

- Use extreme caution when helping others who have been exposed to chemical agents.
- Remove all clothing and other items in contact with the body. Contaminated clothing normally removed over the head should be cut off to avoid contact with the eyes, nose, and mouth. Put contaminated clothing and items into a plastic bag and seal it. Decontaminate hands using soap and water. Remove eyeglasses or contact lenses. Put glasses in a pan of household bleach to decontaminate them, and then rinse and dry.
- Flush eyes with water.
- Gently wash face and hair with soap and water before thoroughly rinsing with water.
- Decontaminate other body areas likely to have been contaminated. Blot (do not swab or scrape) with a cloth soaked in soapy water and rinse with clear water.
- Change into uncontaminated clothes. Clothing stored in drawers or closets is likely to be uncontaminated.
- Proceed to a medical facility for screening and professional treatment.

Nuclear and Radiological Attack

A nuclear blast is an explosion with intense light and heat, a damaging pressure wave, and widespread radioactive material that can contaminate the air, water, and ground surfaces for miles around. A nuclear device can range from a weapon carried by an intercontinental missile launched by a hostile nation or terrorist organization, to a small portable nuclear devise transported by an individual. All nuclear devices cause deadly effects when exploded, including blinding light, intense heat (thermal radiation), initial nuclear radiation, blast, fires started by the heat pulse, and secondary fires caused by the destruction.

Hazards of Nuclear Devices

The extent, nature, and arrival time of these hazards are difficult to predict. The geographical dispersion of hazard effects will be defined by the following:

- Size of the device. A more powerful bomb will produce more distant effects.
- Height above the ground the device was detonated. This will determine the extent of blast effects.
- Nature of the surface beneath the explosion. Some materials are more likely to become radioactive and airborne than others. Flat areas are more susceptible to blast effects.

- Existing meteorological conditions. Wind speed and direction will affect arrival time of fallout; precipitation may wash fallout from the atmosphere.

Radioactive Fallout

Even if individuals are not close enough to the nuclear blast to be affected by the direct impacts, they may be affected by radioactive fallout. Any nuclear blast results in some fallout. Blasts that occur near the earth's surface create much greater amounts of fallout than blasts that occur at higher altitudes. This is because the tremendous heat produced from a nuclear blast causes an up-draft of air that forms the familiar mushroom cloud. When a blast occurs near the earth's surface, millions of vaporized dirt particles also are drawn into the cloud. As the heat diminishes, radioactive materials that have vaporized condense on the particles and fall back to Earth. The phenomenon is called radioactive fallout. This fallout material decays over a long period of time, and is the main source of residual nuclear radiation.

Fallout from a nuclear explosion may be carried by wind currents for hundreds of miles if the right conditions exist. Effects from even a small portable device exploded at ground level can be potentially deadly.

Nuclear radiation cannot be seen, smelled, or otherwise detected by normal senses. Radiation can only be detected by radiation monitoring devices. This makes radiological emergencies different from other types of emergencies, such as floods or hurricanes. Monitoring can project the fallout arrival times, which will be announced through official warning channels. However, any increase in surface buildup of gritty dust and dirt should be a warning for taking protective measures.

In addition to other effects, a nuclear weapon detonated in or above the earth's atmosphere can create an electromagnetic pulse (EMP), a high-density electrical field. An EMP acts like a stroke of lightning but is stronger, faster, and shorter. An EMP can seriously damage electronic devices connected to power sources or antennas. This includes communication systems, computers, electrical appliances, and automobile or aircraft ignition systems. The damage could range from a minor interruption to actual burnout of components. Most electronic equipment within 1,000 miles of a high-altitude nuclear detonation could be affected. Battery-powered radios with short antennas generally would not be affected. Although an EMP is unlikely to harm most people, it could harm those with pacemakers or other implanted electronic devices.

Protection from a Nuclear Blast

The danger of a massive strategic nuclear attack on the United States is predicted by experts to be less likely today. However, terrorism, by nature, is unpredictable.

If there were threat of an attack, people living near potential targets could be advised to evacuate or they could decide on their own to evacuate to an area not considered a likely target. Protection from radioactive fallout would require taking shelter in an underground area or in the middle of a large building.

In general, potential targets include the following:

- Strategic missile sites and military bases.
- Centers of government such as Washington, DC, and state capitals.
- Important transportation and communication centers.
- Manufacturing, industrial, technology, and financial centers.
- Petroleum refineries, electrical power plants, and chemical plants.
- Major ports and airfields.

The three factors for protecting oneself from radiation and fallout are distance, shielding, and time.

- Distance—the more distance between you and the fallout particles, the better. An underground area such as a home or office building basement offers more protection than the first floor of a building. A floor near the middle of a high-rise may be better, depending on what is nearby at that level on which significant fallout particles would collect. Flat roofs collect fallout particles so the top floor is not a good choice, nor is a floor adjacent to a neighboring flat roof.
- Shielding—the heavier and denser the materials—thick walls, concrete, bricks, books and earth—between you and the fallout particles, the better.
- Time—fallout radiation loses its intensity fairly rapidly. In time, you will be able to leave the fallout shelter. Radioactive fallout poses the greatest threat to people during the first two weeks, by which time it has declined to about 1 percent of its initial radiation level.

Remember that any protection, however temporary, is better than none at all, and the more shielding, distance, and time you can take advantage of, the better.

TAKE PROTECTIVE MEASURES

Before a Nuclear Blast

To prepare for a nuclear blast, you should do the following:

- Find out from officials if any public buildings in your community have been designated as fallout shelters. If none have been designated, make your own list of potential shelters near your home, workplace, and school. These places would include basements or the windowless center area of middle floors in high-rise buildings, as well as subways and tunnels.

- If you live in an apartment building or high-rise, talk to the manager about the safest place in the building for sheltering and about providing for building occupants until it is safe to go out.
- During periods of increased threat increase your disaster supplies to be adequate for up to two weeks.

Taking shelter during a nuclear blast is absolutely necessary. There are two kinds of shelters—blast and fallout. The following describes the two kinds of shelters:

- Blast shelters are specifically constructed to offer some protection against blast pressure, initial radiation, heat, and fire. But even a blast shelter cannot withstand a direct hit from a nuclear explosion.
- Fallout shelters do not need to be specially constructed for protecting against fallout. They can be any protected space, provided that the walls and roof are thick and dense enough to absorb the radiation given off by fallout particles.

During a Nuclear Blast

The following are guidelines for what to do in the event of a nuclear explosion.

If an attack warning is issued:

- Take cover as quickly as you can, below ground if possible, and stay there until instructed to do otherwise.
- Listen for official information and follow instructions.

If you are caught outside and unable to get inside immediately:

- Do not look at the flash or fireball—it can blind you.
- Take cover behind anything that might offer protection.
- Lie flat on the ground and cover your head. If the explosion is some distance away, it could take 3 0 seconds or more for the blast wave to hit.
- Take shelter as soon as you can, even if you are many miles from ground zero where the attack occurred—radioactive fallout can be carried by the winds for hundreds of miles. Remember the three protective factors: Distance, shielding, and time.

After a Nuclear Blast

Decay rates of the radioactive fallout are the same for any size nuclear device. However, the amount of fallout will vary based on the size of the device and its proximity to the ground. Therefore, it might be necessary for those in the areas with highest radiation levels to shelter for up to a month.

The heaviest fallout would be limited to the area at or downwind from the explosion, and 80 percent of the fallout would occur during the first 24 hours.

People in most of the areas that would be affected could be allowed to come out of shelter within a few days and, if necessary, evacuate to unaffected areas.

Returning to Your Home

Remember the following:

- Keep listening to the radio and television for news about what to do, where to go, and places to avoid.
- Stay away from damaged areas. Stay away from areas marked "radiation hazard" or "HAZMAT." Remember that radiation cannot be seen, smelled, or otherwise detected by human senses.

Radiological Dispersion Device

Terrorist use of a Radiological Dispersion Device (RDD)—often called "dirty nuke" or "dirty bomb"—is considered far more likely than use of a nuclear explosive device. An RDD combines a conventional explosive device—such as a bomb—with radioactive material. It is designed to scatter dangerous and sub-lethal amounts of radioactive material over a general area. Such RDDs appeal to terrorists because they require limited technical knowledge to build and deploy compared to a nuclear device. Also, the radioactive materials in RDDs are widely used in medicine, agriculture, industry, and research, and are easier to obtain than weapons grade uranium or plutonium.

The primary purpose of terrorist use of an RDD is to cause psychological fear and economic disruption. Some devices could cause fatalities from exposure to radioactive materials. Depending on the speed at which the area of the RDD detonation was evacuated or how successful people were at sheltering-in-place, the number of deaths and injuries from an RDD might not be substantially greater than from a conventional bomb explosion.

The size of the affected area and the level of destruction caused by an RDD would depend on the sophistication and size of the conventional bomb, the type of radioactive material used, the quality and quantity of the radioactive material, and the local meteorological conditions—primarily wind and precipitation. The area affected could be placed off-limits to the public for several months during cleanup efforts.

TAKE PROTECTIVE MEASURES

Before an RDD Event

There is no way of knowing how much warning time there will be before an attack by terrorists using an RDD, so being prepared in advance and knowing what to do and when is important. Take the same protective measures you would for fallout resulting from a nuclear blast.

During an RDD Event

While the explosive blast will be immediately obvious, the presence of radiation will not be known until trained personnel with specialized equipment are on the scene. Whether you are indoors or outdoors, home or at work, be extra cautious. It would be safer to assume radiological contamination has occurred—particularly in an urban setting or near other likely terrorist targets—and take the proper precautions. As with any radiation, you want to avoid or limit exposure. This is particularly true of inhaling radioactive dust that results from the explosion. As you seek shelter from any location (indoors or outdoors) and there is visual dust or other contaminants in the air, breathe though the cloth of your shirt or coat to limit your exposure. If you manage to avoid breathing radioactive dust, your proximity to the radioactive particles may still result in some radiation exposure.

If the explosion or radiological release occurs inside, get out immediately and seek safe shelter. Otherwise, if you are:

After a RDD Event

After finding safe shelter, those who may have been exposed to radioactive material should decontaminate themselves. To do this, remove and bag your clothing (and isolate the bag away from you and others), and shower thoroughly with soap and water. Seek medical attention after officials indicate it is safe to leave shelter.

Contamination from an RDD event could affect a wide area, depending on the amount of conventional explosives used, the quantity and type of radioactive material released, and meteorological conditions. Thus, radiation

Table 9.1 Safe Shelters

Outdoors	*Indoors*
Seek shelter indoors immediately in the nearest undamaged building.	If you have time, turn off ventilation and heating systems, close windows, vents, fireplace dampers, exhaust fans, and clothes dryer vents. Retrieve your disaster supplies kit and a battery-powered radio and take them to your shelter room.
If appropriate shelter is not available, move as rapidly as is safe upwind and away from the location of the explosive blast. Then, seek appropriate shelter as soon as possible.	Seek shelter immediately, preferably underground or in an interior room of a building, placing as much distance and dense shielding as possible between you and the outdoors where the radioactive material may be.
Listen for official instructions and follow directions.	Seal windows and external doors that do not fit snugly with duct tape to reduce infiltration of radioactive particles. Plastic sheeting will not provide shielding from radioactivity nor from blast effects of a nearby explosion.
	Listen for official instructions and follow directions.

dissipation rates vary, but radiation from an RDD will likely take longer to dissipate due to a potentially larger localized concentration of radioactive material.

Follow These Additional Guidelines after an RDD Event:

- Continue listening to your radio or watch the television for instructions from local officials, whether you have evacuated or sheltered-in-place.
- Do not return to or visit an RDD incident location for any reason.
- Follow the instructions for recovering from a disaster in Part 5.

Terrorism Knowledge Check

Answer the following questions.

1. Would you know what to do, if you were at work and . . .
   a. There was an explosion in the building?
   b. You received a package in the mail that you considered suspicious?
   c. You received a telephone call that was a bomb threat?
2. If caught outside during a nuclear blast, what should you do?
3. What are the three key factors for protection from nuclear blast and fallout?

4. If you take shelter in your own home, what kind of room would be safest during a chemical or biological attack?
5. In case of a chemical attack, what extra items should you have in your disaster supplies kit?

If you answered "no" to any of these questions, read the relevant sections of this book so that you are better prepared.

NATURAL HAZARD DISASTERS

Floods

Floods are among the most common hazards in the United States. However, all floods are not alike. Riverine floods develop slowly, sometimes over a period of days. Flash floods can develop quickly, sometimes in just a few minutes, without any visible signs of rain. Flash floods often have a dangerous wall of roaring water that carries a deadly cargo of rocks, mud and other debris and can sweep away most things in its path. Overland flooding occurs outside a defined river or stream, such as when a levee is breached, but still can be destructive. Flooding can also occur from a dam break producing effects similar to flash floods.

Flood effects can be very local, impacting a neighborhood or community, or very large, affecting entire river basins and multiple states. Be aware of flood hazards no matter where you live, but especially if you live in a low-lying area, near water or downstream from a dam. Even very small streams, gullies, creeks, culverts, dry streambeds, or low-lying ground that appear harmless in dry weather can flood. Every state is at risk from this hazard.

What to Do Before a Flood

1. Know the terms used to describe flooding:
   - Flood Watch—Flooding is possible. Stay tuned to NOAA Weather Radio or commercial radio or television for information. Watches are issued 12 to 36 hours in advance of a possible flooding event.
   - Flash Flood Watch—Flash flooding is possible. Be prepared to move to higher ground. A flash flood could occur without any warning. Listen to NOAA Weather Radio or commercial radio or television for additional information.
   - Flood Warning—Flooding is occurring or will occur soon. If advised to evacuate, do so immediately.
   - Flash Flood Warning—A flash flood is occurring. Seek higher ground on foot immediately.

2. Ask local officials whether your property is in a flood-prone or high-risk area. (Remember that floods often occur outside high-risk areas.) Ask about official flood warning signals and what to do when you hear them. Also ask how you can protect your home from flooding.
3. Identify dams in your area and determine whether they pose a hazard to you.
4. Purchase a NOAA Weather Radio with battery backup and a tone-alert feature that automatically alerts you when a Watch or Warning is issued (tone alert is not available in some areas). Purchase a battery-powered commercial radio and extra batteries.
5. Be prepared to evacuate. Learn your community's flood evacuation routes and where to find high ground. See the "Evacuation" sections in this handbook.
6. Talk to your household about flooding. Plan a place to meet your household in case you are separated from one another in a disaster and cannot return home. Choose an out-of-town contact for everyone to call to say they are OK. In some emergencies, calling out-of-state is possible even when local phone lines are down.
7. Determine how you would care for household members who may live elsewhere but might need your help in a flood. Determine any special needs your neighbors might have.
8. Prepare to survive on your own for at least three days. Assemble a disaster supply kit. Keep a stock of food and extra drinking water. See the "Emergency Planning and Disaster Supplies" sections for more information.
9. Know how to shut off electricity, gas and water at main switches and valves. Know where gas pilot lights are located and how the heating system works.
10. Consider purchasing flood insurance:
    - Flood losses are not covered under homeowners' insurance policies.
    - FEMA manages the National Flood Insurance Program, which makes federally backed flood insurance available in communities that agree to adopt and enforce floodplain management ordinances to reduce future flood damage.
    - Flood insurance is available in most communities through insurance agents.
    - There is a 30-day waiting period before flood insurance goes into effect, so don't delay.
    - Flood insurance is available whether the building is in or out of the identified flood-prone area.
11. Consider options for protecting your property:
    - Make a record of your personal property. Take photographs or videotapes of your belongings. Store these documents in a safe place.

- Keep insurance policies, deeds, property records and other important papers in a safe place away from your home.
- Avoid building in a floodplain unless you elevate and reinforce your home.
- Elevate furnace, water heater, and electric panel to higher floors or the attic if they are susceptible to flooding.
- Install "check valves" in sewer traps to prevent floodwater from backing up into the drains of your home.
- Construct barriers, such as levees, berms, and floodwalls to stop floodwater from entering the building.
- Seal walls in basements with waterproofing compounds to avoid seepage.
- Call your local building department or emergency management office for more information.

What to Do During a Flood

1. Be aware of flash flooding. If there is any possibility of a flash flood, move immediately to higher ground. Do not wait for instructions to move.
2. Listen to radio or television stations for local information.
3. Be aware of streams, drainage channels, canyons, and other areas known to flood suddenly. Flash floods can occur in these areas with or without such typical warning signs as rain clouds or heavy rain.
4. If local authorities issue a flood watch, prepare to evacuate:
   - Secure your home. If you have time, tie down or bring outdoor equipment and lawn furniture inside. Move essential items to the upper floors.
   - If instructed, turn off utilities at the main switches or valves. Disconnect electrical appliances. Do not touch electrical equipment if you are wet or standing in water.
   - Fill the bathtub with water in case water becomes contaminated or services cut off. Before filling the tub, sterilize it with a diluted bleach solution.
5. Do not walk through moving water. Six inches of moving water can knock you off your feet. If you must walk in a flooded area, walk where the water is not moving. Use a stick to check the firmness of the ground in front of you.
6. Do not drive into flooded areas. Six inches of water will reach the bottom of most passenger cars causing loss of control and possible stalling. A foot of water will float many vehicles. Two feet of water will wash away almost all vehicles. If floodwaters rise around your car, abandon the car and move to higher ground, if you can do so safely. You and your vehicle can be quickly swept away as floodwaters rise.
7. See the "Evacuation" section for important information.

What to Do After a Flood

1. Avoid floodwaters. The water may be contaminated by oil, gasoline, or raw sewage. The water may also be electrically charged from underground or downed power lines.
2. Avoid moving water. Moving water only six inches deep can sweep you off your feet.
3. Be aware of areas where floodwaters have receded. Roads may have weakened and could collapse under the weight of a car.
4. Stay away from downed power lines and report them to the power company.
5. Stay away from designated disaster areas unless authorities ask for volunteers.
6. Return home only when authorities indicate it is safe. Stay out of buildings if they are surrounded by floodwaters. Use extreme caution when entering buildings. There may be hidden damage, particularly in foundations.
7. Consider your family's health and safety needs:
   - Wash hands frequently with soap and clean water if you come in contact with floodwaters.
   - Throw away food that has come in contact with floodwaters.
   - Listen for news reports to learn whether the community's water supply is safe to drink.
   - Listen to news reports for information about where to get assistance for housing, clothing and food.
   - Seek necessary medical care at the nearest medical facility.
8. Service damaged septic tanks, cesspools, pits, and leaching systems as soon as possible. Damaged sewage systems are serious health hazards.
9. Contact your insurance agent. If your policy covers your situation, an adjuster will be assigned to visit your home. To prepare:
   - Take photos of your belongings and your home or videotape them.
   - Separate damaged and undamaged belongings.
   - Locate your financial records.
   - Keep detailed records of cleanup costs.
10. If your residence has been flooded obtain a copy of "Repairing Your Flooded Home" from the local American Red Cross chapter.
11. See the "Recovering From Disaster" sections for more information.

Hurricanes

A hurricane is a type of tropical cyclone, the generic term for a low-pressure system that generally forms in the tropics. The ingredients for a hurricane include a preexisting weather disturbance, warm tropical oceans, moisture, and relatively light winds aloft. A typical cyclone is accompanied by thunderstorms, and in the Northern Hemisphere, a counterclockwise circulation of winds near the earth's surface. Tropical cyclones are classified as follows:

Tropical Depression—An organized system of clouds and thunderstorms with a defined surface circulation and maximum sustained winds of 38 mph (33 knots) or less. Sustained winds are defined as one-minute average wind measured at about 33 feet (10 meters) above the surface.

Tropical Storm—An organized system of strong thunderstorms with a defined surface circulation and maximum sustained winds of 39–73 mph (34–63 knots).

Hurricane—An intense tropical weather system of strong thunderstorms with a well-defined surface circulation and maximum sustained winds of 74 mph (64 knots) or higher. All Atlantic and Gulf of Mexico coastal areas are subject to hurricanes and tropical storms. Although rarely struck by hurricanes, parts of the Southwest United States and the Pacific Coast experience heavy rains and floods each year from hurricanes spawned off Mexico. The Atlantic hurricane season lasts from June through November with the peak season from mid-August to late October.

As we all have seen, hurricanes can cause catastrophic and deadly damage to coastlines and several hundred miles inland. Winds can exceed 155 miles per hour. Hurricanes and tropical storms can also spawn tornadoes and microbursts, create a surge along the coast, and cause extensive damage due to inland flooding from trapped water.

Tornadoes most often occur in thunderstorms embedded in rain bands well away from the center of the hurricane; however, they also occur near the eyewall. Typically, tornadoes produced by tropical cyclones are relatively weak and short-lived but still pose a threat.

A storm surge is a huge dome of water pushed on-shore by hurricane and tropical storm winds. Storm surges can reach 25 feet high and be 50–100 miles wide. Storm tide is a combination of the storm surge and the normal tide (i.e., a 15-foot storm surge combined with a 2-foot normal high tide over the mean sea level creates a 17-foot storm tide). These phenomena cause severe erosion and extensive damage to coastal areas.

Despite improved warnings and a decrease in the loss of life, property damage continues to rise because an increasing number of people are living or vacationing near coastlines. Those in hurricane-prone areas need to be prepared for hurricanes and tropical storms.

Hurricanes are classified into five categories based on their wind speed, central pressure and damage potential. Category Three and higher are considered major hurricanes, though Category One and Two are still extremely dangerous and warrant your full attention.

Inland/Freshwater Flooding from Hurricanes

Hurricanes can produce widespread torrential rains. Floods are the deadly and destructive result. Excessive rain can also trigger landslides or mud slides, especially in mountainous regions. Flash flooding can occur due to the intense rainfall. Flooding on rivers and streams may persist for several days or more after the storm.

The speed of the storm and the geography beneath the storm are the primary factors regarding the amount of rain produced. Slow moving storms and tropical storms moving into mountainous regions tend to produce more rain.

Between 1970 and 1999, more people lost their lives from freshwater flooding associated with land-falling tropical cyclones than from any other weather hazard related to tropical cyclones.

What to Do Before a Hurricane

1. Know the difference between "Watches" and "Warnings":
   - *Hurricane/Tropical Storm Watch*—Hurricane/tropical storm conditions are possible in the specified area, usually within 36 hours.
   - *Hurricane/Tropical Storm Warning*—Hurricane/tropical storm conditions are expected in the specified area, usually within 24 hours.
   - *Short-Term Watches and Warnings*—warnings provide detailed information on specific hurricane threats, such as flash floods and tornadoes.

Table 9.2 Saffir-Simpson Hurricane Scale

Scale Number (Category)	*Sustained Winds (MPH)*	*Damage*	*Surge*
1	74–95	Minimal: Untied mobile homes, vegetation, and signs.	4–5 feet
2	96–110	Moderate: All mobile homes, roofs, small crafts, flooding.	6–8 feet
3	111–130	Extensive: Small buildings, low-lying roads cut off.	9–12 feet
4	131–155	Extreme: Roofs destroyed, trees down, roads cut off, mobile homes destroyed. Beach homes flooded.	13–18 feet
5	>155	Catastrophic: Most buildings destroyed. Vegetation destroyed. Major roads cut off. Homes flooded.	>18 feet

2. Listen for local radio or television weather forecasts. Purchase a NOAA Weather Radio with battery backup and a tone-alert feature that automatically alerts you when a Watch or Warning is issued (tone alert is not available in some areas). Purchase a battery-powered commercial radio and extra batteries, because the media will be broadcasting information on other events as well.
3. Ask your local emergency management office about community evacuation plans relating to your neighborhood. Learn evacuation routes. Determine where you would go and how you would get there if you needed to evacuate. Sometimes alternate routes are desirable.
4. Talk to your household about hurricane issues. Create a household disaster plan. Plan to meet at a place away from your residence in case you are separated. Choose an out-of-town contact for everyone to call to say they are safe.
5. Determine the needs of your household members who may live elsewhere but need your help in a hurricane. Consider the special needs of neighbors, such as people who are disabled or those with limited sight or vision problems.
6. Prepare to survive on your own for at least three days. Assemble a disaster supply kit. Keep a stock of food and extra drinking water. See the "Emergency Planning and Disaster Supplies" and "Evacuation" sections for more information.
7. Make plans to secure your property. Permanent storm shutters offer the best protection for windows. A second option is to board up windows with 5/8 inch marine plywood, cut to fit and ready to install. Tape does not prevent windows from breaking.
8. Learn how to shut off utilities and where gas pilots and water mains are located.
9. Have your home inspected for compliance with local building codes. Many of the roofs destroyed by hurricanes have not been constructed or retrofitted according to building codes. Installing straps or additional clips to securely fasten your roof to the frame structure will substantially reduce roof damage.
10. Be sure trees and shrubs around your home are well trimmed. Dead limbs or trees could cause personal injury or property damage. Clear loose and clogged rain gutters and downspouts.
11. If you have a boat, determine where to secure it in an emergency.
12. Consider flood insurance. Purchase insurance well in advance—there is a 30-day waiting period before flood insurance takes effect.
13. Make a record of your personal property. Take photographs or videotapes of your belongings. Store these documents in a safe place.

What to Do During a Hurricane Threat

1. Listen to radio or television newscasts. If a hurricane "Watch" is issued, you typically have 24 to 36 hours before the hurricane hits land.
2. Talk with household members. Make sure everyone knows where to meet and who to call, in case you are separated. Consider the needs of relatives and neighbors with special needs.
3. Secure your home. Close storm shutters. Secure outdoor objects or bring them indoors. Moor your boat if time permits.
4. Gather several days' supply of water and food for each household member. Water systems may become contaminated or damaged. Sterilize (with diluted bleach solution of one-part bleach to 10 parts water) and fill the bathtub to ensure a supply of safe water in case you are unable or told not to evacuate. Refer to the "Shelter and Emergency Planning" and "Disaster Supplies" chapters for important information.
5. If you are evacuating, take your disaster supply kit with you to the shelter. Remember that alcoholic beverages and weapons are prohibited within shelters. Also, pets are not usually allowed in a public shelter due to health reasons. See the "Animals in Disasters" chapter and contact your local humane society for additional information.
6. Prepare to evacuate. Fuel your car—service stations may be closed after the storm. If you do not have a car, make arrangements for transportation with a friend or relative. Review evacuation routes. If instructed, turn off utilities at the main valves.
7. Evacuate to an inland location if:
   - Local authorities announce an evacuation and you live in an evacuation zone.
   - You live in a mobile home or temporary structure—they are particularly hazardous during hurricanes no matter how well fastened to the ground.
   - You live in a high-rise. Hurricane winds are stronger at higher elevations.
   - You live on the coast, on a floodplain near a river or inland waterway.
   - You feel you are in danger.
8. When authorities order an evacuation:
   - Leave immediately.
   - Follow evacuation routes announced by local officials.
   - Stay away from coastal areas, riverbanks and streams.
   - Tell others where you are going.
9. If you are not required or are unable to evacuate, stay indoors during the hurricane and away from windows and glass doors. Keep curtains and blinds closed. Do not be fooled if there is a lull, it could be the eye of the storm—winds will pick up again.

- Turn off utilities if told to do so by authorities.
- If not instructed to turn off, turn the refrigerator to its coldest setting and keep closed.
- Turn off propane tanks.

10. In strong winds, follow these rules:
    - Take refuge in a small interior room, closet, or hallway.
    - Close all interior doors. Secure and brace external doors.
    - In a two-story residence, go to an interior first-floor room, such as a bathroom or closet.
    - In a multiple-story building, go to the first or second floors and stay in interior rooms away from windows.
    - Lie on the floor under a table or another sturdy object.
11. Avoid using the phone except for serious emergencies. Local authorities need first priority on telephone lines.
12. See the "Evacuation" sections for important information.

What to Do After a Hurricane

1. Stay where you are if you are in a safe location until local authorities say it is safe to leave. If you evacuated the community, do not return to the area until authorities say it is safe to return.
2. Keep tuned to local radio or television stations for information about caring for your household, where to find medical help, how to apply for financial assistance, and so on.
3. Drive only when necessary. Streets will be filled with debris. Roads may have weakened and could collapse. Do not drive on flooded or barricaded roads or bridges. Closed roads are for your protection. As little as six inches of water may cause you to lose control of your vehicle—two feet of water will carry most cars away.
4. Do not drink or prepare food with tap water until notified by officials that it is safe to do so.
5. Consider your family's health and safety needs. Be aware of symptoms of stress and fatigue. Keep your household together and seek crisis counseling if you have need.
6. Talk with your children about what has happened and how they can help during the recovery. Being involved will help them deal with the situation. Consider the needs of your neighbors. People often become isolated during hurricanes.
7. Stay away from disaster areas unless local authorities request volunteers. If you are needed, bring your own drinking water, food, and sleeping gear.
8. Stay away from riverbanks and streams until potential flooding has passed. Do not allow children, especially under the age of 13, to play

in flooded areas. There is a high risk of injury or drowning in areas that may appear safe.

9. Stay away from moving water. Moving water only six inches deep can sweep you off your feet. Standing water may be electrically charged from underground or downed power lines.
10. Stay away from downed power lines and report them to the power company. Report to local officials any broken gas, sewer or water mains.
11. Don't use candles or other open flames indoors. Use a flashlight to inspect damage.
12. Set up a manageable schedule to repair property.
13. Contact your insurance agent. An adjuster will be assigned to visit your home. To prepare:
    - Take photos of your belongings and your home or videotape them.
    - Separate damaged and undamaged belongings.
    - Locate your financial records.
    - Keep detailed records of cleanup costs.
14. Consider building a "Safe Room or Shelter" to protect your household. See the "Thunderstorms" sections for additional information in the "Tornadoes" section.
15. See the "Recovering From Disaster" section for more important information.

Thunderstorms

Thunderstorms are very common and affect great numbers of people each year. Despite their small size in comparison to hurricanes and winter storms, all thunderstorms are dangerous. Every thunderstorm produces lightning. Other associated dangers of thunderstorms include tornadoes, strong winds, hail, and flash flooding. Flash flooding is responsible for more fatalities—more than 140 annually—than any other thunderstorm-associated hazard.

Some thunderstorms do not produce rain that reaches the ground. These are generically referred to as dry thunderstorms and are most prevalent in the western United States. Known to spawn wildfires, these storms occur when there is a large layer of dry air between the base of the cloud and the ground. The falling raindrops evaporate, but lightning can still reach the ground.

What to Do Before Thunderstorms Approach

1. Know the terms used by weather forecasters:
    - *Severe Thunderstorm Watch*—Tells you when and where severe thunderstorms are likely to occur. Watch the sky and stay tuned to radio or television to know when warnings are issued.

- *Severe Thunderstorm Warning*—Issued when severe weather has been reported by spotters or indicated by radar. Warnings indicate imminent danger to life and property to those in the path of the storm.

2. Know thunderstorm facts:
   - Thunderstorms may occur singly, in clusters, or in lines.
   - Some of the most severe weather occurs when a single thunderstorm affects one location for an extended time.
   - Thunderstorms typically produce heavy rain for a brief period, anywhere from 30 minutes to an hour.
   - Warm, humid conditions are very favorable for thunderstorm development.
   - A typical thunderstorm is 15 miles in diameter and lasts an average of 30 minutes.
   - Of the estimated 100,000 thunderstorms each year in the United States, about 10 percent are classified as severe.
   - A thunderstorm is classified as severe if it produces hail at least three-quarters of an inch in diameter, has winds of 58 miles per hour or higher, or produces a tornado.
3. Know the calculation to determine how close you are to a thunderstorm:
   - Count the number of seconds between a flash of lightning and the next clap of thunder. Divide this number by 5 to determine the distance to the lightning in miles.
4. Remove dead or rotting trees and branches that could fall and cause injury or damage during a severe thunderstorm.
5. When a thunderstorm approaches, secure outdoor objects that could blow away or cause damage. Shutter windows, if possible, and secure outside doors. If shutters are not available, close window blinds, shades, or curtains.

Lightning

The ingredient that defines a thunderstorm is lightning. Since lightning creates thunder, a storm producing lightning is called a thunderstorm. Lightning occurs during all thunderstorms. Lightning results from the buildup and discharge of electrical energy between positively and negatively charged areas. The unpredictability of lightning increases the risk to individuals and property. On average, in the United States, 300 people are injured and 80 people are killed each year by lightning. Although most lightning victims survive, people struck by lightning often report a variety of long-term, debilitating symptoms, including memory loss, attention deficits, sleep disorders, numbness, dizziness, stiffness in joints, irritability, fatigue, weakness, muscle spasms, depression, and an inability to sit for a long period of time.

When thunderstorms threaten your area, get inside a home, building or hard-top automobile (not a convertible) and stay away from metallic objects and fixtures.

1. If you are inside a home:
   - Avoid showering or bathing. Plumbing and bathroom fixtures can conduct electricity.
   - Avoid using a corded telephone, except for emergencies. Cordless and cellular telephones are safe to use.
   - Unplug appliances and other electrical items, such as computers, and turn off air conditioners. Power surges from lightning can cause serious damage.
   - Use your battery-operated NOAA Weather Radio for updates from local officials.
2. If outside, with no time to reach a safe location, follow these recommendations:
   - In a forest, seek shelter in a low area under a thick growth of small trees.
     - In open areas, go to a low place, such as a ravine or valley. Be alert for flash floods.
     - Do not stand under a natural lightning rod, such as a tall, isolated tree in an open area.
     - Do not stand on a hilltop, in an open field, on the beach or in a boat on the water.
     - Avoid isolated sheds or other small structures in open areas.
     - Get away from open water. If you are boating or swimming, get to land and find shelter immediately.
     - Get away from anything metal—tractors, farm equipment, motorcycles, golf carts, golf clubs and bicycles.
     - Stay away from wire fences, clotheslines, metal pipes, rails and other metallic paths that could carry lightning to you from some distance away.
     - If you feel your hair stand on end (which indicates that lightning is about to strike), squat low to the ground on the balls of your feet. Place your hands over your ears and your head between your knees.
3. Make yourself the smallest target possible and minimize your contact with the ground. DO NOT lie flat on the ground.
4. Remember the following facts and safety tips about lightning.

Facts:

- Lightning often strikes outside of heavy rain and may occur as far as 10 miles away from any rainfall.

- Lightning-strike victims carry no electrical charge and should be attended to immediately. If breathing has stopped, begin mouth-to-mouth resuscitation. If the heart has stopped, a trained person should administer CPR. If the victim has a pulse and is breathing, look for other possible injuries. Check for burns where the lightning entered and left the body. Be alert also for nervous system damage, broken bones, and loss of hearing or eyesight. Contact your local emergency management office or American Red Cross chapter for information on CPR and first-aid classes.
- "Heat lightning" is actually lightning from a thunderstorm too far away for thunder to be heard. However, the storm may be moving in your direction!
- Most lightning deaths and injuries occur when people are caught outdoors in the summer months during the afternoon and evening.
- Lightning starts many fires in the western United States and Alaska.
- Lightning can occur from cloud-to-cloud, within a cloud, cloud-to-ground, or cloud-to-air.
- Your chances of being struck by lightning are estimated to be 1 in 600,000 but could be better by following safety tips.

Safety Tips:

- Postpone outdoor activities if thunderstorms are likely.
- Remember the 30/30 lightning safety rule: Go inside if, after seeing lighting, you cannot count to 30 before hearing thunder. Stay indoors for 30 minutes after hearing the last clap of thunder.
- Rubber-soled shoes and rubber tires provide NO protection from lightning. However, the steel frame of a hard-topped vehicle provides increased protection if you are not touching metal.

Although you may be injured if lightning strikes your car, you are much safer inside a vehicle than outside.

Tornadoes

Tornadoes are nature's most violent storms. Spawned from powerful thunderstorms, tornadoes can uproot trees, destroy buildings and turn harmless objects into deadly missiles. They can devastate a neighborhood in seconds. A tornado appears as a rotating, funnel-shaped cloud that extends to the ground with whirling winds that can reach 300 miles per hour. Damage paths can be in excess of one mile wide and 50 miles long. Every state is at some risk from this hazard.

Tornado Facts

1. A tornado is a violently rotating column of air extending from a thunderstorm to the ground.
2. Tornadoes are capable of destroying homes and vehicles and can cause fatalities.
3. Tornadoes may strike quickly, with little or no warning.
4. Tornadoes may appear nearly transparent until dust and debris are picked up or a cloud forms in the funnel. The average tornado moves SW to NE but tornadoes have been known to move in any direction.
5. The average forward speed is 30 mph but may vary from stationary to 70 mph with rotating winds that can reach 300 miles per hour.
6. Tornadoes can accompany tropical storms and hurricanes as they move onto land.
7. Waterspouts are tornadoes that form over water.
8. Tornadoes are most frequently reported east of the Rocky Mountains during spring and summer months, but can occur in any state at any time of year.
9. In the southern states, peak tornado season is March through May, while peak months in the northern states are during the late spring and early summer.
10. Tornadoes are most likely to occur between 3 p.m. and 9 p.m., but can occur at any time of the day or night.

What to Do Before Tornadoes Threaten

1. Know the terms used to describe tornado threats:
   - *Tornado Watch*—Tornadoes are possible. Remain alert for approaching storms. Listen to your battery-operated NOAA Weather Radio or local radio/television outlets for updated reports.
   - *Tornado Warning*—A tornado has been sighted or indicated by weather radar. Take shelter immediately.
2. Ask your local emergency management office or American Red Cross chapter about the tornado threat in your area. Ask about community warning signals.
3. Purchase a NOAA Weather Radio with a battery backup and tone-alert feature that automatically alerts you when a Watch or Warning is issued (tone alert is not available in some areas). Purchase a battery-powered commercial radio and extra batteries as well.
4. Know the county or parish in which you live. Counties and parishes are used in Watches and Warnings to identify the location of tornadoes.

5. Determine places to seek shelter, such as a basement or storm cellar. If an underground shelter is not available, identify an interior room or hallway on the lowest floor.
6. Practice going to your shelter with your household.
7. Know the locations of designated shelters in places where you and your household spend time, such as public buildings, nursing homes and shopping centers. Ask local officials whether a registered engineer or architect has inspected your children's schools for shelter space.
8. Ask your local emergency manager or American Red Cross chapter if there are any public safe rooms or shelters nearby. See the "Safe Room and Shelter" section at the end of this chapter for more information.
9. Assemble a disaster supply kit. Keep a stock of food and extra drinking water. See the "Emergency Planning and Disaster Supplies" and "Evacuation" sections for more information.
10. Make a record of your personal property. Take photographs or videotapes of your belongings. Store these documents in a safe place.

What to Do During a Tornado Watch

1. Listen to NOAA Weather Radio or to commercial radio or television newscasts for the latest information.
2. Be alert for approaching storms. If you see any revolving funnel-shaped clouds, report them immediately by telephone to your local police department or sheriff's office.
3. Watch for tornado danger signs:
   - Dark, often greenish sky
   - Large hail
   - A large, dark, low-lying cloud (particularly if rotating)
   - Loud roar, similar to a freight train

Caution:

- Some tornadoes are clearly visible, while rain or nearby low-hanging clouds obscure others.
- Occasionally, tornadoes develop so rapidly that little, if any, advance warning is possible.
- Before a tornado hits, the wind may die down and the air may become very still.
- A cloud of debris can mark the location of a tornado even if a funnel is not visible.
- Tornadoes generally occur near the trailing edge of a thunderstorm. It is not uncommon to see clear, sunlit skies behind a tornado.

4. Avoid places with wide-span roofs, such as auditoriums, cafeterias, large hallways, supermarkets or shopping malls.
5. Be prepared to take shelter immediately. Gather household members and pets. Assemble supplies to take to the shelter, such as flashlight, battery-powered radio, water, and first-aid kit.

What to Do During a Tornado Warning

When a tornado has been sighted, go to your shelter immediately.

1. In a residence or small building, move to a pre-designated shelter, such as a basement, storm cellar or "Safe Room or Shelter."
2. If there is no basement, go to an interior room on the lower level (closets, interior hallways). Put as many walls as possible between you and the outside. Get under a sturdy table and use arms to protect your head and neck. Stay there until the danger has passed.
3. Do not open windows. Use the time to seek shelter.
4. Stay away from windows, doors and outside walls. Go to the center of the room. Stay away from corners because they attract debris.
5. In a school, nursing home, hospital, factory or shopping center, go to predetermined shelter areas. Interior hallways on the lowest floor are usually safest. Stay away from windows and open spaces.
6. In a high-rise building, go to a small, interior room or hallway on the lowest floor possible.
7. Get out of vehicles, trailers, and mobile homes immediately and go to the lowest floor of a sturdy nearby building or a storm shelter. Mobile homes, even if tied down, offer little protection from tornadoes.
8. If caught outside with no shelter, lie flat in a nearby ditch or depression and cover your head with your hands. Be aware of potential for flooding.
9. Do not get under an overpass or bridge. You are safer in a low, flat location.
10. Never try to outrun a tornado in urban or congested areas in a car or truck; instead, leave the vehicle immediately for safe shelter. Tornadoes are erratic and move swiftly.
11. Watch out for flying debris. Flying debris from tornadoes causes most fatalities and injuries.

What to Do After a Tornado

1. Look out for broken glass and downed power lines.
2. Check for injuries. Do not attempt to move seriously injured people unless they are in immediate danger of death or further injury. If you must move

an unconscious person, first stabilize the neck and back, and then call for help immediately.
- If the victim is not breathing, carefully position the victim for artificial respiration, clear the airway and commence mouth-to-mouth resuscitation.
- Maintain body temperature with blankets. Be sure the victim does not become overheated.
- Never try to feed liquids to an unconscious person.

3. Use caution when entering a damaged building. Be sure that walls, ceiling and roof are in place and that the structure rests firmly on the foundation. Wear sturdy work boots and gloves.
4. See the "Recovering From Disaster" section for more important information.

Wind "Safe Room and Shelter"

Extreme windstorms in many parts of the country pose a serious threat to buildings and their occupants. Your residence may be built "to code," but that does not mean that it can withstand winds from extreme events like tornadoes or major hurricanes.

The purpose of a wind shelter or "Safe Room" is to provide a space where you and your household can seek refuge that provides a high level of protection. You can build a shelter in one of the several places in your home:

- In your basement
- Beneath a concrete slab-on-grade foundation or garage floor
- In an interior room on the first floor

Shelters built below ground level provide the greatest protection, but a shelter built in a first-floor interior room can also provide the necessary protection. Below-ground shelters must be designed to avoid accumulating water during the heavy rains that often accompany severe windstorms.

To protect its occupants, an in-house shelter must be built to withstand high winds and flying debris, even if the rest of the residence is severely damaged or destroyed. Therefore:

- The shelter must be adequately anchored to resist overturning and uplift.
- The walls, ceiling, and door of the shelter must withstand wind pressure and resist penetration by windborne objects and falling debris.
- The connections between all parts of the shelter must be strong enough to resist the wind.
- If sections of either interior or exterior residence walls are used as walls of the shelter, they must be separated from the structure of the residence, so that damage to the residence will not cause damage to the shelter.

If you are concerned about wind hazards where you live, especially if you live in a high-risk area, you should consider building a shelter. Publications are available from FEMA to assist in determining if you need a shelter and how to construct one. Contact the FEMA distribution center for a copy of Taking Shelter from the Storm (L-233 for the brochure and FEMA-320 for the booklet with complete construction plans).

Winter Storms and Extreme Cold

Heavy snowfall and extreme cold can immobilize an entire region. Even areas that normally experience mild winters can be hit with a major snowstorm or extreme cold. The impacts include flooding, storm surge, closed highways, blocked roads, downed power lines, and hypothermia. You can protect yourself and your household from the many hazards of winter by planning ahead.

What to Do Before a Winter Storm Threatens

1. Know the terms used by weather forecasters:
   - *Freezing rain*—Rain that freezes when it hits the ground, creating a coating of ice on roads, walkways, trees, and power lines
   - *Sleet*—Rain that turns to ice pellets before reaching the ground (Sleet also causes roads to freeze and become slippery)
   - *Winter Storm Watch*—A winter storm is possible in your area
   - *Winter Storm Warning*—A winter storm is occurring, or will soon occur in your area
   - *Blizzard Warning*—Sustained winds or frequent gusts to 35 miles per hour or greater and considerable falling or blowing snow (reducing visibility to less than a quarter mile) are expected to prevail for a period of three hours or longer
   - *Frost/Freeze Warning*—Below-freezing temperatures are expected
2. Prepare to survive on your own for at least three days. Assemble a disaster supply kit. Be sure to include winter-specific items, such as rock salt to melt ice on walkways, sand to improve traction, snow shovels and other snow removal equipment. Keep a stock of food and extra drinking water. See the "Emergency Planning and Disaster Supplies" and "Evacuation" sections for more information. Prepare for possible isolation in your home:
   - Have sufficient heat*ing fuel; regular fuel sources may be cut off.*
   - *Have emergency heating equipment and fuel (a gas fireplace or a wood-burning stove or fireplace) so you can keep at least one room of your residence liv*able. (Be sure the room is well ventilated.) If a

thermostat controls your furnace and your electricity is cut off by a storm, you will need emergency heat.
   - Kerosene heaters are another emergency heating option.
   - Store a good supply of dry, seasoned wood for your fireplace or wood-burning stove.
   - Keep fire extinguishers on hand, and make sure your household knows how to use them.
   - Never burn charcoal indoors.
3. Winterize your home to extend the life of your fuel supply:
   - Insulate walls and attics.
   - Caulk and weather-strip doors and windows.
   - Install storm windows or cover windows with plastic.
4. Maintain several days' supply of medicines, water, and food that needs no cooking or refrigeration.
5. What to do during a winter storm:
   - Listen to the radio or television for weather reports and emergency information.
   - Eat regularly and drink ample fluids, but avoid caffeine and alcohol.
6. Dress for the season:
   - Wear several layers of loose fitting, lightweight and warm clothing rather than one layer of heavy clothing. The outer garments should be tightly woven and water repellent.
   - Mittens are warmer than gloves.
   - Wear a hat; most body heat is lost through the top of the head if the head is uncovered.
   - Cover your mouth with a scarf to protect your lungs.
7. Be careful when shoveling snow. Overexertion can bring on a heart attack—a major cause of death in the winter. If you must shovel snow, stretch before going outside and don't overexert yourself.
8. Watch for signs of frostbite: loss of feeling and white or pale appearance in extremities, such as fingers, toes, ear lobes or the tip of the nose. If symptoms are detected, get medical help immediately.
9. Watch for signs of hypothermia: uncontrollable shivering, memory loss, disorientation, incoherence, slurred speech, drowsiness and apparent exhaustion. If symptoms of hypothermia are detected, get the victim to a warm location, remove any wet clothing, warm the center of the body first, and give warm, non-alcoholic beverages if the victim is conscious. Get medical help as soon as possible.
10. When at home:
    - Conserve fuel if necessary by keeping your residence cooler than normal. Temporarily "close off" heat to some rooms.

- When using kerosene heaters, maintain ventilation to avoid buildup of toxic fumes. Refuel kerosene heaters outside and keep them at least three feet from flammable objects.

Winter Driving

About 70 percent of winter deaths related to snow and ice occur in automobiles. Consider public transportation if you must travel. If you travel by car, travel in the day, don't travel alone, and keep others informed of your schedule. Stay on main roads; avoid back-road shortcuts.

1. Winterize your car. This includes a battery check, antifreeze, wipers and windshield-washer fluid, ignition system, thermostat, lights, flashing hazard lights, exhaust system, heater, brakes, defroster, oil level, and tires. Consider snow tires, snow tires with studs, or chains. Keep your car's gas tank full.
2. Carry a "winter car kit" in the trunk of your car. The kit should include the following:
   - Shovel
   - Windshield scraper
   - Battery-powered radio
   - Flashlight
   - Extra batteries
   - Water
   - Snack food
   - Mittens
   - Hat
   - Blanket
   - Tow chain or rope
   - Tire chains
   - Bag of road salt and sand
   - Fluorescent distress flag
   - Booster cables
   - Roadmaps
   - Emergency flares
   - Cellular telephone or two-way radio, if available
3. If a blizzard traps you in your car:
   - Pull off the highway. Turn on hazard lights and hang a distress flag from the radio aerial or window.
   - Remain in your vehicle where rescuers are most likely to find you. Do not set out on foot unless you can see a building close by where you know you can take shelter. Be careful: distances are distorted by blowing snow. A building may seem close but be too far to walk to in deep snow.

- Run the engine and heater about 10 minutes each hour to keep warm. When the engine is running, open a window slightly for ventilation. This will protect you from possible carbon monoxide poisoning. Periodically clear snow from the exhaust pipe.
- Exercise to maintain body heat, but avoid overexertion. In extreme cold, use roadmaps, seat covers and floor mats for insulation. Huddle with passengers and use your coat for a blanket.
- Take turns sleeping. One person should be awake at all times to look for rescue crews.
- Drink fluids to avoid dehydration.
- Be careful not to waste battery power. Balance electrical energy needs—the use of lights, heat and radio—with supply.
- At night, turn on the inside light so work crews or rescuers can see you.
- If stranded in a remote area, spread a large cloth over the snow to attract attention of rescue personnel who may be surveying the area by airplane.
- Once the blizzard passes, you may need to leave the car and proceed on foot.

Extreme Heat (Heat Wave)

Heat kills by pushing the human body beyond its limits. Under normal conditions, the body's internal thermostat produces perspiration that evaporates and cools the body. However, in extreme heat and high humidity, evaporation is slowed and the body must work extra hard to maintain a normal temperature. Most heat disorders occur because the victim has been overexposed to heat or has over-exercised for his or her age and physical condition. The elderly, young children and those who are sick or overweight are more likely to succumb to extreme heat.

Conditions that can induce heat-related illnesses include stagnant atmospheric conditions and poor air quality. Consequently, people living in urban areas may be at greater risk from the effects of a prolonged heat wave than those living in rural areas. Also, asphalt and concrete store heat longer and gradually release heat at night, which can produce higher nighttime temperatures known as the "urban heat island effect."

What to Do Before an Extreme Heat Emergency

1. Know the terms associated with extreme heat:
   - Heat wave—Prolonged period of excessive heat, often combined with excessive humidity.
   - Heat index—A number in degrees Fahrenheit (F) that tells how hot it feels when relative humidity is added to the air temperature. Exposure to full sunshine can increase the heat index by 15 degrees.

- Heat cramps—Muscular pains and spasms due to heavy exertion. Although heat cramps are the least severe, they are often the first signal that the body is having trouble with the heat.
- Heat exhaustion—Typically occurs when people exercise heavily or work in a hot, humid place where body fluids are lost through heavy sweating. Blood flow to the skin increases, causing blood flow to decrease to the vital organs. This results in a form of mild shock. If not treated, the victim's condition will worsen. Body temperature will keep rising and the victim may suffer heat stroke.
- Heat stroke—Heat stroke is life-threatening. The victim's temperature-control system, which produces sweating to cool the body, stops working. The body's temperature can increase so high that brain damage and death may result if the body is not cooled quickly.
- Sun stroke—Another term for heat stroke.

2. Consider the following preparedness measures when faced with the possibility of extreme heat:
   - Install window air conditioners snugly, insulate if necessary.
   - Close any floor heat registers nearby, and use a circulating or box fan to spread cool air.
   - Check air-conditioning ducts for proper insulation.
   - Install temporary reflectors, such as aluminum foil covered cardboard, to reflect heat back outside and be sure to weather-strip doors and sills to keep cool air in.
   - Cover windows that receive morning or afternoon sun with drapes, shades, awnings or louvers.
   - Outdoor awnings or louvers can reduce the heat that enters a home by up to 80 percent. Consider keeping storm windows up all year.
3. See the "Emergency Planning and Disaster Supplies" section for more information.

What to Do During Extreme Heat or a Heat-wave Emergency

1. Stay indoors as much as possible:
   - If air conditioning is not available, stay on the lowest floor out of the sunshine.
   - Remember that electric fans do not cool; they just blow hot air around.
2. Eat well-balanced, light and regular meals. Avoid using salt tablets unless directed to do so by a physician.
3. Drink plenty of water regularly even if you do not feel thirsty:
   - People who have epilepsy or heart, kidney, or liver disease, are on fluid-restrictive diets, or have a problem with fluid retention should consult a doctor before increasing liquid intake.

4. Limit intake of alcoholic beverages:
   - Although beer and alcoholic beverages appear to satisfy thirst, they actually cause further body dehydration.
5. Never leave children or pets alone in closed vehicles.
6. Dress in loose clothes that cover as much skin as possible.
   - Lightweight, light-colored clothing reflects heat and sunlight and helps maintain normal body temperature.
7. Protect your face and head by wearing a wide-brimmed hat.
8. Avoid too much sunshine:
   - Sunburn slows the skin's ability to cool itself. Use a sunscreen lotion with a high SPF (sun protection factor) rating (i.e., 15 or greater).
9. Avoid strenuous work during the warmest part of the day. Use a buddy system when working in extreme heat and take frequent breaks.
10. Spend at least two hours per day in an air-conditioned place. If your home is not air conditioned, consider spending the warmest part of the day in public buildings, such as libraries, schools, movie theaters, shopping malls, and other community facilities.
11. Check on family, friends and neighbors who do not have air conditioning and who spend much of their time alone.

First Aid for Heat-induced Illnesses

1. Sunburn
   - Symptoms: Skin redness and pain, possible swelling, blisters, fever, headaches.
   - First aid: Take a shower, using soap, to remove oils that may block pores, preventing the body from cooling naturally. If blisters occur, apply dry, sterile dressings and get medical attention.
2. Heat cramps
   - Symptoms: Painful spasms, usually in leg and abdominal muscles. Heavy sweating.
   - First aid: Get the victim to a cooler location. Lightly stretch and gently massage affected muscles to relieve spasm. Give sips of up to a half glass of cool water every 15 minutes. Do not give liquids with caffeine or alcohol. If nauseous, discontinue liquids.
3. Heat exhaustion
   - Symptoms: Heavy sweating and skin may be cool, pale or flushed. Weak pulse. Normal body temperature is possible but temperature will likely rise. Fainting or dizziness, nausea or vomiting, exhaustion and headaches are possible.
   - First aid: Get victim to lie down in a cool place. Loosen or remove clothing. Apply cool, wet cloths. Fan or move victim to air-conditioned

place. Give sips of water if victim is conscious. Be sure water is consumed slowly. Give half a glass of cool water every 15 minutes. If nausea occurs, discontinue. If vomiting occurs, seek immediate medical attention.

4. Heat stroke (sun stroke)
   - Symptoms: High body temperature (105+). Hot, red, dry skin. Rapid, weak pulse; and rapid, shallow breathing. Possible unconsciousness. Victim will likely not sweat unless victim was sweating from recent strenuous activity.
   - First aid: Heat stroke is a severe medical emergency. Call 911 or emergency medical services or get the victim to a hospital immediately. Delay can be fatal. Move victim to a cooler environment. Remove clothing. Try a cool bath, sponging or wet sheet to reduce body temperature. Watch for breathing problems. Use extreme caution. Use fans and air conditioners.

Emergency Water Shortage

A prolonged drought, poor water supply management, or contamination of a surface water supply source or aquifer can cause an emergency water shortage.

A drought is a period of abnormally dry weather that persists long enough to produce serious effects (crop damage, water supply shortages, etc.). The severity of the drought depends upon the degree of moisture deficiency, the duration, and the size of the affected area. Drought can affect vast territorial regions and large population numbers. In effect, drought is a silent but very damaging phenomenon that is rarely lethal but enormously destructive. Drought can ruin local and regional economies that are agricultural and tourism based. Drought also creates environmental conditions that increase risk of other hazards, such as fire, flash flood, and possible landslides/debris flow.

Poor water-quality management can result in the demand for water exceeding the available supply. This can be exacerbated by fluctuations in regional precipitation, excessive water demand, or rapid residential development.

Emergency water shortages can also be caused by contamination of a water supply. A major spill of a petroleum product or hazardous chemical on a major river can force communities to shut down water treatment plants. Although typically more localized, the contamination of ground water or an aquifer can also disrupt the use of well water.

Water Conservation

Conserving water is very important during emergency water shortages. Water saved by one user may be enough to protect the critical needs of others.

Irrigation practices can be changed to use less water, and crops that use less water can be planted. Cities and towns can ration water, factories can change manufacturing methods, and individuals can practice water-saving measures to reduce consumption. If everyone reduces water use during a drought, more water will be available to share.

1. Practice indoor water conservation.
   General:
   - Never pour water down the drain when there may be another use for it. Use it to water your indoor plants or garden.
   - Repair dripping faucets by replacing washers. One drop per second wastes 2,700 gallons of water per year!

   Bathroom:
   - Check all plumbing for leaks. Have leaks repaired by a plumber.
   - Install a toilet displacement device to cut down on the amount of water needed to flush. Place a one-gallon plastic jug of water into the tank to displace toilet flow (do not use a brick, it may dissolve and loose pieces may cause damage to the internal parts). Be sure installation does not interfere with the operating parts.
   - Consider purchasing a low-volume toilet that uses less than half the water of older models. NOTE: The law requires low-volume units in many areas.
   - Replace your showerhead with an ultra-low-flow version.
   - Do not take baths—take short showers—only turn on water to get wet and lather and then again to rinse off.
   - Place a bucket in the shower to catch excess water for watering plants.
   - Don't let the water run while brushing your teeth, washing your face or shaving.
   - Don't flush the toilet unnecessarily. Dispose of tissues, insects, and other similar waste in the trash rather than the toilet.

   Kitchen:
   - Operate automatic dishwashers only when they are fully loaded. Use the "light wash" feature if available to use less water.
   - Hand-wash dishes by filling two containers—one with soapy water and the other with rinse water containing a small amount of chlorine bleach.
   - Most dishwashers can clean soiled dishes very well, so dishes do not have to be rinsed before washing. Just remove large particles of food, and put the soiled dishes in the dishwasher.
   - Store drinking water in the refrigerator. Don't let the tap run while you are waiting for water to cool.

- Do not waste water waiting for it to get hot. Capture it for other uses, such as plant watering or heat it on the stove or in a microwave.
- Do not use running water to thaw meat or other frozen foods. Defrost food overnight in the refrigerator, or use the defrost setting on your microwave.
- Clean vegetables in a pan filled with water rather than running water from the tap.
- Kitchen sink disposals require a lot of water to operate properly. Start a compost pile as an alternate method of disposing of food waste, or simply dispose of food in the garbage.

Laundry:

- Operate automatic clothes washers only when they are fully loaded or set the water level for the size of your load.

Long-term indoor water conservation:

- Retrofit all household faucets by installing aerators with flow restrictors.
- Consider installing an instant hot water heater on your sink.
- Insulate your water pipes to reduce heat loss and prevent them from breaking if you have a sudden and unexpected spell of freezing weather.
- If you are considering installing a new heat pump or air-conditioning system, the new air-to-air models are just as efficient as the water-to-air type, and they do not waste water.
- Install a water-softening system only when the minerals in the water would damage your pipes. Turn the softener off while on vacation.
- When purchasing a new appliance, choose one that is more energy and water efficient.

2. Practice outdoor water conservation.

General:

- If you have a well at home, check your pump periodically. If the automatic pump turns on and off while water is not being used, you have a leak.

Car washing:

- Use a shut-off nozzle on your hose that can be adjusted down to a fine spray, so that water flows only as needed.
- Consider using a commercial car wash that recycles water. If you wash your own car, park on the grass so that you will be watering it at the same time.

Lawn-care:

- Don't over water your lawn. A heavy rain eliminates the need for watering for up to two weeks. Most of the year, lawns only need 1 inch of water per week.

- Water in several short sessions rather than one long one in order for your lawn to better absorb moisture.
- Position sprinklers so water lands on the lawn and shrubs and not on paved areas.
- Avoid sprinklers that spray a fine mist. Mist can evaporate before it reaches the lawn. Check sprinkler systems and timing devices regularly to be sure they operate properly.
- Raise the lawn-mower blade to at least three inches, or to its highest level. A higher cut encourages grass roots to grow deeper, shades the root system, and holds soil moisture.
- Plant drought-resistant lawn seed.
- Avoid over-fertilizing your lawn. Applying fertilizer increases the need for water. Apply fertilizers that contain slow-release, water-insoluble forms of nitrogen.
- Use a broom or blower instead of a hose to clean leaves and other debris from your driveway or sidewalk.
- Do not leave sprinklers or hoses unattended. A garden hose can pour out 600 gallons or more in only a few hours.

Pool:

- Consider installing a new water-saving pool filter. A single back flushing with a traditional filter uses 180–250 gallons of water.
- Cover pools and spas to reduce evaporation of water.

Long-term outdoor conservation:

- Plant native and/or drought-tolerant grasses, ground covers, shrubs and trees. Once established, they do not need water as frequently and usually will survive a dry period without watering. Small plants require less water to become established. Group plants together based on similar water needs.
- Install irrigation devices that are the most water efficient for each use. Micro and drip irrigation and soaker hoses are examples of efficient devices.
- Use mulch to retain moisture in the soil. Mulch also helps control weeds that compete with landscape plants for water.
- Avoid purchasing recreational water toys that require a constant stream of water.
- Avoid installing ornamental water features (such as fountains) unless they use recycled water.

Participate in public water-conservation programs of your local government, utility or water management district. Follow water conservation and water shortage rules in effect. Remember, you are included in the restrictions even if your water comes from a private well. Be sure to support community efforts that help develop and promote a water-conservation ethic. Contact your local water authority, utility district, or local emergency management agency for information specific to your area.

Earthquakes

An earthquake is a sudden shaking of the earth caused by the breaking and shifting of rock beneath the earth's surface. Earthquakes can cause buildings and bridges to collapse, telephone and power lines to fall, and result in fires, explosions, and landslides. Earthquakes can also cause huge ocean waves, called tsunamis, which travel long distances over water until they crash into coastal areas.

The following information includes general guidelines for earthquake preparedness and safety. Because injury-prevention techniques may vary from state to state, it is recommended that you contact your local emergency management office, health department, or American Red Cross chapter.

What to Do Before an Earthquake

1. Know the terms associated with earthquakes.
   - *Earthquake*—a sudden slipping or movement of a portion of the earth's crust accompanied and followed by a series of vibrations.
   - *Aftershock*—an earthquake of similar or lesser intensity that follows the main earthquake.
   - *Fault*—the earth's crust slips along a fault, an area of weakness where two sections of crust have separated. The crust may only move a few inches to a few feet in a severe earthquake.
   - *Epicenter*—the area of the earth's surface directly above the origin of an earthquake.
   - *Seismic Waves*—vibrations that travel outward from the center of the earthquake at speeds of several miles per second. These vibrations can shake some buildings so rapidly that they collapse.
   - *Magnitude*—indicates how much energy was released. This energy can be measured on a recording device and graphically displayed through lines on a Richter Scale. A magnitude of 7.0 on the Richter Scale would indicate a very strong earthquake. Each whole number on the scale represents an increase of about thirty times the energy released. Therefore, an earthquake measuring 6.0 is about thirty times more powerful than one measuring 5.0.
2. Look for items in your home that could become a hazard in an earthquake:
   - Repair defective electrical wiring, leaky gas lines, and inflexible utility connections.
   - Bolt down water heaters and gas appliances (have an automatic gas shut-off device installed that is triggered by an earthquake).

- Place large or heavy objects on lower shelves. Fasten shelves to walls. Brace high and top-heavy objects.
- Store bottled foods, glass, china and other breakables on low shelves or in cabinets that can fasten shut.
- Anchor overhead lighting fixtures.
- Check and repair deep plaster cracks in ceilings and foundations. Get expert advice, especially if there are signs of structural defects.
- Be sure the residence is firmly anchored to its foundation.
- Install flexible-pipe fittings to avoid gas or water leaks. Flexible fittings are more resistant to breakage.

3. Know where and how to shut off electricity, gas and water at main switches and valves. Check with your local utilities for instructions.
4. Hold earthquake drills with your household:
   - Locate safe spots in each room under a sturdy table or against an inside wall. Reinforce this information by physically placing yourself and your household in these locations.
   - Identify danger zones in each room—near windows where glass can shatter, bookcases or furniture that can fall over, or under ceiling fixtures that could fall down.
5. Develop a plan for reuniting your household after an earthquake. Establish an out-of-town telephone contact for household members to call to let others know that they are OK.
6. Review your insurance policies. Some damage may be covered even without specific earthquake insurance. Protect important home and business papers.
7. Prepare to survive on your own for at least three days. Assemble a disaster supply kit. Keep a stock of food and extra drinking water. See the "Emergency Planning and Disaster Supplies" and "Evacuation" sections for more information.

What to Do During an Earthquake

Stay inside until the shaking stops and it is safe to go outside. Most injuries during earthquakes occur when falling objects hit people as they enter or exit buildings.

1. Drop, Cover, and Hold On! Minimize your movements during an earthquake to a few steps to a nearby safe place. Stay indoors until the shaking has stopped and you are sure exiting is safe.
2. If you are indoors, take cover under a sturdy desk, table or bench, or against an inside wall, and hold on. Stay away from glass, windows, outside doors or walls and anything that could fall, such as lighting fixtures or furniture. If you are in bed, stay there, hold on and protect your head with a pillow, unless you are under a heavy light fixture that could fall.

3. If there isn't a table or desk near you, cover your face and head with your arms and crouch in an inside corner of the building. Doorways should only be used for shelter if they are in close proximity to you and if you know that they have a strongly supported load-bearing doorway.
4. If you are outdoors, stay there. Move away from buildings, streetlights, and utility wires.
5. If you live in an apartment building or other multi-household structure with many levels, consider the following:
   - Get under a desk and stay away from windows and outside walls.
   - Stay in the building (many injuries occur as people flee a building and are struck by falling debris from above).
   - Be aware that the electricity may go out and sprinkler systems may come on.
   - DO NOT use the elevators.
6. If you are in a crowded indoor public location:
   - Stay where you are. Do not rush for the doorways.
   - Move away from tall shelves, cabinets and bookcases containing objects that may fall.
   - Take cover and grab something to shield your head and face from falling debris and glass.
   - Be aware that the electricity may go out or the sprinkler systems or fire alarms may turn on.
   - DO NOT use elevators.
7. In a moving vehicle, stop as quickly as safety permits, and stay in the vehicle. Avoid stopping near or under buildings, trees, overpasses or utility wires. Then, proceed cautiously, watching for road and bridge damage.
8. If you become trapped in debris:
   - Do not light a match.
   - Do not move about or kick up dust.
   - Cover your mouth with a handkerchief or clothing.
   - Tap on a pipe or wall so rescuers can locate you. Use a whistle if one is available. Shout only as a last resort—shouting can cause you to inhale dangerous amounts of dust.
9. Stay indoors until the shaking has stopped and you are sure exiting is safe.

What to Do After an Earthquake

1. Be prepared for aftershocks. These secondary shock waves are usually less violent than the main quake but can be strong enough to do additional damage to weakened structures.
2. Check for injuries. Do not attempt to move seriously injured people unless they are in immediate danger of death or further injury. If you

must move an unconscious person, first stabilize the neck and back, and then call for help immediately.

- If the victim is not breathing, carefully position the victim for artificial respiration, clear the airway and start mouth-to-mouth resuscitation.
- Maintain body temperature with blankets. Be sure the victim does not become overheated.
- Never try to feed liquids to an unconscious person.

3. If the electricity goes out, use flashlights or battery-powered lanterns. Do not use candles, matches or open flames indoors after the earthquake because of possible gas leaks.
4. In areas covered with fallen debris and broken glass, wear sturdy shoes.
5. Check your home for structural damage. If you have any doubts about safety, have your home inspected by a professional before entering.
6. Check chimneys for visual damage; however, have a professional inspect the chimney for internal damage before lighting a fire.
7. Clean up spilled medicines, bleaches, gasoline and other flammable liquids. Evacuate the building if gasoline fumes are detected and the building is not well ventilated.
8. Visually inspect utility lines and appliances for damage.
   - If you smell gas or hear a hissing or blowing sound, open a window and leave. Shut off the main gas valve. Report the leak to the gas company from the nearest working phone or cell phone available.
   - Stay out of the building. If you shut off the gas supply at the main valve, you will need a professional to turn it back on.
   - Switch off electrical power at the main fuse box or circuit breaker if electrical damage is suspected or known.
   - Shut off the water supply at the main valve if water pipes are damaged.
   - Do not flush toilets until you know that sewage lines are working properly.
9. Open cabinets cautiously. Beware of objects that can fall off shelves.
10. Use the phone only to report life-threatening emergencies.
11. Listen to news reports for the latest emergency information.
12. Stay off the streets. If you must go out, watch for fallen objects, downed electrical wires, weakened walls, bridges, roads and sidewalks.
13. Stay away from damaged areas unless your assistance has been specifically requested by police, fire or relief organizations.
14. If you live in coastal areas, be aware of possible tsunamis, sometimes mistakenly called tidal waves. When local authorities issue a tsunami warning, assume that a series of dangerous waves is on the way. Stay away from the beach. See the "Tsunamis" section for more information.

Volcanoes

A volcano is a vent through which molten rock escapes to the earth's surface. When pressure from gases within the molten rock becomes too great, an eruption occurs. Some eruptions are relatively quiet, producing lava flows that creep across the land at two to 10 miles per hour.

Explosive eruptions can shoot columns of gases and rock fragments tens of miles into the atmosphere, spreading ash hundreds of miles downwind. Lateral blasts can flatten trees for miles. Hot, sometimes poisonous, gases may flow down the sides of the volcano.

Lava flows are streams of molten rock that either pour from a vent quietly through lava tubes or by lava fountains. Because of their intense heat, lava flows are also great fire hazards. Lava flows destroy everything in their path, but most move slowly enough that people can move out of the way.

Fresh volcanic ash, made of pulverized rock, can be harsh, acidic, gritty, glassy, and odorous. While not immediately dangerous to most adults, the combination of acidic gas and ash could cause lung damage to small infants, very old people or those suffering from severe respiratory illnesses. Volcanic ash can also damage machinery, including engines and electrical equipment. Ash accumulations mixed with water become heavy and can collapse roofs.

Volcanic eruptions can be accompanied by other natural hazards: earthquakes, mudflows and flash floods, rock falls and landslides, acid rain, fire, and (under special conditions) tsunamis. Active volcanoes in the United States are found mainly in Hawaii, Alaska and the Pacific Northwest.

What to Do Before an Eruption

1. Make evacuation plans. If you live in a known volcanic hazard area, plan a route out and have a backup route in mind.
2. Develop a household disaster plan. In case household members are separated from one another during a volcanic eruption (a real possibility during the day when adults are at work and children are at school), have a plan for getting back together. Ask an out-of-town relative or friend to serve as the "household contact" because, after a disaster, it's often easier to call long distance. Make sure everyone knows the name, address, and phone number of the contact person.
3. Assemble a disaster supply kit (see "Emergency Planning and Disaster Supplies" section).
4. Get a pair of goggles and a throwaway breathing mask for each member of the household in case of ash-fall.
5. Do not visit an active volcano site unless officials designate a safe-viewing area.

What to Do During an Eruption

1. If close to the volcano, evacuate immediately away from the volcano to avoid flying debris, hot gases, lateral blast, and lava flow.
2. Avoid areas downwind from the volcano to avoid volcanic ash.
3. Be aware of mudflows. The danger from a mudflow increases as you approach a stream channel and decreases as you move away from a stream channel toward higher ground. In some parts of the world (Central and South America, Indonesia, the Philippines), this danger also increases with prolonged heavy rains. Mudflows can move faster than you can walk or run. Look upstream before crossing a bridge, and do not cross if the mudflow is approaching. Avoid river valleys and low-lying areas.
4. Stay indoors until the ash has settled unless there is danger of the roof collapsing.
5. During an ash-fall, close doors, windows, and all ventilation in the house (chimney vents, furnaces, air conditioners, fans and other vents).
6. Avoid driving in heavy dust unless absolutely required. If you do drive in dense dust, keep speed down to 35 mph or slower.
7. Remove heavy ash from flat or low-pitched roofs and rain gutters.
8. Volcanic ash is actually fine, glassy fragments and particles that can cause severe injury to breathing passages, eyes, and open wounds, and irritation to skin. Follow these precautions to keep yourself safe from ash-fall:
   - Wear long-sleeved shirts and long pants.
   - Use goggles and wear eyeglasses instead of contact lenses.
   - Use a dust mask or hold a damp cloth over your face to help breathing.
   - Keep car or truck engines off. Driving can stir up volcanic ash that can clog engines and stall vehicles. Moving parts can be damaged from abrasion, including bearings, brakes and transmissions.

What to Do After the Eruption

1. Avoid ash-fall areas if possible. If you are in an ash-fall area, cover your mouth and nose with a mask, keep skin covered and wear goggles to protect the eyes.
2. Clear roofs of ash-fall because it is very heavy and can cause buildings to collapse. Exercise great caution when working on a roof.
3. Avoid driving through ash-fall that is easily stirred up and can clog engines, causing vehicles to stall.
4. If you have a respiratory ailment, avoid contact with any amount of ash. Stay indoors until local health officials advise it is safe to go outside.

Landslides and Debris Flow

(Mudslide)

Landslides occur in all U.S. states and territories and occur when masses of rock, earth, or debris move down a slope. Landslides may be small or large, and can move at slow or very high speeds. They are activated by storms, earthquakes, volcanic eruptions, fires and by human modification of the land.

Debris and mudflows are rivers of rock, earth, and other debris saturated with water. They develop when water rapidly accumulates in the ground, during heavy rainfall or rapid snowmelt, changing the earth into a flowing river of mud or "slurry." They can flow rapidly down slopes or through channels, and can strike with little or no warning at avalanche speeds. They can also travel several miles from their source, growing in size as they pick up trees, large boulders, cars, and other materials along the way.

Landslide, mudflow, and debris-flow problems are occasionally caused by land mismanagement. Improper land-use practices on ground of questionable stability, particularly in mountain, canyon and coastal regions, can create and accelerate serious landslide problems. Land-use zoning, professional inspections, and proper design can minimize many landslide, mudflow, and debris-flow problems.

What to Do Before a Landslide or Debris Flow

1. Contact your local emergency management office or American Red Cross chapter for information on local landslide and debris flow hazards.
2. Get a ground assessment of your property:
   - County or state geological experts, local planning department or departments of natural resources may have specific information on areas vulnerable to landslides. Consult an appropriate professional expert for advice on corrective measures you can take.
3. Minimize home hazards by having flexible-pipe fittings installed to avoid gas or water leaks. Flexible fittings are more resistant to breakage. Only the gas company or its professionals should install gas fittings.
4. Familiarize yourself with your surrounding area:
   - Small changes in your local landscape could alert you to the potential of greater future threat.
   - Observe the patterns of storm-water drainage on slopes and especially the places where runoff water converges.
   - Watch for any sign of land movement, such as small slides, flows, or progressively leaning trees, on the hillsides near your home.
5. Be particularly observant of your surrounding area before and during intense storms that could heighten the possibility of landslide or debris

flow from heavy rains. Many debris-flow fatalities occur when people are sleeping.

6. Talk to your insurance agent. Debris flow may be covered by flood insurance policies from the National Flood Insurance Program (NFIP).
7. Learn to recognize landslide-warning signs.
   - Doors or windows stick or jam for the first time.
   - New cracks appear in plaster, tile, brick, or foundations.
   - Outside walls, walks or stairs begin pulling away from the building.
   - Slowly developing, widening cracks appear on the ground or on paved areas, such as streets or driveways.
   - Underground utility lines break.
   - Bulging ground appears at the base of a slope.
   - Water breaks through the ground surface in new locations.
   - Fences, retaining walls, utility poles, or trees tilt or move.
   - You hear a faint rumbling sound that increases in volume as the landslide nears.
   - The ground slopes downward in one specific direction and may begin shifting in that direction under your feet.

What to Do During a Heightened Threat (intense storm) of Landslide or Debris Flow

1. Listen to radio or television for warnings of intense rainfall:
   - Be prepared to evacuate if instructed by local authorities or if you feel threatened.
   - Should you remain at home, move to a second story if possible to distance yourself from the direct path of debris flow and landslide debris.
2. Be alert when intense, short bursts of rain follow prolonged heavy rains or damp weather, which increases the risk of debris flows.
3. Listen for any unusual sounds that might indicate moving debris, such as trees cracking or boulders knocking together. A trickle of flowing or falling mud or debris may precede larger landslides. Moving debris can flow quickly and sometimes without warning.
4. If you are near a stream or channel, be alert for sudden increases or decreases in water flow and for a change from clear to muddy water. Such changes may indicate landslide activity upstream. Be prepared to move quickly.
5. Be especially alert when driving. Embankments along roadsides are particularly susceptible to landslides. Watch for collapsed pavement, mud, fallen rocks, and other indications of possible debris flows.
6. Evacuate when ordered by local authorities. See the "Evacuation" section for more information.

What to Do During a Landslide or Debris Flow

1. Quickly move away from the path of a landslide or debris flow.
2. Areas generally considered safe include the following:
   - Areas that have not moved in the past.
   - Relatively flat-lying areas away from drastic changes in slope.
   - Areas at the top of or along ridges set back from the tops of slopes.
3. If escape is not possible, curl into a tight ball and protect your head.

What to Do After a Landslide or Debris Flow

1. Stay away from the slide area. There may be danger of additional slides.
2. Check for injured and trapped people near the slide, without entering the direct slide area, and direct rescuers to them.
3. Help a neighbor who may require special assistance—large families, children, elderly people, and people with disabilities.
4. Listen to local radio or television stations for the latest emergency information.
5. Landslides and flows can provoke associated dangers, such as broken electrical, water, gas, and sewage lines, and disrupt roadways and railways:
   - Look for and report broken utility lines to appropriate authorities. Reporting potential hazards will get the utilities turned off as quickly as possible, preventing further hazard and injury.
   - Check the building foundation, chimney, and surrounding land for damage. Damage to foundations, chimneys, or surrounding land may help you assess the safety of the area.
6. Watch for flooding, which may occur after a landslide or debris flow. Floods sometimes follow landslides and debris flows because they may both be started by the same event.
7. Replant damaged ground as soon as possible since erosion caused by loss of ground cover can lead to flash flooding and additional landslides in the near future.
8. Seek the advice of a geotechnical expert for evaluating landslide hazards or designing corrective techniques to reduce landslide risk. A professional will be able to advise you of the best ways to prevent or reduce landslide risk, without creating further hazard.
9. See the "Recovering From Disaster" section for more information.

Tsunamis

Tsunami (pronounced soo-ná-mee), sometimes mistakenly called a tidal wave, is a series of enormous waves created by an underwater disturbance, such as an earthquake. A tsunami can move hundreds of miles per hour in the open ocean and smash into land with waves as high as 100 feet or more, although most waves are less than 18 feet high.

From the area where the tsunami originates, waves travel outward in all directions much like the ripples caused by throwing a rock into a pond. In deep water, the tsunami wave is not noticeable. Once the wave approaches the shore it builds in height. All tsunamis are potentially dangerous, even though they may not damage every coastline they strike. A tsunami can strike anywhere along most of the U.S. coastline. The most destructive tsunamis have occurred along the coasts of California, Oregon, Washington, Alaska, and Hawaii.

Earthquake-induced movement of the ocean floor most often generates tsunamis. Landslides, volcanic eruptions, and even meteorites can also generate tsunamis. If a major earthquake or landslide occurs close to shore, the first wave in a series could reach the beach in a few minutes, even before a warning is issued.

Areas are at greater risk if less than 25 feet above sea level and within a mile of the shoreline. Drowning is the most common cause of death associated with a tsunami. Tsunami waves and the receding water are very destructive to structures in the run-up zone. Other hazards include flooding, contamination of drinking water and fires from gas lines or ruptured tanks.

What to Do Before a Tsunami

1. Know the terms used by the West Coast/Alaska Tsunami Warning Center (WC/ATWC—responsible for tsunami warnings for California, Oregon, Washington, British Columbia, and Alaska) and the Pacific Tsunami Warning Center (PTWC—responsible for tsunami warnings to international authorities, Hawaii, and the U.S. territories within the Pacific basin).
   - *Advisory*—An earthquake has occurred in the Pacific basin, which might generate a tsunami.
     WC/ATWC and PTWC will issue hourly bulletins advising of the situation.
   - *Watch*—A tsunami was or may have been generated, but is at least two hours travel time to the area in Watch status.
   - *Warning*—A tsunami was or may have been generated, which could cause damage; therefore, people in the warning area are strongly advised to evacuate.
2. Listen to radio or television for more information and follow the instructions of your local authorities.
3. Immediate warning of tsunamis sometimes comes in the form of a noticeable recession in water away from the shoreline. This is nature's tsunami

warning, and it should be heeded by moving inland to higher ground immediately.
4. If you feel an earthquake in a coastal area, turn on your radio to learn if there is a tsunami warning.
5. Know that a small tsunami at one beach can be a giant wave a few miles away. The topography of the coastline and the ocean floor will influence the size of the wave.
6. A tsunami may generate more than one wave. Do not let the modest size of one wave allow you to forget how dangerous a tsunami is. The next wave may be bigger.
7. Prepare for possible evacuation. Learn evacuation routes. Determine where you would go and how you would get there if you needed to evacuate. See the "Evacuation" and "Emergency Planning and Disaster Supplies" sections for more information.

What to Do During a Tsunami

1. If you are advised to evacuate, do so immediately.
2. Stay away from the area until local authorities say it is safe. Do not be fooled into thinking that the danger is over after a single wave—a tsunami is not a single wave but a series of waves that can vary in size.
3. Do not go to the shoreline to watch for a tsunami. When you can see the wave, it is too late to escape.

What to Do After a Tsunami

1. Avoid flooded and damaged areas until officials say it is safe to return.
2. Stay away from debris in the water; it may pose a safety hazard to boats and people.
3. See the "Recovering From Disaster" section for more information.

Fire

Each year more than 4,000 Americans die and more than 25,000 are injured in fires, many of which could be prevented. Direct property loss due to fires is estimated at $8.6 billion annually.

To protect yourself, it's important to understand the basic characteristics of fire. Fire spreads quickly. There is no time to gather valuables or make a phone call. In just two minutes a fire can become life threatening. In five minutes a residence can be engulfed in flames.

Heat and smoke from fire can be more dangerous than the flames. Inhaling the super-hot air can sear your lungs. Fire produces poisonous gases that make you disoriented and drowsy. Instead of being awakened by a fire, you may fall into a deeper sleep. Asphyxiation is the leading cause of fire deaths, exceeding burns by a three-to-one ratio.

What to Do Before Fire Strikes

1. Install smoke alarms. Working smoke alarms decrease your chances of dying in a fire by half:
   - Place smoke alarms on every level of your residence—outside bedrooms on the ceiling or high on the wall, at the top of open stairways or at the bottom of enclosed stairs and near (but not in) the kitchen.
   - Test and clean smoke alarms once a month and replace batteries at least once a year. Replace smoke alarms once every 10 years.
2. With your household, plan two escape routes from every room in the residence. Practice with your household escaping from each room:
   - Make sure windows are not nailed or painted shut. Make sure security gratings on windows have a fire safety-opening feature so that they can be easily opened from the inside.
   - Consider escape ladders if your home has more than one level and ensure that burglar bars and other anti-theft mechanisms that block outside window entry are easily opened from inside.
   - Teach household members to stay low to the floor (where the air is safer in a fire) when escaping from a fire.
   - Pick a place outside your home for the household to meet after escaping from a fire.
3. Clean out storage areas. Don't let trash, such as old newspapers and magazines, to accumulate.
4. Check the electrical wiring in your home:
   - Inspect extension cords for frayed or exposed wires or loose plugs.
   - Outlets should have cover plates and no exposed wiring.
   - Make sure wiring does not run under rugs, over nails, or across high-traffic areas.
   - Do not overload extension cords or outlets. If you need to plug in two or three appliances, get a UL-approved unit with built-in circuit breakers to prevent sparks and short circuits.
   - Make sure home insulation does not touch electrical wiring.
   - Have an electrician check the electrical wiring in your home.
5. Never use gasoline, benzine, naptha, or similar liquids indoors:
   - Store flammable liquids in approved containers in well-ventilated storage areas.
   - Never smoke near flammable liquids.
   - After use, safely discard all rags or materials soaked in flammable material.

6. Check heating sources. Many home fires are started by faulty furnaces or stoves, cracked or rusted furnace parts and chimneys with creosote buildup. Have chimneys, wood stoves and all home-heating systems inspected and cleaned annually by a certified specialist.
7. Insulate chimneys and place spark arresters on top. The chimney should be at least three feet higher than the roof. Remove branches hanging above and around the chimney.
8. Take care when using alternative heating sources, such as wood, coal and kerosene heaters and electrical space heaters:
   - Check with your local fire department on the legality of using kerosene heaters in your community. Be sure to fill kerosene heaters outside after they have cooled.
   - Place heaters at least three feet away from flammable materials. Make sure the floor and nearby walls are properly insulated.
   - Use only the type of fuel designated for your unit and follow manufacturer's instructions.
   - Store ashes in a metal container outside and away from the residence.
   - Keep open flames away from walls, furniture, drapery, and flammable items. Keep a screen in front of the fireplace.
   - Have chimneys and wood stoves inspected annually and cleaned, if necessary.
   - Use portable heaters only in well-ventilated rooms.
9. Keep matches and lighters up high, away from children, and if possible, in a locked cabinet.
10. Do not smoke in bed, or when drowsy or medicated. Provide smokers with deep, sturdy ashtrays. Douse cigarette and cigar butts with water before disposal.
11. Safety experts recommend that you sleep with your door closed.
12. Know the locations of the gas valve and electric fuse or circuit-breaker box and how to turn them off in an emergency. If you shut off your main gas line for any reason, allow only a gas company representative to turn it on again.
13. Install A-B-C type fire extinguishers in the home and teach household members how to use them (Type A—wood or papers fires only; Type B—flammable liquid or grease fires; Type C—electrical fires; Type A-B-C—rated for all fires and recommended for the home).
14. Consider installing an automatic fire sprinkler system in your home.
15. Ask your local fire department to inspect your residence for fire safety and prevention.
16. Teach children how to report a fire and when to use 911.
17. To support insurance claims in case you do have a fire, conduct an inventory of your property and possessions and keep the list in a separate location. Photographs are also helpful.

18. See the "Emergency Planning and Disaster Supplies" section for additional information.

What to Do During a Fire

1. Use water or a fire extinguisher to put out small fires. Do not try to put out a fire that is getting out of control. If you're not sure if you can control it, get everyone out of the residence and call the fire department from a neighbor's residence.
2. Never use water on an electrical fire. Use only a fire extinguisher approved for electrical fires.
3. Smother oil and grease fires in the kitchen with baking soda or salt, or put a lid over the flame if it is burning in a pan. Do not attempt to take the pan outside.
4. If your clothes catch fire, stop, drop, and roll until the fire is extinguished. Running only makes the fire burn faster.
5. If you are escaping through a closed door, use the back of your hand to feel the top of the door, the doorknob, and the crack between the door and doorframe before you open it. Never use the palm of your hand or fingers to test for heat—burning those areas could impair your ability to escape a fire (i.e., ladders and crawling):
   - If the door is cool, open slowly and ensure fire and/or smoke is not blocking your escape route. If your escape route is blocked, shut the door immediately and use an alternate escape route, such as a window. If clear, leave immediately through the door. Be prepared to crawl. Smoke and heat rise. The air is clearer and cooler near the floor.
   - If the door is warm or hot, do not open. Escape through a window. If you cannot escape, hang a white or light-colored sheet outside the window, alerting fire fighters to your presence.
6. If you must exit through smoke, crawl low under the smoke to your exit—heavy smoke and poisonous gases collect first along the ceiling.
7. Close doors behind you as you escape to delay the spread of the fire.
8. Once you are safely out, stay out and call 911.

What to Do After a Fire

1. Give first aid where needed. After calling 911 or your local emergency number, cool and cover burns to reduce chance of further injury or infection.
2. Do not enter a fire-damaged building unless authorities say it is OK.
3. If you must enter a fire-damaged building, be alert for heat and smoke. If you detect either, evacuate immediately.

4. Have an electrician check your household wiring before the current is turned on.
5. Do not attempt to reconnect any utilities yourself. Leave this to the fire department and other authorities.
6. Beware of structural damage. Roofs and floors may be weakened and need repair.
7. Contact your local disaster relief service, such as the American Red Cross or Salvation Army, if you need housing, food or a place to stay.
8. Call your insurance agent:
   - Make a list of damage and losses. Pictures are helpful.
   - Keep records of cleanup and repair costs. Receipts are important for both insurance and income tax claims.
   - Do not throw away any damaged goods until an official inventory has been taken. Your insurance company takes all damages into consideration.
9. If you are a tenant, contact the landlord. It's the property owner's responsibility to prevent further loss or damage to the site.
10. Secure personal belongings or move them to another location.
11. Discard food, beverages and medicines that have been exposed to heat, smoke or soot. Refrigerators and freezers left closed hold their temperature for a short time. Do not attempt to refreeze food that has thawed.
12. If you have a safe or strong box, do not try to open it. It can hold intense heat for several hours. If the door is opened before the box has cooled, the contents could burst into flames.
13. If a building inspector says the building is unsafe and you must leave your home:
    - Ask local police to watch the property during your absence.
    - Pack identification, medicines, glasses, jewelry, credit cards, checkbooks, insurance policies, and financial records, if you can reach them safely.
    - Notify friends, relatives, police and fire departments, your insurance agent, the mortgage company, utility companies, delivery services, employers, schools, and the Post Office of your whereabouts.
14. See the "Shelter" and "Recovering From Disaster" sections for more information.

Wildland Fires

If you live on a remote hillside, in a valley, prairie or forest where flammable vegetation is abundant, your residence could be vulnerable to wildland fire. These fires are usually triggered by lightning or accidents.

1. Fire facts about rural living:
   - Once a fire starts outdoors in a rural area, it is often hard to control. Firefighters dealing with such fires are trained to protect natural resources, not homes and buildings.
   - Many homes are located far from fire stations. The result is longer emergency response times. Within a matter of minutes, an entire home may be destroyed by fire.
   - Limited water supply in rural areas can make fire suppression difficult.
   - Homes may be secluded and surrounded by woods, dense brush and combustible vegetation that fuel fires.
2. Ask fire authorities for information about wildland fires in your area. Request that they inspect your residence and property for hazards.
3. Be prepared and have a fire safety and evacuation plan:
   - Practice fire escape and evacuation plans.
   - Mark the entrance to your property with address signs that are clearly visible from the road.
   - Know which local emergency services are available and have those numbers posted near telephones.
   - Provide emergency-vehicle access through roads and driveways at least 12 feet wide with adequate turnaround space.
4. Tips for making your property fire resistant:
   - Keep lawns trimmed, leaves raked, and the roof and rain gutters free from debris, such as dead limbs and leaves.
   - Stack firewood at least 30 feet away from your home.
   - Store flammable materials, liquids and solvents in metal containers outside the home, at least 30 feet away from structures and wooden fences.
   - Create defensible space by thinning trees and brush within 30 feet around your home. Beyond 30 feet, remove dead wood, debris and low tree branches.
   - Landscape your property with fire-resistant plants and vegetation to prevent fire from spreading quickly. For example, hardwood trees are more fire resistant than pine, evergreen, eucalyptus, or fir trees.
   - Make sure water sources, such as hydrants, ponds, swimming pools and wells, are accessible to the fire department.
5. Protect your home:
   - Use fire resistant, protective roofing and materials like stone, brick and metal to protect your home. Avoid using wood materials. They offer the least fire protection.
   - Cover all exterior vents, attics and eaves with metal mesh screens no larger than six millimeters or 1/4 inch to prevent debris from collecting and to help keep sparks out.

- Install multi-pane windows, tempered safety glass or fireproof shutters to protect large windows from radiant heat.
- Use fire-resistant draperies for added window protection.
- Have chimneys, wood stoves and all home-heating systems inspected and cleaned annually by a certified specialist.
- Insulate chimneys, and place spark arresters on top. Chimney should be at least three feet above the roof.
- Remove branches hanging above and around the chimney.

6. Follow local burning laws:
   - Do not burn trash or other debris without proper knowledge of local burning laws, techniques and the safest times of day and year to burn.
   - Before burning debris in a wooded area, make sure you notify local authorities and obtain a burning permit.
   - Use an approved incinerator with a safety lid or covering with holes no larger than 3/4 inch.
   - Create at least a 10-foot clearing around the incinerator before burning debris.
   - Have a fire extinguisher or garden hose on hand when burning debris.
7. If wildfire threatens your home and time permits, consider the following:

   Inside
   - Shut off gas at the meter. Turn off pilot lights.
   - Open fireplace damper. Close fireplace screens.
   - Close windows, vents, doors, blinds or noncombustible window coverings and heavy drapes.
   - Remove flammable drapes and curtains.
   - Move flammable furniture into the center of the home away from windows and sliding-glass doors.
   - Close all interior doors and windows to prevent drafts.
   - Place valuables that will not be damaged by water in a pool or pond.
   - Gather pets into one room. Make plans to care for your pets if you must evacuate.
   - Back your car into the garage or park it in an open space facing the direction of escape. Shut doors, and roll up windows. Leave the key in the ignition and the car doors unlocked. Close garage windows and doors, but leave them unlocked. Disconnect automatic garage door openers.

   Outside
   - Seal attic and ground vents with pre-cut plywood or commercial seals.
   - Turn off propane tanks.
   - Place combustible patio furniture inside.
   - Connect garden hose to outside taps. Place lawn sprinklers on the roof and near above-ground fuel tanks. Wet the roof.

- Wet or remove shrubs within 15 feet of the home.
- Gather fire tools, such as a rake, axe, handsaw or chainsaw, bucket, and shovel.

8. If advised to evacuate, do so immediately. Choose a route away from the fire hazard. Watch for changes in the speed and direction of fire and smoke.
9. See the "Evacuation" section for detailed information about evacuation preparedness. Also see the "Recovering from Disaster" and "Shelters" sections for additional information.

TECHNOLOGICAL AND MAN-MADE HAZARDS

There are thousands of chemicals in our homes and places of work. Some are dangerous and even more so when they get into the wrong hands. A child can die from drinking a chemical stored in a bottle under the sink; thousands could die in a bioterrorist attack on a crowded subway or a packed indoor concert.

Most hazardous chemicals have warning properties that provide a practical means for detecting a hazard and initiating protective actions. Such warning properties make chemicals perceptible; for example, vapors or gases can be perceived by the human senses (i.e., smell, sight, taste, or irritation of the eyes, skin, or respiratory tract) before serious effects occur. The distinction between perceptible and imperceptible agents is not an exact one. The concentrations at which a person can detect an odor vary from person to person, and these thresholds also vary relative to the concentration that can produce immediate, injurious effects.

Most industrial chemicals and chemical warfare agents are readily detectable by smell. Soldiers in World Wars I and II were taught to identify, by smell, such agents as mustard, phosgene, and chlorine, and this detection method proved effective for determining when to put on and take off a gas mask. An exception is the chemical warfare agent Sarin, which is odorless and colorless in its pure form and, therefore, imperceptible. Japanese terrorists used Sarin when they attacked morning rush-hour commuters on several Tokyo subway trains in 1995. Twelve people died and hundreds more were injured, many of them permanently. Two years ago, police in London foiled a plot to release Sarin among the spectators of a major soccer game.

Among the most common toxic industrial chemicals, carbon monoxide is one of the few that is imperceptible. Because it is odorless and colorless, it causes many deaths in buildings each year.

Biological agents are also imperceptible and there are no detection devices that can determine their presence in the air in real time, which is why a bioterrorist attack could be so devastating. Current methods for detecting bacterial spores, such as anthrax, require a trained operator and expensive equipment. It is not currently possible to base protective responses to biological agents on detection.

Chemical incidents are characterized by the rapid onset (minutes to hours) of medical symptoms and easily observed indicators (e.g., colored residue, dead foliage, pungent odor, and dead animals, birds, fish, or insects).

In the case of a biological incident, the onset of symptoms takes days to weeks and, typically, there will be no characteristic indicators. Because of the delayed onset of symptoms in a biological incident, the area affected may be greater due to the migration of infected individuals.

Hazardous-Materials Incidents

From industrial chemicals and toxic waste to household detergents and air fresheners, hazardous materials are part of our everyday lives. Affecting urban, suburban and rural areas, hazardous-materials incidents can range from a chemical spill on a highway to groundwater contamination by naturally occurring methane gas.

Hazardous materials are substances that, because of their chemical nature, pose a potential risk to life, health or property if they are released. Hazards can exist during production, storage, transportation, use or disposal.

Chemical plants are one source of hazardous materials, but there are many others. Your local service station stores gasoline and diesel fuel, hospitals store a range of radioactive and flammable materials, and there are about 30,000 hazardous-materials waste sites in the country.

Many communities have Local Emergency Planning Committees (LEPCs) that identify industrial hazardous materials and keep the community informed of the potential risk. All companies that have hazardous chemicals must report annually to the LEPC. The public is encouraged to participate in the process. Contact your local emergency management office to find out if your community has an LEPC and how you can participate.

What to Do Before a Hazardous-Materials Incident

1. Ask your fire or police department about warning procedures. These could include the following:
   - Outdoor warning sirens or horns
   - Emergency Alert System (EAS)—Information provided via radio and television

- "All-Call" telephoning—An automated system for sending recorded messages
- News media—Radio, television and cable
- Residential route alerting—Messages announced to neighborhoods from vehicles equipped with public-address systems

2. Ask your LEPC or emergency management office about community plans for responding to a hazardous-materials accident at a plant or other facility, or a transportation accident involving hazardous materials.
3. Ask your LEPC about storage and usage of hazardous chemicals in your local area.
4. Use the information gathered from LEPC and local emergency management offices to evaluate risks to your household. Determine how close you are to factories, freeways, or railroads that may produce or transport toxic waste.
5. Be prepared to evacuate. An evacuation could last for a few hours or several days. See the "Evacuation" and "Emergency Planning and Disaster Supplies" sections for important information.
6. Be prepared to shelter in place; that is, to seek safety in your home or any other building you might be in at the time of a chemical release. At home you should select a room to be used as a shelter. The shelter room for use in case of a hazardous-material incident should be above ground, large enough to accommodate all household members and pets, and should have the fewest possible exterior doors and windows. You should also assemble a shelter kit to be used to seal the shelter room during a chemical release. The kit should include plastic sheeting, duct tape, scissors, a towel, and modeling clay or other material to stuff into cracks.

What to Do During a Hazardous-Materials Incident

1. If you witness (or smell) a hazardous materials accident, call 911, your local emergency notification number or the fire department as soon as safely possible.
2. If you hear a warning signal, listen to local radio or television stations for further information. Follow instructions carefully.
3. Stay away from the incident site to minimize the risk of contamination.
4. If you are caught outside during an incident, remember that gases and mists are generally heavier than air. Try to stay upstream, uphill and upwind—hazardous materials can quickly be transported by water and wind. In general, try to go at least one-half mile (10 city blocks) from the danger area; for many incidents you will need to go much further.
5. If you are in a motor vehicle, stop and seek shelter in a permanent building if possible. If you must remain in your car, keep car windows and vents closed, and shut off the air conditioner and heater.

6. If asked to evacuate your home, do so immediately:
   - If authorities indicate there is enough time, close all windows, shut vents and turn off attic, heating and air-conditioning fans to minimize contamination.
   - See the "Evacuation" section for more information.
7. If you are requested to stay indoors—shelter in place—rather than evacuate:
   - Follow all instructions given by emergency authorities.
   - Get household members and pets inside as quickly as possible.
   - Close and lock all exterior doors and windows. Close vents, fireplace dampers, and as many interior doors as possible.
   - Turn off air conditioners and ventilation systems. In large buildings, the building superintendent should set all ventilation systems to 100-percent recirculation, so that no outside air is drawn into the building. If this is not possible, ventilation systems should be turned off.
   - Go into the pre-selected shelter room (the above-ground room with the fewest openings to the outside). Take a battery-powered radio, water, sanitary supplies, a flashlight, and the shelter kit containing plastic sheeting, duct tape, scissors, a towel, and modeling clay or other materials to stuff into cracks.
   - Close doors and windows in the room. Stuff a towel tightly under each door and tape around the sides and top of the door. Cover each window and vent in the room with a single piece of plastic sheeting, taping all around the edges of the sheeting to provide a continuous seal. If there are any cracks or holes in the room, such as those around pipes entering a bathroom, fill them with modeling clay or other similar material.
   - Remain in the room, listening to emergency broadcasts on the radio, until authorities advise you to leave your shelter.
   - If authorities warn of the possibility of an outdoor explosion, close all drapes, curtains, and shades in the room. Stay away from windows to prevent injury from breaking glass.
   - When authorities advise people in your area to leave their shelters, open all doors and windows and turn on air-conditioning and ventilation systems. These measures will flush out any chemicals that infiltrated into the building.
   - See the "Shelter" section for more information.
8. Schools and other public buildings may institute procedures to shelter in place. If there is a hazardous materials incident and your children are at school, you will probably not be permitted to drive to the school to pick up your children. Even if you go to the school, the doors will probably be locked to keep your children safe. Follow the directions of your local emergency officials.

9. Avoid contact with spilled liquids, airborne mists or condensed solid chemical deposits. Keep your body fully covered to provide some protection. Wear gloves, socks, shoes, pants and long-sleeved shirts.
10. Do not eat or drink food or water that may have been contaminated.
11. If indoors, fill the bathtub (first sterilize it with a diluted bleach solution—one-part bleach to ten parts water) and large containers with water for drinking, cooking, and dishwashing. Be prepared to turn off the main water intake valve in case authorities advise you to do so.

What to Do After an Incident

1. Do not return home until local authorities say it is safe.
2. Upon returning home, open windows, vents, and turn on fans to provide ventilation.
3. A person or item that has been exposed to a hazardous chemical may be contaminated and could contaminate other people or items. If you have come in contact with or have been exposed to hazardous chemicals, you should:
   - Follow decontamination instructions from local authorities. (Depending on the chemical, you may be advised to take a thorough shower, or you may be advised to stay away from water and follow another procedure.)
   - Seek medical treatment for unusual symptoms as soon as possible.
   - If medical help is not immediately available and you think you might be contaminated, remove all of your clothing and shower thoroughly (unless local authorities say the chemical is water reactive and advise you to do otherwise). Change into fresh, loose clothing and seek medical help as soon as possible.
   - Place exposed clothing and shoes in tightly sealed containers. Do not allow them to contact other materials. Call local authorities to find out about proper disposal.
   - Advise everyone who comes in contact with you that you may have been exposed to a toxic substance.
4. Find out from local authorities how to clean up your land and property.
5. Report any lingering vapors or other hazards to your local emergency services office.
6. See the "Recovering from Disaster" and "Shelter" sections for more information.

Household Chemical Emergencies

Nearly every household uses products containing hazardous materials. Although the risk of a chemical accident is slight, knowing how to handle these products and how to react during an emergency can reduce the risk of injury.

How to Prepare for Household Chemical Emergencies

1. Contact agencies with expertise on hazardous household materials, such as your local public health department or the Environmental Protection Agency, for information about potentially dangerous household products and their antidotes. Ask about the advisability of maintaining antidotes in your home for cleaners and germicides, deodorizers, detergents, drain and bowl cleaners, gases, home medications, laundry bleaches, liquid fuels, paint removers and thinners.
2. Follow instructions on the product label for proper disposal of chemicals. Proper disposal will ensure environmental and public health, as well as household well-being. If you have additional questions on chemical disposal, call your local environmental or recycling agency.
   - Small amounts of the following products can be safely poured down the drain with plenty of water: bathroom and glass cleaner, bleach, drain cleaner, household disinfectant, laundry and dishwashing detergent, rubbing alcohol, rug and upholstery cleaner, and toilet bowl cleaner.
   - Small amounts of the following products should be disposed by wrapping the container in newspaper and plastic and placing it in the trash: brake fluid, car wax or polish, dish and laundry soap, fertilizer, furniture and floor polish, insect repellent, nail polish, oven cleaner, paint thinners and strippers, pesticides, powder cleansers, water-based paint and wood preservatives.
   - Dispose of the following products at a recycling center or a collection site: kerosene, motor or fuel oil, car battery or battery acid, diesel fuel, transmission fluid, large amounts of paint, thinner or stripper, power-steering fluid, turpentine, gun cleaning solvents and tires.
   - Empty spray cans completely before placing in the trash. Do not place spray cans into a burning barrel, incinerator, or trash compactor because they may explode.
   - Flush outdated and unused medicines down the toilet and place the empty container in the trash. Outdated medicines can cause ill effects. Flushing will eliminate the risk of people or animals picking them out of the trash.
3. Read directions before using a new chemical product, and be sure to store household chemicals according to the instructions on the label.
4. Store chemicals in a safe, secure location, preferably up high and always out of the reach of children.

5. Avoid mixing household chemical products. Deadly toxic fumes can result from the mixture of chemicals, such as chlorine bleach and ammonia.
6. Never smoke while using household chemicals. Avoid using hair spray, cleaning solutions, paint products, or pesticides near an open flame, pilot light, lighted candle, fireplace, wood-burning stove, etc. Although you may not be able to see or smell them, vapor particles in the air could catch fire or explode.
7. If you spill a chemical, clean it up immediately with rags. Be careful to protect your eyes and skin, wear gloves and eye protection. Allow the fumes in the rags to evaporate outdoors, and then dispose of the rags by wrapping them in a newspaper and placing them in a sealed plastic bag in your trashcan.
8. Buy only as much of a chemical as you think you will use. If you have product left over, try to give it to someone who will use it. Storing hazardous chemicals increases the risk of chemical emergencies.
9. Keep an A-B-C rated fire extinguisher in the home and car, and get training from your local fire department on how to use it.
10. Post the number of the nearest poison control center by all telephones. In an emergency situation you may not have time to look up critical phone numbers.
11. Learn to detect hazardous materials. Many hazardous materials do not have a taste or an odor, and some can be detected because they cause physical reactions, such as watering eyes or nausea. Other hazardous materials exist beneath the ground and can be recognized by an oil or foam-like appearance.
12. Learn to recognize the symptoms of toxic poisoning:
    - Difficulty breathing
    - Irritation of the eyes, skin, throat or respiratory tract
    - Changes in skin color
    - Headache or blurred vision
    - Dizziness
    - Clumsiness or lack of coordination
    - Cramps or diarrhea

What to Do During a Household Chemical Emergency

1. If your child should eat or drink a non-food substance, find any containers immediately and take them to the phone. Medical professionals may need specific information from the container to give you the best emergency advice.

2. Call the poison control center, emergency medical services (EMS), 911, hospital emergency room, county health department, fire department or your local pharmacy. They will give you emergency advice while you wait for professional help. You should have such numbers on hand for easy access and use.
3. Follow the emergency operator or dispatcher's instructions carefully. The first-aid advice found on containers may not be appropriate. Do not give anything by mouth until medical professionals have advised you.
4. Take immediate action if the chemical gets into the eyes. Delaying first aid can greatly increase the likelihood of injury. Flush the eye with clear, water for a minimum of 15 minutes, unless authorities instruct you not to use water on the particular chemical involved. Continue the cleansing process even if the victim indicates he or she is no longer feeling any pain, and then seek medical attention.
5. Get out of the residence immediately if there is danger of a fire or explosion. Do not waste time collecting items or calling the fire department when you are in danger.
6. If there is a fire or explosion, call the fire department from outside (a cellular phone or a neighbor's phone) once you are safely away from danger.
7. Stay upwind and away from the residence to avoid breathing toxic fumes.
8. Wash hands, arms, or other exposed body parts that may have been exposed to the chemical. Chemicals may continue to irritate the skin until they are washed off.
9. Discard clothing that may have been contaminated. Some chemicals may not wash out completely. Discarding clothes will prevent potential future exposure.
10. Administer first-aid treatment to victims of chemical burns:
    - Call 911 for emergency help.
    - Remove clothing and jewelry from around the injury.
    - Pour clean, cool water over the burn for 15 to 30 minutes.
    - Loosely cover the burn with a sterile or clean dressing. Be sure that the dressing will not stick to the burn.
    - Refer the victim to a medical professional for further treatment.

Nuclear Power Plants

Nuclear power plants operate in most states in the country and produce about 20 percent of the nation's power. Nearly three million Americans live within 10 miles of an operating nuclear power plant. Although the construction and operation of these facilities are closely monitored and regulated by the Nuclear Regulatory Commission (NRC), accidents at these plants are possible. An accident could result in dangerous levels of radiation that could affect the health and safety of the public living near the nuclear power plant.

Local and state governments, federal agencies and the electric utilities have emergency response plans in the event of a nuclear power plant incident. The plans define two "emergency planning zones." One covers an area within a 10-mile radius of the plant where it is possible that people could be harmed by direct radiation exposure. The second zone covers a broader area, usually up to a 50-mile radius from the plant, where radioactive materials could contaminate water supplies, food crops, and livestock.

Understanding Radiation

Radioactive materials are composed of atoms that are unstable. An unstable atom gives off its excess energy until it becomes stable. The energy emitted is radiation. Each of us is exposed to radiation daily from natural sources, including the sun and earth. Small traces of radiation are present in food and water. Radiation also is released from man-made sources such as X-ray machines, television sets, and microwave ovens. Nuclear power plants use the heat generated from nuclear fission in a contained environment to convert water to steam, which powers generators to produce electricity.

Radiation has a cumulative effect. The longer a person is exposed to radiation, the greater the risk. A high exposure to radiation can cause serious illness or death. The potential danger from an accident at a nuclear power plant is exposure to radiation. This exposure could come from the release of radioactive material from the plant into the environment, usually characterized by a plume (cloud-like) formation of radioactive gases and particles. The area that the radioactive release may affect is determined by the amount released from the plant, wind direction and speed, and weather conditions. The major hazards to people in the vicinity of the plume are radiation exposure to the body from the cloud and particles deposited on the ground, inhalation of radioactive materials and ingestion of radioactive materials.

If an accident at a nuclear power plant were to release radiation in your area, local authorities would activate warning sirens or another approved alerting method. They would also instruct you through the EAS on local television and radio stations on how to protect yourself.

The three ways to minimize radiation exposure are distance, shielding and time:

- *Distance.* The more distance between you and the source of the radiation the better. In a serious nuclear power plant accident, local authorities will call for an evacuation to increase the distance between you and the radiation.
- *Shielding.* As with distance, the more heavy, dense material between you and the source of the radiation the better. This is why local authorities could advise you to remain indoors if an accident occurs at a nearby nuclear

power plant. In some cases, the walls in your home would be sufficient shielding to protect you.

- *Time.* Usually, radioactivity loses its strength fairly quickly. In a nuclear power plant accident, local authorities will monitor any release of radiation and determine when the threat has passed.

What to Do Before a Nuclear Power Plant Emergency

1. Know the terms used to describe a nuclear emergency:
   - *Notification of Unusual Event*—A small problem has occurred at the plant. No radiation leak is expected. Federal, state and county officials will be told right away. No action on your part will be necessary.
   - *Alert*—A small problem has occurred, and small amounts of radiation could leak inside the plant. This will not affect you. You should not have to do anything.
   - *Site Area Emergency*—A more serious problem. Small amounts of radiation could leak from the plant. If necessary, state and county officials will act to ensure public safety. Area sirens may be sounded. Listen to your radio or television for safety information.
   - General Emergency—The most serious problem. Radiation could leak outside the plant and off the plant site. The sirens will sound. Tune to your local radio or television station for reports. State and county officials will act to protect the public. Be prepared to follow instructions promptly.
2. Learn your community's warning system. Nuclear power plants are required to install sirens and other warning systems (flash warning lights) to cover a 10-mile area around the plant.
   - Find out when the warning systems will be tested next.
   - When tested in your area, determine whether you can hear and/or see sirens and flash warning lights from your home.
3. Obtain public emergency information materials from the power company that operates your local nuclear power plant or your local emergency services office. If you live within 10 miles of the power plant, you should receive these materials yearly from the power company or your state or local government.
4. Learn the emergency plans for schools, day-care centers, nursing homes and other places where members of your household frequent. Learn where people would go in case of evacuation. Stay tuned to your local radio and television stations.
5. Be prepared to evacuate:
   - Prepare an emergency evacuation supply kit (see the "Emergency Planning and Disaster Supplies" section).

- Consider your transportation options. If you do not own or drive a car, ask your local emergency manager about plans for people without private vehicles. (See the "Evacuation" section for important details.)

What to Do During a Nuclear Power Plant Emergency

1. Listen to the warning. Not all incidents result in the release of radiation. The incident could be contained inside the plant and pose no danger to the public.
2. Stay tuned to local radio or television. Local authorities will provide specific instructions and information:
   - The advice given will depend on the nature of the emergency, how quickly it is evolving and how much radiation, if any, is likely to be released.
   - Local instructions should take precedence over any advice given in this handbook.
   - Review the public information materials you received from the power company or government officials.
3. Evacuate if you are advised to do so:
   - Close and lock doors and windows.
   - Keep car windows and vents closed; use re-circulating air.
   - Listen to radio for evacuation routes and other instructions.
   - See the "Evacuation" section for important details.
4. If you are not advised to evacuate, remain indoors:
   - Close doors and windows.
   - Turn off the air conditioner, ventilation fans, furnace and other air intakes.
   - Go to a basement or other underground area, if possible.
   - Keep a battery-powered radio with you at all times.
5. Shelter livestock and give them stored feed, if time permits.
6. Do not use the telephone unless absolutely necessary. Lines will be needed for emergency calls.
7. If you suspect exposure, take a thorough shower:
   - Change clothes and shoes.
   - Put exposed clothing in a plastic bag.
   - Seal the bag and place it out of the way.
8. Put food in covered containers or in the refrigerator. Food not previously covered should be washed before being put in containers.

What to Do After a Nuclear Power Plant Emergency

1. If told to evacuate, do not return home until local authorities say it is safe.

2. If you were advised to stay in your home, do not go outside until local authorities indicate it is safe.
3. Seek medical treatment for any unusual symptoms, such as nausea, that may be related to radiation exposure.
4. See the "Shelter" and "Recovering from Disaster" sections for more information.

Recovering from a Disaster

Here is some general advice on steps to take after disaster strikes to begin putting your home, your community, and your life back to normal.

Health and Safety

Your first concern after a disaster is your household's health and safety.

1. Be aware of new hazards created by the disaster. Watch for washed-out roads, contaminated buildings, contaminated water, gas leaks, broken glass, damaged wires and slippery floors.
2. Be aware of exhaustion. Don't try to do too much at once. Set priorities and pace yourself.
3. Drink plenty of clean water. Eat well and get enough rest.
4. Wear sturdy work boots and gloves. Wash your hands thoroughly and often with soap and clean water when working in debris.
5. Inform local authorities about health and safety hazards, including chemical releases, downed power lines, washed-out roads, smoldering insulation or dead animals.

Returning to a Damaged Home

Returning to a damaged home can be both physically and mentally challenging. Above all, use caution.

1. Keep a battery-powered radio with you so you can listen for emergency updates.
2. Wear sturdy work boots and gloves.
3. Before going inside, walk carefully around the outside of your home and check for loose power lines, gas leaks and structural damage. If you smell gas, do not enter the home and leave immediately. Do not enter if floodwaters remain around the building. If you have any doubts about safety, have your home inspected by a professional before entering.
4. If your home was damaged by fire, do not enter until authorities say it is safe.

5. Check for cracks in the roof, foundation and chimneys. If it looks like the building may collapse, leave immediately.
6. A battery-powered flashlight is the best source of light for inspecting a damaged home.

   CAUTION: The flashlight should be turned on outside before entering a damaged home—the battery may produce a spark that could ignite leaking gas, if present.
7. Do not use oil, gas lanterns, candles or flashlights for lighting inside a damaged home. Leaking gas or other flammable materials may be present. Do not smoke. Do not turn on the lights until you're sure they're safe to use.
8. Enter the home carefully and check for damage. Be aware of loose boards and slippery floors.
9. Watch out for animals, especially poisonous snakes. Use a stick to poke through debris.
10. If you smell gas or hear a hissing or blowing sound, open a window and leave immediately. Turn off the main gas valve from the outside, if you can. Call the gas company from a neighbor's residence. If you shut off the gas supply at the main valve, you will need a professional to turn it back on.
11. Check the electrical system where visible and accessible. If you see sparks, broken or frayed wires, or if you smell hot insulation, turn off the electricity at the main fuse box or circuit breaker. If, however, you are wet, standing in water or unsure of your safety, do not touch anything electrical. Instead, leave the building and call for help.
12. Check appliances. If appliances are wet, turn off the electricity at the main fuse box or circuit breaker. Then unplug appliances and let them dry out. Have appliances checked by a professional before using them again. Also have the electrical system checked by an electrician before turning the power back on.
13. Check the water and sewage systems. If pipes are damaged, turn off the main water valve.
14. Clean up spilled medicines, bleaches and gasoline. Open cabinets carefully. Be aware of objects that may fall.
15. Try to protect your home from further damage. Open windows and doors to get air moving through.
16. Clean and disinfect everything that got wet. Mud left behind by floodwaters can contain sewage and chemicals.
17. If your basement has flooded, pump it out gradually (about one-third of the water per day) to avoid damage. The walls may collapse and the floor may buckle if the basement is pumped out while the surrounding ground is still waterlogged.

18. Check with local authorities before using any water; it could be contaminated. Wells should be pumped out and the water tested by authorities before drinking.
19. Throw out fresh food, cosmetics, and medicines that have come into contact with floodwaters.
20. Check refrigerated food for spoilage—your power supply may have been disrupted during the emergency. Throw out all spoiled food and any food that you suspect might be spoiled.
21. Call your insurance agent. Take pictures of damage. Keep good records of repair and cleaning costs.

Getting Disaster Assistance

Throughout the recovery period, it's important to monitor local radio or television reports and other media sources for information about where to get emergency housing, food, first aid, clothing, and financial assistance. Following is general information about the kinds of assistance that may be available.

Direct assistance to individuals and families may come from any number of organizations. The American Red Cross is often stationed right at the scene to help people with their most immediate medical, food and housing needs. Other voluntary organizations, such as the Salvation Army, may also provide food, shelter and supplies, and assist in cleanup efforts. Church groups and synagogues are often involved as well. In addition, social service agencies from local or state governments may be available to help people in shelters or provide direct assistance to families.

In the most severe disasters, the federal government is also called in to help individuals and families with temporary housing, counseling (for post-disaster trauma), low-interest loans and grants, and other assistance.

Small businesses and farmers are also eligible. Most federal assistance becomes available when the president of the United States declares a "Major Disaster" for the affected area at the request of a state governor. When this happens, FEMA may establish a Disaster Recovery Center (DRC). A DRC is a facility established in, or near to, the community affected by the disaster, where people can meet face-to-face with represented federal, state, local, and volunteer agencies to:

- Discuss their disaster-related needs
- Obtain information about disaster assistance programs
- Teleregister for assistance
- Update registration information
- Learn about measures for rebuilding that can eliminate or reduce the risk of future loss

- Learn how to complete the Small Business Administration (SBA) loan application, which is also the form used to qualify all individuals for low-cost loans or grants, including those for repair or replacement of damaged homes and furnishings
- Request the status of their Disaster Housing Application

People can apply for assistance by telephone without going to a DRC by dialing 1-800-621-FEMA (3362).

Mental Health and Crisis Counseling

The emotional toll that a disaster brings can sometimes be even more devastating than the financial strains of damage and loss of home, business or personal property. Children and the elderly are special concerns in the aftermath of disasters. Even individuals who experience a disaster "second hand" through exposure to extensive media coverage can be affected.

Crisis-counseling programs often include community outreach, consultation, and education. FEMA and the state and local governments of the affected area may provide crisis-counseling assistance to help people cope with and recover from disaster. If you feel you need assistance, get help.

Coping with Disasters

You need to be aware of signs that you need help in coping with the stress of a disaster.

1. Things to remember when trying to understand disaster events:
   - No one who sees a disaster is untouched by it.
   - It is normal to feel anxious about your own safety and that of your family and close friends.
   - Profound sadness, grief and anger are normal reactions to an abnormal event.
   - Acknowledging your feelings helps you recover.
   - Focusing on your strengths and abilities will help you to heal.
   - Accepting help from community programs and resources is healthy.
   - We each have different needs and different ways of coping.
   - It is common to want to strike back at people who have caused great pain, but nothing good is accomplished by hateful language or actions.
2. Signs that adults need crisis-counseling/stress-management assistance:
   - Difficulty communicating thoughts
   - Difficulty sleeping
   - Difficulty maintaining balance

- Easily frustrated
- Increased use of drugs/alcohol
- Limited attention span
- Poor work performance
- Headaches/stomach problems
- Tunnel vision/muffled hearing
- Colds or flu-like symptoms
- Disorientation or confusion
- Difficulty concentrating
- Reluctance to leave home
- Depression, sadness
- Feelings of hopelessness
- Mood-swings and crying easily
- Overwhelming guilt and self-doubt
- Fear of crowds, strangers, or being alone.

3. Ways to ease disaster-related stress:
   - Talk with someone about your feelings—anger, sorrow, and other emotions—even though it may be difficult.
   - Seek help from professional counselors who deal with post-disaster stress.
   - Don't hold yourself responsible for the disastrous event or be frustrated because you feel that you cannot help directly in the rescue work.
   - Take steps to promote your own physical and emotional healing by staying active in your daily life patterns or by adjusting them. This healthy outlook will help you and your household (e.g., healthy eating, rest, exercise, relaxation, meditation).
   - Maintain a normal household and daily routine, limiting demanding responsibilities of yourself and your household.
   - Spend time with family and friends.
   - Participate in memorials, rituals, and use of symbols as a way to express feelings.
   - Use existing support groups of family, friends and church.
   - Establish a family emergency plan. Feeling there is something you can do, can be very comforting.

Helping Children Cope with a Disaster

Disasters can leave children feeling frightened, confused and insecure. Whether a child has personally experienced trauma, has merely seen the event on television or heard it discussed by adults, it is important for parents and teachers to be informed and ready to help if reactions to stress begin to occur.

Children respond to trauma in many different ways. Some may have reactions very soon after the event; others may seem to be doing fine for weeks or months and then begin to show worrisome behavior. Knowing the signs that are common at different ages can help parents and teachers recognize problems and respond appropriately.

Reassurance is the key to helping children through a traumatic time. Very young children need a lot of cuddling, as well as verbal support. Answer questions about the disaster honestly, but don't dwell on frightening details or allow the subject to dominate family or classroom time indefinitely. Encourage children of all ages to express emotions through conversation, drawing or painting and to find a way to help others who were affected by the disaster. Also, limit the amount of disaster-related material (television, etc.) your children are seeing or hearing and pay careful attention to how graphic it is. Try to maintain a normal household or classroom routine and encourage children to participate in recreational activity. Reduce your expectations temporarily about performance in school or at home, perhaps by substituting less-demanding responsibilities for normal chores.

Additional information about how to communicate with children can be found on the FEMA for Kids website at www.fema.gov/kids.

Helping Others

The compassion and generosity of the American people is never more evident than after a disaster. People want to help. Here are some general guidelines on helping others after a disaster.

1. In addition to the people you care for on a day-to-day basis, consider the needs of your neighbors and people with special needs.
2. If you want to volunteer, check with local organizations or listen to local news reports for information about where volunteers are needed. Until volunteers are specifically requested, stay away from disaster areas.
3. If you are needed in a disaster area, bring your own food, water and emergency supplies. This is especially important in cases where a large area has been affected and emergency items are in short supply.
4. Do not drop off food, clothing or any other item to a government agency or disaster relief organization unless a particular item has been requested. Normally these organizations do not have the resources to sort through the donated items.
5. You can give a check or money order to a recognized disaster relief organization. These groups are organized to process checks, purchase what is needed and get it to the people who need it most.

6. If your company wants to donate emergency supplies, donate a quantity of a given item or class of items (such as nonperishable food) rather than a mix of different items. Also, determine where your donation is going, how it's going to get there, who is going to unload it and how it's going to be distributed. Without sufficient planning, much needed supplies will be left unused.

Chapter 10

Staying Safe at School, Work, and Play

SAFE SCHOOLS

How Safe Is Your Child's School?

Every weekday a quarter of the population of the United States is in school, either as students, teachers, or ancillary staff. Thankfully, the vast majority of them are safe.

On average there is one shooting event at a school somewhere in the United States but the overwhelming majority of these incidents involve just one person and in most of these cases, it is a successful or failed suicide. Many of the other cases involve the accidental discharge of a firearm with no injuries.

In the last twenty years there have been five mass casualty shootings in schools—Columbine High School in April 1999 with fifteen fatalities, Virginia Tech in April 2017 with thirty-three fatalities, Sandy Hook Elementary in December 2012 with twenty-eight fatalities, Umpqua Community College in October 2015 with ten fatalities, and Parkland in February 2018 with seventeen fatalities.

To date, no US school has been the target of an international terrorist attack. However, schools could be considered viable terrorist targets for the following reasons:

- An attack at a school would instill fear and panic nationwide. Consider the effects of the deadly student attacks at Columbine and Red Lake High Schools.
- An attack on a school could promote the reputation and power of a terrorist/ group.

- Schools provide an essential community service. About one-quarter of the nation's population attends school each school day.
- Domestic terrorists have previously attacked schools and education facilities.
- An attack at a school would warrant national media coverage.
- Schools symbolize America and America's future.
- Many schools remain easy targets due to their accessibility, vulnerability, and prominence in the community.

Although the likelihood of a direct attack against a school is very low, recent terrorist events have increased interest in the vulnerability of all types of buildings to CBR threats. Of particular concern are building HVAC systems, because they can become an entry point and distribution system for airborne hazardous contaminants.

Even without special protective systems, buildings can provide protection in varying degrees against airborne hazards that originate outdoors. Conversely, the hazards produced by a release inside a building can be much more severe than a similar release outdoors. Because buildings allow only a limited exchange of air between indoors and outdoors, not only can higher concentrations occur when there is a release inside, but hazards may persist longer indoors.

In an ideal world, one would like to prevent all school crime, violence, and bullying and ensure the safety of each staff member and student. The reality is so many things are outside the control of the school that such a task is virtually impossible. However, there are many measures and steps that can be taken to make our schools safer.

Ensuring School Safety

Local law enforcement agencies have been asked to play an increasingly prominent role in ensuring school safety and following Parkland, there is currently a debate about whether teachers should be armed on campus.

Youth violence, gangs, and drugs are important issues not only for federal and local policy makers, but for school administrators, parents, and students as well. The community policing philosophy has become a critical element to making schools safer in the twenty-first century. State and local law enforcement agencies are bringing community policing to schools:

- Hiring school resource officers (SROs) who utilize a community oriented policing philosophy.
- Engaging in student training programs proven to reduce violence and drug use and reinforce student conflict resolution skills.

- Developing early warning systems that can alert school administrators and SROs to potentially problematic students and geographic areas within schools.
- Using problem-solving skills to develop innovative solutions to local school problems, including gangs, drugs, loitering, disorderly conduct, fights, larceny, and vandalism.
- Collecting and analyzing data using a wide variety of traditional and non-traditional data sources (such as suspensions, dropout rates, calls for service, GPA, attendance, and parent and counselor information) to study school safety problems.
- Developing partnerships with parents, local businesses, school bus drivers, school officials, and students to develop outcome-based solutions to school safety problems.
- Mentoring students and exposing them to positive role models.

The US Department of Justice's Community Oriented Policing Services (COPS) has played an instrumental role in supplying more than 2,300 law enforcement agencies with over 4,900 School Resource Officers through its popular COPS in Schools (CIS) program. This program is designed to assist law enforcement agencies in hiring new, additional SROs to engage in community policing in and around primary and secondary schools. This program provides an incentive for law enforcement agencies to build collaborative partnerships with the school community and to use community policing to combat school violence. These officers have helped to improve school safety and student development around the country.

In addition, through the School Based Partnership Program (SBP) the COPS office has provided funding specifically to develop problem-solving partnerships between local police, schools, and community-based organizations. These partnerships focus on developing cooperative solutions to specific crime problems through the use of the SARA problem-solving model.

The COPS Office, along with other US Department of Justice components, recognizes the challenges local law enforcement and schools face when confronting school safety issues. The federal government is therefore involved in providing Guides & Reports and Training & Technical Assistance to community members, school officials, and local law enforcement that are interested in increasing school safety. The COPS Office has also provided Links to Other Resources to assist in this important effort.

The aim of the program is to identify all possible threats, develop a plan to prevent them, or mitigate their effects in order to get back to normal as quickly as possible after an incident. The components of the program are listed below.

MITIGATION AND PREVENTION

The goal of mitigation is to decrease the need for response as opposed to simply increasing response capability.

- Connect with community emergency responders to identify local hazards.
- Review the last safety audit to examine school buildings and grounds.
- Determine who is responsible for overseeing violence prevention strategies in your school.
- Encourage staff to provide input and feedback during the crisis planning process.
- Review incident data.
- Determine major problems in your school with regard to student crime and violence.
- Assess how the school addresses these problems.
- Conduct an assessment to determine how these problems—as well as others—may impact your vulnerability to certain crises.

Preparedness

Good planning will facilitate a rapid, coordinated, effective response when a crisis occurs.

- Determine what crisis plans exist in the district, school, and community.
- Identify all stakeholders involved in crisis planning.
- Develop procedures for communicating with staff, students, families, and the media.
- Establish procedures to account for students during a crisis.
- Gather information about the school facility, such as maps and the location of utility shutoffs.
- Identify the necessary equipment that needs to be assembled to assist staff in a crisis.

RESPONSE

A crisis is the time to follow the crisis plan and make use of your preparations.

- Determine if a crisis is occurring.
- Identify the type of crisis that is occurring and determine the appropriate response.
- Activate the incident management system.

- Ascertain whether an evacuation, reverse evacuation, lockdown, or shelter-in-place needs to be implemented.
- Maintain communication among all relevant staff at officially designated locations.
- Establish what information needs to be communicated to staff, students, families, and the community.
- Monitor how emergency first aid is being administered to the injured.
- Decide if more equipment and supplies are needed.

RECOVERY

During recovery, return to learning and restore the infrastructure as quickly as possible.

- Strive to return to learning as quickly as possible.
- Restore the physical plant as well as the school community.
- Monitor how staff members are assessing students for the emotional impact of the crisis.
- Identify what follow-up interventions are available to students, staff, and first responders.
- Conduct debriefings with staff and first responders.
- Assess curricular activities that address the crisis.
- Allocate appropriate time for recovery.
- Plan how anniversaries of events will be commemorated.
- Capture "lessons learned" and incorporate them into revisions and trainings.

There are many things that parents and students can do to make both schools and getting to and from them safer.

TRAVELING TO AND FROM SCHOOL

Riding the Bus

The school day begins and ends with a trip on a school bus for some 22 million students nationwide. Unfortunately, each year many children are injured and several are killed in school bus incidents.

However, school bus transportation is safe. In fact, buses are safer than cars! Even so, last year, some 26 students were killed and another 9,000 were injured in incidents involving school buses.

Over the past six years, about 70 percent of the deaths in fatal school bus-related crashes were occupants of vehicles other than the school bus and 20 percent were pedestrians. About 4 percent were school bus passengers and 2 percent were school bus drivers. Of the pedestrians killed in school bus related crashes over this period, approximately 77 percent were struck by the school bus. Of the people injured in school bus related crashes from 1994 through 1999, about 44 percent were school bus passengers, 9 percent were school bus drivers, and another 43 percent were occupants of other vehicles.

Many of the accidents occurred as the pupils were entering and exiting the bus. Although drivers of all vehicles are required to stop for a school bus when it is stopped to load or unload passengers, children should not rely on them to do so. The National Safety Council encourages parents to teach their children these rules for getting on and off the school bus.

RULES FOR GETTING ON AND OFF THE SCHOOL BUS

Getting on the School Bus

- Have a safe place to wait for your bus.
- Stay away from traffic and avoid roughhousing or other behavior that can lead to carelessness. Do not stray onto streets, alleys, or private property.
- Line up away from the street or road as the school bus approaches.
- Stay away from the bus until it comes to a complete stop, the doors have opened, and the driver signals you to enter.
- When being dropped off, exit the bus and walk ten giant steps away from the bus. Keep a safe distance between you and the bus. Also, remember that the bus driver can see you best when you are back away from the bus.
- Use the handrail to enter and exit the bus.
- Stay away from the bus until the driver gives his/her signal that it's OK to approach.

Be aware of the street traffic around you. Drivers *are* required to follow certain rules of the road concerning school buses, however, not all do. Protect yourself and watch out!

Behavior on the Bus

- When on the bus, find a seat and sit down.
- Loud talking or other noise can distract the bus driver and is not allowed.
- Never put head, arms, or hands out of the window.
- Keep aisles clear—books or bags are tripping hazards and can block the way in an emergency.

- Before you reach your stop, get ready to leave by getting your books and belongings together.
- At your stop, wait for the bus to stop completely before getting up from your seat. Then, walk to the front door and exit using the hand rail.

Getting off the School Bus

- If you have to cross the street in front of the bus, walk at least 10 feet ahead of the bus along the side of the road, until you can turn around and see the driver.
- Make sure that the driver can see you.
- Wait for a signal from the driver before beginning to cross.
- When the driver signals, walk across the road, keeping an eye out for sudden traffic changes.
- Do not cross the center line of the road until the driver has signaled that it is safe for you to begin walking.
- Stay away from the bus's rear wheels at all times.

Correct Way to Cross the Street

- Children should always stop at the curb or the edge of the road and look left, then right, and then left again before crossing.
- They should continue looking in this manner until they are safely across.
- If students' vision is blocked by a parked car or other obstacle, they should move out to where drivers can see them and they can see other vehicles—then stop, and look left-right-left again.

Walking and Biking to School

Even if you don't ride in a motor vehicle, you still have to protect yourself. Because of minimal supervision, young pedestrians face a wide variety of decisions making situations and dangers while walking to and from school. Here are a few basic safety tips to follow:

- Mind all traffic signals and/or the crossing guard—never cross the street against a light, even if you don't see any traffic coming.
- Walk your bike through intersections.
- Walk with a buddy.
- Wear reflective material—it makes you more visible to street traffic.

Riding in a Car

- You might have heard before that most traffic crashes occur close to home—they do.
- Safety belts are the best form of protection passengers have in the event of a crash. They can lower your risk of injury by 45 percent.
- You are four times more likely to be seriously injured or killed if ejected from the vehicle in a crash.
- Everyone needs to be buckled up properly. That means older kids in seat belts, younger kids in booster seats, and little kids in child safety seats.

Safety in Schools

School safety is *everyone's* business. The best safety effort will involve all aspects of your child's school, including classroom instruction, school services, and the school climate.

If you're like many parents, the very idea of "school safety" is overwhelming and frightening. Do your kids feel safe at school? What's the school doing to keep its students safe? Use this list of questions to find out if you school could be doing more.

Getting Started

Take the checklist below with you to your next PTO meeting and review it with other parents.

Ask a group of parents to fill it out. You can then present it to the principal.

Fill it out with your child.

Send it to your local school board for its consideration.

The List

1. Is student safety a priority for your school and your community?
2. Do parents have access to reports that include information about the number of violent or other unsafe incidents at the school?
3. Does your school have procedures for responding quickly to unsafe situations?
4. Is your school addressing ways to prevent, as well as respond, to crises?
5. Are the school board, school principal, school superintendent, teachers, school staff, parents, students, and community professionals all involved in these efforts?
6. Has someone been designated to coordinate activities to maintain a safe and healthy environment?

7. Are counselors and psychologists available to work with students who are troubled or disruptive?
8. Do students in all grades participate in classes to help them develop conflict resolution and other life skills?
9. Do school health service providers help or refer students who come to them with concerns about safety?
10. Does the school emphasize promoting self-esteem and respect for others in all aspects of the school program?
11. Are parents and students involved in activities that promote school safety?
12. Are teachers encouraged to set examples of positive physical, social, and emotional health?
13. Does the school have fair, firm, consistent discipline policies?
14. Are school facilities attractive and hazard-free?
15. Is safety addressed in all aspects of the school program—the cafeteria, physical education, classrooms, playgrounds, after-school programs, and so on?
16. Do school staff, parents, and community members working on school safety collaborate?

Once you have the answers to these questions, you'll know what else your child's school should be doing.

Coping with Fear

The return to school is a time filled with both excitement and reluctance for most children. There may be an unusual kind of anxiety added to the mixture: memories and fears of schoolyard shootings. Virtually every child in the country has heard—or seen—stories about students like themselves being gunned down by classmates.

How Do Children Experience the Media's Coverage of These Events?

How children respond to memories, or news stories, of schoolyard murders will depend on their age, temperament, and experience. Some children may be openly frightened. Some may fear that it could happen to them. Some may feel distanced from the possibility of violence in their own lives. Some may be unable to grasp that this really happened to real people. Some may take a protective stance of cynicism and apathy. However children respond, all of them need help and guidance from the adults who care for them.

How Can Parents Help with Children's Fears?

Helping children feel safe about returning to school means talking with them and *listening* to them. Ask how they're feeling about school starting. Ask what they're excited about. Ask if they have any particular worries. Listen carefully for clues about their feelings.

Encourage them to tell you about anything that worries them. "Boy, the kid in that television program was really mad—do any kids you know get that angry? What would you do if you heard a kid talking about doing something violent? Or if you knew that someone in your class brought a gun to school?"

Assure them that you will be working to make their school safe.

WHAT YOU CAN DO

What Can Parents Do to Help Ensure School Safety?

Support and become involved in violence-prevention programs. There are many effective programs available, but they don't work if they're not used. Find out what your school district offers, or whether a local youth group or your religious community is interested. Then offer your time.

Keep guns out of the home. If this is not possible, keep them secure from children. The easiest place for children to get hold of weapons is in their own homes or those of family members. This was the case in all of the reported schoolyard killings.

Take the threat of children who threaten violence seriously. Again, all of the boys in recent school shootings had made verbal threats prior to acting on them. Clearly, an aggressive battle with action figures does not automatically signal horror in the making. But a kid who talks about wanting to "blow someone away" should never be dismissed. Urge school districts to increase counseling for depressed and angry students. One of the ironies of these terrible events has been the willingness of schools to mobilize counseling after the fact to help students deal with their experiences. A preventative mobilization of these resources could be much more effective, and would perhaps have averted the loss of young lives. School budgets are always under great pressure, but our children's safety is not a "frill."

The students whose schoolyards exploded with gunfire are not the only ones who were affected. When children return to school, we need to help them come to terms with their fears and do our best to prevent future violence.

Bullying

Kids are bullying each other, assaulting teachers, and bringing guns to school. What are schools doing to stop youth violence? A growing number are turning to conflict-resolution programs.

Violence prevention comes in a variety of shapes and sizes: It can be peer mediation, life-skills training, or anger management. But at its core, each program attempts to teach children one thing—how to resolve their conflicts peacefully.

One Success Story

The staff at Kenwood Heights Elementary School in urban Springfield, OH, has found a way to keep the peace. Its top-notch Kenwood Peace Program is a model for other schools. Principal Marni Gochenour began the program out of desperation: "Many of our children come from violent neighborhoods. We had a lot of assaults on teachers and other children. We literally had first-graders being led out of school in handcuffs. We were not able to teach."

Today, a peace flag flies over Kenwood Heights. In nearly every nook and cranny are reminders of the program's mission: Peace banners hang in the hallways; a scoreboard in the cafeteria records each peaceful day; and students, teachers, custodians, and parents all work together to maintain the school's award-winning gardens. Gochenour starts each day by announcing over the PA a hypothetical problem for students and teachers to solve. And classrooms are equipped with "peace tables" where students go to resolve their disagreements.

A Unique Approach

They tend to do things a little differently at Kenwood Heights. Most schools that adopt violence-prevention programs target 4th or 5th graders because that's often the age when behavior problems become unmanageable. "We chose initially to implement our program in the kindergarten and first grade," says Gochenour. "We wanted to reach kids before the problems began." Kenwood Heights also chose not to use "peer mediation," a common violence-prevention method that involves training a few students to diffuse conflicts among their peers. Instead the faculty's goal is to teach all students to be mediators. The results have been impressive: dramatic drops in fights and emergency removals; increased attendance, and more students making honor roll.

What It Takes to Be Great

Not all violence-prevention programs are as effective as Kenwood's. According to the Drug Strategies' *Safe Schools, Safe Students* study, high-quality violence-prevention programs do make a difference. But the majority of programs studied did not make the grade. Only ten out of the eighty-four reviewed received an A. The study states that to be successful, a violence-prevention program must:

- Encourage school attitudes against violence, aggression, and bullying.
- Provide teacher training.
- Create a positive school climate.
- Include families, peers, media, and the community.

A Culture of Denial

School shootings have become sickeningly routine. While school officials wrangle with security issues and psychologists offer up "profiles" to identify would-be killers, a few observers suggest a new approach. Instead of asking experts to pinpoint the motives of "kids who kill," why not examine the "culture of denial" that keeps good people from doing something about warnings they've been given by a troubled teen?

The Burden's on the Bystanders

No school metal-detector can sense feelings that may lead to ongoing cycles of violence. But since motives and personalities vary, perhaps the best violence-prevention efforts will focus not on the behavior of killers but on that of bystanders, those who hear or see signs of potential trouble, and must then choose whether or not to act.

If you see a boy who is so scared, or angry, or bullied, and he's talking about bringing a gun to school, that's enough for you to begin talking with the boy's parents, first, and then possibly the police or school guidance counselors.

Violent Kids: Warning Signs

Communities across the country have searched for reasons why some kids become violent. Major stressors can send a child over the brink: domestic violence, alcohol and drugs, and social and academic pressures.

Here are some warning signs that parents and educators should keep in mind when considering a troubled child:

- Detachment—a lack of bonding and "connectedness" to others
- Withdrawal or feelings of hopelessness
- Threats of violence, and efforts to establish the means and opportunity to carry them out
- Significant changes in behavior, appearance, and so on.

- Disciplinary problems in school, and/or delinquent, criminal activity in school or the community
- Unusual interest in, or preoccupation with, weapons, bombs, and violent entertainment (music, movies, etc.).
- Abuse of animals, suicide threats or attempts, self-mutilation, and so on.

SCHOOL SAFETY

Some Facts

Violent Deaths

According to the National Center for Education Statistics youths aged between five and nineteen were at least seventy times more likely to be murdered away from school than at school. Violent deaths in schools are tragic events that affect not only the individuals and families directly involved, but also many others in the schools and communities where they occur.

Deaths that occurred while the victim was on the way to or from regular sessions at school, or while the victim was attending or traveling to or from an official school-sponsored event, were also considered a school-associated violent death. Victims of school-associated violent deaths include students, staff members, and other non-students.

Victimization

Theft and violence at school and while going to and from school can lead to a disruptive and threatening environment, physical injury, and emotional stress, and can be an obstacle to student achievement.

Students ages between twelve and eighteen were more likely to be victims of theft at school than away from school. In 2015–2016 (the latest data available), about 840,000 thefts occurred at school, and about 545,000 occurred away from school. However, thefts, violent victimizations, and serious violent victimizations both at and away from school all declined between 1992 and 2015.

The rates of theft and serious violence at school are higher for urban and suburban students than for rural students. However, no difference is detected in the likelihood of theft and serious violent victimization away from school between students living in urban, suburban, and rural areas.

Victimization rates vary according to student characteristics. Male students are more likely than female students to report being the victim of a violent crime at school (2 vs. 1%) but no difference is detected in their likelihood of reporting theft (4% for both). Students in urban schools are more likely than their rural school counterparts to report being victims of violent crime (2 vs. 1%). Nonetheless, less than 3 percent of students report being a victim of violent crime, regardless of students' characteristic.

Sexual minority youth (those who identify as gay, lesbian, and bisexual and those who are not sure about their sexual orientation) are at a greater risk of harassment, victimization, and social isolation, compared to heterosexual youth. These experiences can lead to more depression symptoms, suicidal thoughts, and problem behaviors among sexual minority youth, as well as lower academic outcomes and increased unexcused absences from school (Burton et al. 2013; Kosciw et al. 2013; Robinson and Espelage 2011). These hostile experiences, combined with their resulting negative outcomes, were also found to be factors contributing to sexual minority youth's higher rates of substance abuse (Goldbach et al. 2014).

Injuries from Weapons

Every year, some students are threatened or injured with a weapon while they are on school property. The percentage of students victimized in this way provides an important measure of how safe our schools are and how this has changed over time.

In 2015, about 6 percent of 9th–12th-grade students reported that they had been threatened or injured with a weapon on school property during the previous twelve months. In each survey year from 1993 to 2015, a lower percentage of female students than of male students in grades 9–12 reported being threatened or injured with a weapon on school property during the previous twelve months.

The percentage of students who reported being threatened or injured with a weapon on school property differed by race/ethnicity and grade level. In 2015, lower percentages of Asian students (4%) and White students (5%) than of Black students (8%) and Pacific Islander students (20%) reported being threatened or injured with a weapon on school property during the previous twelve months. In addition, the percentage of students who reported being threatened or injured with a weapon on school property was lower for White students than for Hispanic students (5 vs. 7%). In 2015, lower percentages of 12th (4%) and 11th graders (5%) than of 9th graders (7%) reported being threatened or injured with a weapon on school property.

The likelihood of being threatened or injured with a weapon on school property also varies by student characteristics. Males are more likely than females to report being threatened or injured with a weapon on school property. For example, a recent survey found that 12 percent of male students reported being threatened or injured in the past year, compared with 6 percent of female students. Among 9th–12th-grade students, those in lower grades were more likely to be threatened or injured with a weapon on school property than were those in higher grades. Twelve percent of 9th graders reported that they were threatened or injured with a weapon on school property, compared with 9 percent of 10th graders, 7 percent of 11th graders, and 6 percent of 12th graders.

Physical Fights

The percentage of 9th–12th-grade students who reported being in a physical fight on school property has declined in recent years.

Schools at which there are numerous physical fights may not be able to maintain a focused learning environment for students. Further, students who are involved in fights on school property may have difficulty succeeding in their studies. Thirty-three percent of 9th–12th-grade students reported being in a fight anywhere, and 13 percent said they had been in a fight on school property. The percentage of students who reported being in a fight anywhere declined. Similarly, the percentages of students who reported fighting on school property in these years also declined.

The percentage of public schools that recorded violent incidents and serious violent incidents varied by school characteristics. For example, primary schools recorded lower percentages of violent incidents (53%) than middle schools (88%) and high schools and combined elementary/secondary schools (referred to as high/combined schools) (78%). Similarly, a lower percentage of primary schools recorded serious violent incidents (9%) than middle or high/combined schools (18% and 19%, respectively).

In all survey years, males were more likely than females to have been in a fight anywhere and on school property—41 percent of males said they had been in a fight anywhere, and 17 percent said they had been in a fight on school property while 25 percent of females reported they had been in a fight anywhere, and 8 percent said they had been in a fight on school property. When looking at different grade levels, students in lower grades reported being in fights more frequently than students in higher grades, both anywhere and on school property.

The percentages of students engaging in fights anywhere varied according to students' race/ethnicity. Specifically, Black and Hispanic students were more likely than White students and American Indian students were more likely than Asian and White students to report being in a fight anywhere and on school property—24 percent of American Indian students, 17 percent each of Black and Hispanic students, 13 percent of Asian students, and 10 percent of White students reported being in a fight on school property.

Bullying and Cyberbullying

Between 2005 and 2015, the percentage of students ages between twelve and eighteen who reported being bullied at school during the school year decreased from 28 to 21 percent. A higher percentage of female than of male students reported being bullied at school during the school year in 2015 (23 vs. 19%).

In 2015, a higher percentage of female than of male students ages 12–18 reported being bullied at school during the school year (23 vs. 19%), as well as being the subject of rumors (15 vs. 9%). In contrast, a higher percentage of male than of female students reported being threatened with harm (5 vs. 3%). Higher percentages of Black students (25%) and White students (22%) than of Hispanic students (17%) reported being bullied at school in 2015. The percentage of students who reported being made fun of, called names, or insulted was also higher for Black students (17%) and White students (14%) than for Hispanic students (9%). The percentage of students who reported being the subject of rumors was higher for Black students (14%), White students (13%), and Hispanic students (10%) than for Asian students (5%). A higher percentage of students in grade 6 than of students in grades eight through twelve reported being bullied at school during the school year. In 2015, about 31 percent of 6th graders reported being bullied at school, compared with 22 percent of 8th graders, 19 percent of 9th graders, 21 percent of 10th graders, 16 percent of 11th graders, and 15 percent of 12th graders. In addition, a higher percentage of 7th graders (25%) than of 11th and 12th graders reported being bullied at school. The percentage was also higher for 8th and 10th graders than for 12th graders. No measurable differences were observed in the percentage of students who reported being bullied at school by urbanicity or between those in public and private schools.

Violence and Other Incidents Reported to the Police

The survey found that 71 percent of public schools experienced one or more violent incidents, while 36 percent reported one or more such incidents to the police.

This indicator provides the percentage of schools that experienced one or more crimes and the total number of crimes reported by schools. In the School Survey on Crime and Safety, principals of public schools were asked to provide the number of violent incidents, serious violent incidents, thefts, and other incidents that occurred at their school, as well as the number of incidents that were reported to the police. Violent incidents include rape, sexual battery other than rape, physical attack or fight with or without a weapon, threat of physical attack with or without a weapon, and robbery with or without a weapon.

Figures suggest an estimated 1.5 million violent incidents occur in public elementary and secondary schools. Seventy-one percent of public schools experience one or more violent incidents and 36 percent of schools report one or more such incidents to the police. Twenty percent of schools experience one or more serious violent incidents (which are a subset of violent incidents and include rape, sexual battery other than rape, physical attack or fight with a weapon, threat of physical attack with a weapon, and robbery with or without a weapon) and 46 percent of public schools experience one or more thefts. These translate into an estimated 61,000 serious violent incidents and 218,000 thefts at public schools. When looking at reports to police, 15 percent of public schools reported one or more serious violent incidents to the police and 28 percent reported one or more thefts to the police.

The prevalence of violent incidents and those reported to the police varied by the school level. Schools with students in higher grades were more likely to experience a violent incident than those with students in lower grades. Specifically, secondary schools were more likely to have a violent incident than elementary, middle, or combined schools (92 vs. 61–87% for the other school levels). A similar pattern was observed for those incidents that were reported to the authorities: 71 percent of secondary schools reported a violent incident, compared with 20 percent of elementary schools, 56 percent of middle schools, and 51 percent of combined schools.

When examining violent incidents by the location of public schools, city schools were more likely than urban fringe schools to experience or report to the police at least one violent incident. Seventy-seven percent of city schools had one or more violent incidents and 44 percent reported one or more incidents to the police, compared with 67 and 35 percent, respectively, of urban fringe schools. Rural schools were the least likely to report one or more violent incidents to the police (29 vs. 35–44% for public schools in urban fringe and city areas).

Serious Disciplinary Action

During the year, 3.4 million public school students in the United States received in-school suspensions and 3.2 million received out-of-school suspensions. The percentage of Black students receiving out-of-school suspensions (15%) was higher than the percentages for students of any other racial/ethnic group.

Less than 1 percent of students received each of the following disciplinary actions: referral to law enforcement, corporal punishment, expulsion, and school-related arrest.

The percentage of Black students receiving out-of-school suspensions (15%) was higher than the percentages for students of all other racial/ethnic groups. In contrast, a lower percentage of Asian students (1%) received out-of-school suspensions than students from any other racial/ethnic group.

A higher percentage of male students (9%) than female students (4%) received an out-of-school suspension in 2011–2012. This pattern of higher percentages of male than female students being suspended held across all racial/ethnic groups. In addition, differences by race/ethnicity for male and female students were similar to the overall differences by race/ethnicity. Among males, the percentage of Black students who received an out-of-school suspension (20%) was almost twice the percentage of American Indian/Alaska Native students (10%), and more than twice the percentages of students of two or more races (9%), Hispanic students (8%), White students (6%), Pacific Islander students (5%), and Asian students (2%). Similarly, the percentage of Black female students who received an out-of-school suspension (11%) was more than twice the percentages of female students of any other race/ethnicity. The pattern of greater percentages of Black males and females receiving disciplinary actions than males and females of any other race/ethnicity was also evident for student expulsions.

Teachers Threatened

Most public schools experience one or more violent incidents every year, while 36 percent report one or more such incidents to the police.

This indicator provides the percentage of schools that experienced one or more crimes and the total number of crimes reported by schools. In the School Survey on Crime and Safety, principals of public schools were asked to provide the number of violent incidents, serious violent incidents, thefts, and other incidents that occurred at their school, as well as the number of incidents that were reported to the police. Violent incidents include rape, sexual battery other than rape, physical attack or fight with or without a weapon, threat of physical attack with or without a weapon, and robbery with or without a weapon.

An estimated 1.5 million violent incidents occur in public elementary and secondary schools every year. Seventy-one percent of public schools experience one or more violent incidents and 36 percent of schools report one or more such incidents to the police. Twenty percent of schools experience one or more serious violent incidents (which are a subset of violent incidents and include rape, sexual battery other than rape, physical attack or fight with a weapon, threat of physical attack with a weapon, and robbery with or without a weapon) and 46 percent of public schools experience one or more thefts. These translate into an estimated 61,000 serious violent incidents and 218,000 thefts at public schools. When looking at reports to police, 15 percent of public schools report one or more serious violent incidents to the police and 28 percent report one or more thefts to the police.

The prevalence of violent incidents and those reported to the police varies by the school level. Schools with students in higher grades are more likely to experience a violent incident than those with students in lower grades. Specifically, secondary schools are more likely to have a violent incident than elementary, middle, or combined schools (92 vs. 61–87% for the other school levels). A similar pattern is observed for those incidents that are reported to the authorities: 71 percent of secondary schools report a violent incident, compared with 20 percent of elementary schools, 56 percent of middle schools, and 51 percent of combined schools.

When examining violent incidents by the location of public schools, city schools are more likely than urban fringe schools to experience or report to the police at least one violent incident during the school year. Seventy-seven percent of city schools had one or more violent incidents and 44 percent reported one or more incidents to the police, compared with 67 and 35 percent, respectively, of urban fringe schools. Rural schools are the least likely to report one or more violent incidents to the police (29 vs. 35–44% for public schools in urban fringe and city areas).

Students Carrying Weapons

Between 1993 and 2015, the percentage of 9th–12th-grade students who reported carrying a weapon anywhere during the previous thirty days decreased from 22 to 16 percent, and the percentage of students who reported carrying a weapon on school property during the previous thirty days decreased from 12 to 4 percent.

In 2015, about 16 percent of students reported that they had carried a weapon anywhere at least one day during the previous thirty days: 8 percent reported carrying a weapon anywhere on six or more days, 5 percent reported carrying a weapon on two to five days, and 3 percent reported carrying a weapon on one day. Also in 2015, about 4 percent of students reported carrying a weapon on school property at least 1 day during the previous thirty days. This percentage included 2 percent of students who reported carrying a weapon on six or more days; 1 percent of students who reported carrying a weapon on two to five days; and 1 percent of students who reported carrying a weapon on one day during the previous thirty days.

In every survey year from 1993 to 2015, a higher percentage of male students than of female students reported that they had carried a weapon, both anywhere and on school property, during the previous thirty days. In 2015, for example, 24 percent of male students reported carrying a weapon anywhere, compared with 8 percent of female students. In addition, 6 percent of male students reported carrying a weapon on school property, compared with 2 percent of female students.

Student's Perceptions of Personal Safety at School and Away from School

The percentage of students who reported being afraid of attack or harm at school decreased from 12 percent in 1995 to 3 percent in 2015, and the percentage of students who reported being afraid of attack or harm away from school decreased from 6 percent in 1999 to 2 percent in 2015.

In 2015, a higher percentage of female students than of male students reported being afraid of attack or harm at school (4 vs. 3%) and away from school (3 vs. 1%). In general, the percentages of students who reported being afraid of attack or harm at school and away from school were not measurably different across racial/ethnic groups. However, a higher percentage of Hispanic students (5%) than of White students (3%) reported being afraid of attack or harm at school in 2015. Similarly, a higher percentage of Hispanic students (3%) than of White students (2%) reported being afraid of attack or harm away from school.

Higher percentages of 6th graders (5%) and 7th and 8th graders (4 percent each) reported being afraid of attack or harm at school than did 10th- and 12th graders (2 percent each) in 2015. The percentage of students who reported being afraid of attack or harm away from school was higher for 8th graders (3%) than for 10th graders (1%). In 2015, higher percentages of students in urban (3%) and suburban areas (2%) than of students in rural areas (1%) reported being afraid of attack or harm away from school. However, no measurable differences by urbanicity were observed in the percentage of students who reported being afraid of attack or harm at school.

Avoiding Places

In 2015, about 5 percent of students reported that they avoided at least one school activity or class or one or more places in school during the previous school year because they thought someone might attack or harm them.

In 2015, about 1 percent each of students reported that they avoided any activities, avoided any classes, and stayed home from school because of fear of attack or harm. With respect to avoiding specific places in school, 2 percent each of students reported that they avoided the hallways or stairs in school and any school restrooms, and 1 percent each reported that they avoided parts of the school cafeteria, the entrance to the school, and other places inside the school building.

Students' reports of avoiding one or more places in school because of fear of attack or harm varied by grade. In 2015, a higher percentage of 6th graders (6%) than of 10th (3%), 11th (2%), and 12th graders (3%) reported avoiding one or more places in school.

The percentage of students who reported avoiding one or more places in school was also higher for 7th graders (5%) than for 10th- and 11th graders, and it was higher for 9th graders (4%) than for 11th graders. There were no measurable differences by sex and race/ethnicity in the percentage of students reporting avoiding one or more places in school because of fear of attack or harm.

In 2015, higher percentages of students in urban (5%) and suburban areas (4%) reported avoiding one or more places in school than did students in rural areas (2%). In addition, a higher percentage of public school students than of private school students reported avoiding one or more places in school (4 vs. 2%).

Hate

In 2015, about 7 percent of students ages between twelve and eighteen reported being the target of hate-related words and 27 percent reported seeing hate-related graffiti at school during the school year. The percentage of students who reported seeing hate-related graffiti at school was higher in 2015 than in 2013 (25%). The percentage of students who reported being the target of hate-related words at school in 2015 was not measurably different from the percentage in 2013.

The percentage of male students who reported being called a hate-related word during the school year did not differ measurably from the percentage for female students in any survey year from 2001 to 2015. During this period, the percentage of male students who reported being called a hate-related word decreased from 13 to 8 percent and the percentage for female students decreased from 12 to 7 percent.

However, for both male and female students, there were no measurable differences in the percentage of students who reported being called a hate-related word between 2013 and 2015. The percentage of male students who reported seeing hate-related graffiti at school during the school year did not measurably differ from the percentage for female students in most survey years from 1999 to 2015. During this period, the percentage of male students who reported seeing hate-related graffiti at school decreased from 34 to 26 percent and the percentage for female students decreased from 39 to 28 percent. However, for both male and female students, no measurable differences were observed between the two most recent survey years (2013 and 2015) in the percentage of students who reported seeing hate-related graffiti at school.

In 2015, lower percentages of White (6%) and Hispanic (7%) students than of Black (9%) students reported being called a hate-related word at school during the school year. Also in 2015, a lower percentage of Asian students than students of any other race/ethnicity reported seeing hate-related graffiti at school during the school year. About 17 percent of Asian students reported seeing hate-related graffiti at school, compared with 25 percent of Black students, 27 percent of Hispanic students, and 29 percent of White students. The percentages of White, Black, and Hispanic students who reported being called a hate-related word at school decreased between 2001 and 2015. Similarly, the percentages of White, Black, and Hispanic students who reported seeing hate-related graffiti at school also decreased between 1999 and 2015.

Gangs

Between 2001 and 2015, the percentage of students ages between twelve and eighteen who reported that gangs were present at their school decreased from 20 to 11 percent. The percentage who reported gangs were present at their school was also lower in 2015 than in 2013 (12%). A higher percentage of students from urban areas (15%) reported a gang presence than of students from suburban (10%) and rural areas (4%) in 2015.

The percentage of students from urban areas who reported a gang presence at their school was lower in 2015 than in every survey year between 2001 (29%) and 2011 (23%). However, there was no measurable change in this percentage between 2013 and 2015. The same pattern was observed for students from suburban and rural areas, with lower percentages of students reporting a gang presence in 2015 than in all years from 2001 to 2011, but no measurable change between 2013 and 2015.

A higher percentage of students attending public schools (11%) than of students attending private schools (2%) reported that gangs were present at their school in 2015. The percentage of public school students who reported a gang presence was lower in 2015 than in 2013 (13%). However, the percentage of private school students reporting a gang presence at their school in 2015 was not measurably different from the percentage in 2013.

In 2015, higher percentages of Black (17%) and Hispanic (15%) students reported the presence of gangs at their school than of White (7%) and Asian (4%) students (figure 8.2 and table 8.1). In addition, a higher percentage of White students than of Asian students reported a gang presence. The percentage of students who reported a gang presence was lower in 2015 than in 2013 for both Hispanic (15 vs. 20%) and Asian (4 vs. 9%) students, while the percentages reported in 2015 by White and Black students and students of other racial/ethnic groups were not measurably different from the percentages reported in 2013.

Alcohol

The percentage of 9th–12th grade students who reported consuming alcohol on at least one day during the previous thirty days decreased from 48 to 33 percent between 1993 and 2015.

In 2015, the percentage of students who reported consuming alcohol generally increased with grade level. About 42 percent of 12th graders reported consuming alcohol on at least one day during the previous thirty days. This percentage was higher than the percentages for 9th graders (23%) and 10th graders (29%), although it was not measurably different from the percentage for 11th graders.

The percentage of students who reported consuming alcohol also varied by race/ethnicity. In 2015, higher percentages of American Indian/Alaska Native students (46%), students of two or more races (40%), White students (35%), and Hispanic students (34%) than of Black students (24%) and Asian students (13%) reported consuming alcohol on at least 1 day during the previous thirty days. The percentage of Asian students who reported consuming alcohol on at least one day was also lower than the percentages reported by Pacific Islander students (37%) and Black students.

The consumption of alcohol by students on school property, a crime in itself, may lead to other crimes and misbehavior on the part of students. It may also lead to a school environment that is harmful to students, teachers, and staff.

Drugs

The percentage of 9th–12th-grade students who reported that illegal drugs were made available to them on school property decreased from 32 percent in 1995 to 22 percent in 2015.

In every survey year from 1993 to 2015, a lower percentage of female than of male students reported that illegal drugs were offered, sold, or given to them on school property. For instance, in 2015, about 19 percent of female students reported that illegal drugs were made available to them on school property, compared with 24 percent of male students who reported so. In 2015, lower percentages of Asian students (15%), White students (20%), and Black students (21%) than of Hispanic students (27%) reported that illegal drugs were made available to them on school property. In addition, the percentage of Asian students who reported that illegal drugs were made available to them on school property was lower than that of students of two or more races (25%).

The percentage of Asian students who reported that illegal drugs were offered, sold, or given to them on school property was lower in 2015 than in 2013 (15 vs. 23%); however, no measurable differences were found between the 2013 and 2015 percentages for students of any other racial/ethnic groups.

In 2015, public school students' reports of the availability of illegal drugs on school property varied across the thirty-two states for which data were available. Among these states, the percentages of students reporting that illegal drugs were offered, sold, or given to them on school property ranged from 15 percent in Maine and Oklahoma to 30 percent in Nevada.

Marijuana

In 2015, some 22 percent of 9th–12th-grade students reported using marijuana at least one time during the previous thirty days, which was higher than the percentage reported in 1993 (18%) but not measurably different from the percentage reported in 2013.

Specifically, in 2015 about 7 percent of 9th–12th-grade students reported using marijuana one or two times during the previous thirty days, 10 percent reported using marijuana three to thirty-nine times during the previous thirty days, and 4 percent reported using marijuana forty or more times during the previous thirty days.

In every survey year between 1993 and 2011, higher percentages of male students than of female students reported using marijuana at least one time during the previous thirty days; in 2013 and 2015, however, there were no measurable differences in the percentages reported by male and female students. In 2015, a higher percentage of males (5%) than of females (3%) reported using marijuana 40 or more times during the previous thirty days.

In 2015, some differences in the percentages of students who reported marijuana use were observed by race/ethnicity and grade level. The percentage of Asian students (8%) who reported using marijuana at least one time during the previous thirty days was lower than the percentages reported by White students (20%), students of two or more races (23%), Hispanic students (24%), American Indian/Alaska Native students (27%), and Black students (27%). The percentage for White students was also lower than the percentages for Hispanic and Black students. In addition, the percentage of students in 9th grade (15%) who reported using marijuana at least one time during the previous thirty days was lower than the percentages of students in 10th grade (20%), 11th grade (25%), and 12th grade (28%) who reported doing so. The percentage for students in 10th grade was also lower than the percentages for students in 11th and 12th grade.

SAFETY IN THE WORKPLACE

Workplace Violence

Violence in the workplace has become an epidemic. Not only is workplace violence increasingly common in those workplaces where violence is expected—for example, corrections, law enforcement, and mental health—but in almost every occupation that deals with the public.

According to the Bureau of Labor Statistics, homicides are the second leading cause of death in the workplace and the leading cause of death for women. Although the press focuses on "worker-on-worker-type violence," where a berserk worker kills his supervisor or co-workers, such incidents make up only 4 to 6 percent of all workplace homicides. The rest are the result of robberies, or assaults by residents, patients, or customers against workers.

Violence-related fatalities are only the tip of the iceberg. The Department of Justice estimates there are 2 million cases of violence in the workplace each year. Health care and social services lead the list of occupations with the highest rate of on-the-job injury resulting from violence. Injuries and deaths related to workplace violence should not be tolerated. Most incidents are predictable and preventable. And like any other workplace hazard, it is the responsibility of the employer to take reasonable measures to minimize the likelihood of workplace violence.

Workplace violence is not just limited to physical assault. It can also include near misses, verbal abuse, and sexual harassment. Even the fear of assault or witnessing an assault on a co-worker can have serious health effects on workers.

Effects of Violence

Abusive or threatening incidents, as well as injuries in the workplace, often result in serious and disabling psychological damage. Victims of workplace violence also have an increased risk of post-traumatic stress disorder (PTSD). PSTD is a disorder common among combat veterans and victims of terrorism, crimes, rape, and other violent incidents.

Psychological trauma is a common outcome of violent incidents. Emotional problems include self-doubt; depression; fear; post-traumatic stress syndrome; loss of sleep; irritability; disturbed relationships with family, friends, and co-workers; decreased ability to function at work; and increased absenteeism. Workers often blame themselves when they are injured in an assault, and management often encourages this self-blame.

It is rare that these issues are dealt with effectively even in the short term. But there is increasing evidence that victims and witnesses of violent incidents need long-term treatment to fully overcome these problems.

Preventing Violence

There is a very dangerous common myth that workplace violence is essentially random and unpredictable. According to this way of thinking, because violence cannot be predicted, we can't do anything about it.

The reality, however, is that violence can be predicted and prevented. Violence can be prevented even in workplaces that serve groups of the public who tend to be aggressive and violent. In almost every situation, there are solutions. Some are easy, painless, and cheap, while others are more difficult and more expensive. Solutions differ greatly from occupation to occupation, and from workplace to workplace. Nevertheless, there are measures that can make the work environment less dangerous, even in the most dangerous workplaces.

1. Identify the Conditions That Increase the Chances of Violence (Risk Factors).

Identifying risk factors involves looking at the work environment, work practices, and victim and perpetrator characteristics.

Environmental risk factors that predict violence include:

- a violent society
- a violence-prone neighborhood
- the large number of weapons in circulation
- early release of mental patients who have a history of violence
- hospitalization instead of incarceration of violent criminals

Work practices associated with workplace violence include:

- low staffing levels
- working alone
- working late at night or early in the morning
- working with money or prescription drugs
- long waits for services by customers, clients or patients
- the lack of available services

Victim characteristics include:

- working in homes or in the community;
- handling money or prescription drugs;
- working in correctional institutions or institutions for the mentally ill or developmentally disabled without adequate training in violence avoidance or self-defense;
- providing care, advice or information, such as health care workers, mental health workers, emergency room and admission workers, and social services workers;
- dealing with complaints, such as social service, child welfare and unemployment workers; and/or
- having the authority to act against the public, inspect premises and enforce laws, such as inspectors, child welfare workers, law enforcement/corrections officers and security guards.

Characteristics of those who commit violence (perpetrator characteristics) include:

- persons with a history of violent behavior
- gang members
- relatives of injured persons
- drug users

2. Eliminate or Reduce Risks.

As with other health and safety problems, first try to eliminate the problem. If this is not possible, then attempt to engineer or build the problem out of the workplace. Finally, change administrative procedures.

Try to remove the problem:
Mental health and social service workers are frequently assaulted by patients or residents in health care or social service facilities who should be in jails or forensic facilities.

Engineering controls that can provide protection:

- metal detectors (stationary or hand-held)
- changing office design to provide escape routes for employees
- panic alarms
- bullet-proof glass
- entrance controls in certain parts of the building
- closed-circuit TV cameras
- restricting entrance to a facility after dark
- mobile phones for field personnel

Administrative controls that can reduce risks:

- additional staffing
- a ban on working alone
- recording accidents, verbal abuse and "near misses"
- training in diffusing violent situations or in self-defense

Post-Incident Procedures

Persons trained to treat people exposed to violent incidents should be called in immediately after an incident has taken place. Procedures, such as critical incident debriefing, and sometimes long-term counseling, may be appropriate—not just for victims of assaults, but also for witnesses or people doing similar jobs.

Worker-on-Worker Violence

It is important to attempt to discover the causes of threatening behavior by a worker. Possible causes are:

- a "toxic" workplace (high stress levels, abusive management, threatened layoffs, etc.);
- drug or alcohol abuse;
- domestic problems (relationship, money, etc.); and/or
- mental health problems.

It is useful to attempt to resolve these problems before the discipline and grievance process starts. Discipline and grievance procedures lead to winners and losers. The solution lies in getting help for the troubled worker, if possible. Employee assistance programs (EAPs) can often be useful.

Workplace Violence Policies

Every workplace should have a workplace violence policy that contains a management statement that violence or threatening behavior will not be tolerated, whether it comes from clients, customers, patients, inmates, co-workers, or supervisors.

An Action List

- **Workplace violence is a leading cause of workplace fatalities and injuries**
  - Most homicides are a result of robberies.
  - A large number of non-fatal assaults are aimed at health care and social service workers.
  - Only a small number of workplace homicides result from "worker-on-worker" violence.
- **Preventing workplace violence:**
  - Identify conditions that put workers at risk of violence.
  - Take steps to reduce risks.
  - Help workers after there has been an incident.
- **Preventing worker-on-worker violence:**
  - Find the causes.
  - Address the problem early.
  - Don't put yourself at risk.

- **Workplace violence policies should include:**
  - A strong statement that violence is a serious hazard that will be addressed by the employer.
- **Workplace violence policies should avoid:**
  - Profiling workers.
  - "Zero tolerance" positions.
  - Procedures that violate workers' contractual or other due-process rights.

Shift Work

Millions of people work shifts or work more than one job with subsequent risks to their health and wellbeing.

Shift work (working hours outside the traditional work schedule) and rotating shifts place a great deal of stress (both physical and mental) on the worker. Shift work disrupts the twenty-four-hour "circadian" rhythm that regulates all body functions.

HEALTH EFFECTS OF SHIFT WORK

Shift work can cause a variety of health problems:

- **Disturbed sleep:** Many workers have difficulties maintaining sleep after the night (graveyard) shift and going to sleep before the morning shift. Shift workers may also suffer more fatigue than day workers may. Being overly tired makes it difficult to concentrate, which increases the possibility of errors or injuries. This can be a risk both to the worker and to the public.
- **Interference with social and family life:** Shift work can disrupt family and social life, resulting in psychological harm to the worker.
- **Behavioral effects:** There is more use of alcohol, coffee, and tobacco among shift workers. The stress of shift work can cause irritability, nervousness, and bad tempers, and may play a role in causing ulcers and heart disease.
- **Digestive problems:** Some research has suggested that shift workers have more upset stomachs, constipation, and stomach ulcers than day workers do.
- **Heart problems** also have been noted more often among shift workers than day workers.

Rotating Shifts

There are ways to make rotating shifts less harmful. Because of the human body's natural "biological clock" or "circadian rhythms," the least disruptive is to change to the next later shift after a day or two off. In other words, a day-afternoon-night shift rotation is better than a day-night-afternoon rotation. A fast rotation (every two days, for example) should be avoided because it does not allow enough time to get used to night work.

Even where shift rotations are properly scheduled, prolonged shift work can place great strain on personal relationships for workers whose waking and working hours are "out of synch" with those of their family and friends. In addition, shift workers may find that the quality of time that they spend with family and friends is unsatisfying because the worker's fatigue (from poor sleep or lack of sleep) prevents normal social activity. Some workers just never adjust to shift work. Whenever possible, working the night shift should be voluntary and/or include provisions for employees who simply cannot adjust.

Improving Shift Work Schedules

- Minimize a permanent (fixed or non-rotating) night shift. Most workers never get used to night shift because they go back to a daytime schedule on their days off. Also, some workers on fixed night shifts lose contact with the rest of the workers in the organization.
- Keep consecutive night shifts to a minimum.
- Avoid quick shift changes. Some researchers suggest that forty-eight hours should be the minimum between shifts.
- Schedule as many weekends off as possible.
- Avoid several days of work followed by four- to seven-day "mini- vacations."
- Keep long work shifts and overtime to a minimum.
- Consider different lengths for shifts. Heavier, more boring work should be done during shorter shifts and lighter, more interesting work should be moved to longer shifts.
- Flextime can be useful for those with childcare needs or a long commute time. Start and end times can be moved away from rush hour.
- Keep the schedules regular and predictable. Workers should know their schedule well ahead of time, so they can plan their rest, childcare and contact with family and friends.
- Provide enough rest breaks for jobs requiring repetitive physical work.
- Brief rest breaks each hour seem to be the best for recovery from muscle fatigue.

Stress

Stress is a major health and safety problem. Most workers experience stress whether they are on the shop floor or the boardroom. While not as obvious as safety problems, noise or exposure to toxic chemicals, stress can cause serious health problems.

Stress is the body's natural reaction where there is an imbalance between the demands of the environment and the ability of the worker to respond to those demands. Stress increases when the worker has little control over his or her work. A little stress is not bad, but continuous stress can have serious health effects on workers.

The National Institute of Occupational Safety and Health (NIOSH) reports that psychological disorders, many of which result from stress, are among the ten leading causes of work-related disease. The US Office of Technology Assessment predicts that stress-related illness may be the greatest public health problem faced by office workers of the future.

People have very different abilities to deal with stress. What may be stressful or very upsetting to one person may not be stressful to another. Nevertheless, it is important for people who work under stressful conditions to realize that stress on the job is the body's natural response to poor working conditions. It is not caused by an individual's shortcomings and cannot be cured by being "tough."

HEALTH EFFECTS

Workers suffering from repeated, prolonged, or continuous job stress may experience:

- frequent headaches
- sleeplessness
- loss of appetite
- depression
- short temper
- backache
- stomach problems
- ulcers
- high blood pressure
- heart disease and heart attacks

Stress can also increase the risk of being injured on the job, weaken the body's resistance to disease, and lead to marital problems, drug abuse, and alcoholism.

Burnout is a severe reaction to job stress. Burnout is common for workers in occupations, such as social work, day care, and health care who are in constant contact with clients or patients. These workers are not only exposed to stressful conditions, but also suffer from emotional over-stimulation. The resulting "burnout" leads to a sense of emptiness and loss of feelings, in addition to the other physical and psychological symptoms of stress.

FACTORS THAT CAUSE STRESS (STRESSORS)

Factors that cause stress at work include:

- **Physical and environmental conditions:** poor, unsanitary, or unsafe working conditions; lack of privacy; improper lighting; and poor ventilation and temperature control.
- **Psychological:** harassment, inadequate job skills, role conflicts, conflict with co-workers and supervisors, and lack of adequate breaks.
- **Organizational:** policies and procedures, schedules, deadlines, repetitive work, unreasonable workloads, inadequate pay and benefits.
- **Sociocultural:** lack of support from supervisors or co-workers, overly authoritative or crisis-centered management style.
- **Interpersonal:** lack of respect, lifestyle, financial, or family problems.
- **Work setting:** lack of job security, lack of control over one's job.

REDUCING STRESS AT WORK

Poor working conditions are the primary cause of workplace stress. "Stress management" techniques (such as meditation, muscle relaxation, etc.) are useful in treating the symptoms, not the causes of stress-related illnesses. The best way to reduce stress is to reduce the causes of stress in the workplace. The following steps should be taken:

- **Do a survey** or talk to workers to define what aspects of the job cause stress.
- **Negotiate strong contract language** to reduce the causes of stress. These may include increasing staff levels, reducing overtime, providing more rest time, reducing the frequency of shift rotations, improving physical working conditions.
- **Use (EAPs** to help workers deal with stress and other problems, whether at home or at work.
- **Stress Training and Reduction Workshops** can teach techniques that can reduce the symptoms, but not the causes of stress-related illnesses. These techniques can be helpful to reduce some of the symptoms while the causes of stress are being addressed. Physical exercise, relaxation exercises, and making more time for families, hobbies, and outside interests are important methods of stress reduction.

Health Hazards in the Workplace

- **Types of health hazards in the workplace:**
  - Chemical (asbestos, solvents, chlorine)
  - Biological (tuberculosis, HIV, hepatitis, molds)

  - Physical (noise, heat and cold, radiation, vibration)
  - Ergonomics or Repetitive Strain Injuries (carpal tunnel syndrome, back injuries)
  - Psychological (stress)
- **How health hazards enter your body:**
  - Breathing (inhalation)
  - Swallowing (ingestion)
  - Skin (absorption)
  - Cuts (injection)
- **The harm caused by health hazards depends on:**
  - Strength, or potency, of the agent
  - Amount of the agent that is present
  - How long you are exposed to the agent
  - Part of your body that is exposed
- **Types of health effects:**
  - Acute: the effect shows up right away.
  - Chronic: problems show up after a long period of exposure and/or long after the exposure ends.
  - Local: only the part of the body that was exposed is affected.
  - Systemic: an agent enters the body and affects other parts of the body.
- **Cancer**
  - Cancer is a term for many diseases in different parts of the body.
  - Carcinogens are agents that cause cancer.
  - There is no totally safe level of exposure to something that causes cancer.
  - Cancer from a workplace exposure may develop ten, twenty, or more years after the exposure.
- **Reproductive effects**
  - Both men and women can be affected by reproductive hazards at work.
  - Reproductive hazards cause miscarriages and birth defects.
- **Safety Hazards**
  - Working in traffic zones
  - Confined spaces
  - Trenching
  - Violence
  - Machines
  - Fire and explosions
  - Slips and falls

HOW HAZARDS ENTER THE BODY

To prevent exposure to health hazards, it is important to understand first how harmful agents enter the body. The way an agent enters your body also can make a difference in what happens to you. For example, breathing asbestos fibers causes cancer or other harm to the lungs. Swallowing asbestos fibers can cause damage in the stomach, colon, or other digestive organs.

There are four main ways that harmful agents can enter your body.

Breathing (inhalation)

You can breathe in chemicals, some that you cannot even see or smell. Some diseases, such as tuberculosis (TB), are spread by breathing in germs that are in the air.

Swallowing (ingestion)

Workers can swallow chemicals in different ways. You can swallow chemicals along with your food if you do not wash your hands after handling chemicals. In dusty areas you can swallow chemicals, such as lead or asbestos particles, that are in the air or that land on food or in drinks.

Some diseases, such as hepatitis A, are spread by eating food that is contaminated with the hepatitis A virus. Other germs can be swallowed if you eat or touch your mouth without proper hand washing after touching soiled linens or other items.

Skin (absorption)

Many chemicals and some types of radiation can pass right through your skin. Some types of chemicals, germs, or radiation will cause illness or symptoms just by coming into contact with the skin. For example, polychlorinated biphenyls (PCBs), a group of toxic chemicals used in transformers, can cause a painful skin rash called chloracne.

Cuts (injection)

Chemicals or germs can also pass through your skin if you get a cut or get stuck with a sharp object, such as a needle.

KEY POINTS

- State "Right-To-Know" laws and the OSHA Hazard Communication Standard (29 CFR 1910.1200) give employees the right to know about chemicals in their workplace. Employers must inform workers about chemicals by:
  - Keeping a list of all the hazardous products that are used or stored in the workplace;

- Making sure that all chemical containers have labels that:
  - can be read;
  - show what the product is and who makes it;
  - warn about the dangers of the product.
- Having Material Safety Data Sheets (MSDS) for each chemical in the workplace; and
- Training workers about chemical hazards and how to protect themselves.

- A permissible exposure limit (PEL) is OSHA's legally allowed amount of chemical exposure to workers.
- *LEGAL DOES NOT MEAN SAFE!!!* A legal exposure to a chemical does not mean that workers will not get sick.
- There are tests to find out if workers are exposed to chemicals at work.
  - Testing the air in the work area—air monitoring.
  - Testing dust, building materials or other substances for toxins—bulk or wipe samples.
  - Blood tests, X-rays, or other medical tests to check for exposure—biological monitoring.
- Steps to prevent workers' exposure to chemicals should be taken in the following order:
  - *Substitution:* Do jobs without chemicals or use less toxic products.
  - *Enclosure:* Isolate the job.
  - *Ventilation:* Keep harmful chemicals out of the air that workers breathe.
  - *Housekeeping:* Keep the workplace and workers clean.
  - *Personal Protective Equipment (PPE):* PPE is needed if there is no other solution, or until a better control is used. If PPE fails, the workers get no protection. PPE is often used with other protections.

Workers come into contact with chemicals every day at work. Thousands of different chemicals can be found in the workplace. The chemicals are delivered in fifty-five-gallon drums, small plastic containers, boxes containing powders and in other forms.

Some dangerous chemicals in your workplace are not in containers. Asbestos fibers can be in pipe insulation. Lead can be in dust that contains lead paint or in welding fumes from metal coatings. Road crews may breathe in silica when breaking up rocks.

Chemicals can be in different forms. The way a chemical might enter your body depends on the form of a chemical:

- **Dusts, fumes,** and **certain fibers** are small solids (particulates) which may be inhaled or swallowed.
- **Liquids** can harm skin and also be absorbed through the skin.
- **Mists** are very small droplets of liquids that are inhaled.
- **Gases** and **vapors** are inhaled.

GETTING INFORMATION ABOUT CHEMICALS ON THE JOB—RIGHT-TO-KNOW LAWS

The ***Right-to-Know*** refers to workers' rights to information about chemicals in their workplaces. The federal law that provides these rights is the ***OSHA Hazard Communication Standard*** **(29 CFR 1910.1200)**. Private-sector employers must provide chemical information to their workers under the OSHA standard.

Most states have their own Right-to-Know laws that cover public-sector workers. In some of these states, workers have stronger rights than under the federal Hazard Communication Standard.

Your Right-To-Know Rights

Right-to-Know laws give workers the right to know about dangerous chemicals at work. You have rights to get information about chemicals in four main ways.

1. **Employers must maintain a list of all hazardous products known to be in the workplace.**
2. **Chemical containers must have labels.**
   - Containers must have a label that workers can read.
   - The label must show what the product is and who makes it.
   - The label must warn about the dangers of the chemical.
3. **Material Safety Data Sheets (MSDS) that describe the dangers of a chemical and how to prevent exposure must be provided.**
   - Workers must have access to MSDSs for the chemicals they work with during their shift.
   - MSDSs must not have blank spaces.
   - The name of the product on the MSDS must match the name on the container's label.
4. **Workers must be trained about chemical hazards.**
   - What are the health hazards of a chemical spill.
   - How to handle and store chemicals.
   - How to prevent exposure to chemicals.
   - Employees' rights and employers' duties under the law.

WORKERS' RIGHTS TO INFORMATION UNDER HAZARD COMMUNICATION

Under the Hazard Communication Standard, employers must provide their workers with information about chemicals by:

- a written program to inform workers, including a list of hazardous chemicals;
- labels;
- MSDS; and
- training.

Hazardous Chemicals List

Employers must keep an updated list of hazardous chemicals they use and store. A hazardous chemical is a product that can cause harm to employees.

Labels

The OSHA Hazard Communication Standard requires that manufacturers must label all chemical containers. At a minimum, the OSHA standard requires that the label must contain at least three pieces of information:

- All labels must identify the chemical product. Either the chemical name or the brand name of the product may be used.
- All labels must include the manufacturer's name and address.
- All labels must carry a warning about the specific health and safety hazards of the product. For example, "Absorbed through Skin—Causes Kidney Damage" tells what part of the body may be harmed. This is better than a label that advises to "Avoid contact with skin." OSHA requires that the label lists the organs that are affected by the chemical *(target organ).*

The OSHA Hazard Communication Standard also requires the following:

- Labels must be placed on smaller containers when workers transfer material from one container to another. For example, if a worker transfers one gallon from a fifty-five-gallon drum into a smaller container, the smaller container must also have a label. **EXCEPTION:** The smaller container does not have to have a label if only that worker will use up all the material on his/her shift. AFSCME recommends that employers label ALL chemical containers.
- All labels must be written in English. Labels in other languages are acceptable as long as there is an English label as well.
- Labels can be read. Damaged labels must be replaced before the product may be used.

Chemicals do not always have to be labeled. For example, employers do not have to label pipes or piping systems. Employers must still inform employees about hazardous chemicals in pipes, such as gas or steam. This can be done by color-coding pipes, such as painting all gas pipes yellow.

MATERIAL SAFETY DATA SHEETS AND WORKERS' ACCESS

The law requires the chemical manufacturer to send Material Safety Data Sheets, or MSDS along with the chemicals. The MSDS must not have any blank spaces. If there is no information for a section, the space must say unknown or not applicable (N/A).

Workers have the right to review the MSDS. Right-to-Know laws require that workers have access to an MSDS for each hazardous chemical in their work area during their shift! For example, an MSDS cannot be kept in a day-shift supervisor's office if the office is locked and workers on the evening or night shift cannot get the sheets.

Information Required on Material Safety Data Sheets

An MSDS does not have to be on a specific form. However, Right-to-Know laws require that an MSDS contain certain types of information.

Section 1. Product Identity

- The name of the chemical on the MSDS must match the name on the label.
- The name and location of the manufacturer and when the MSDS was produced.

Section 2. Hazardous Ingredients and Exposure Limits

Products are usually a mixture of different chemicals. Many chemical products contain water or other substances that are not dangerous. These do not have to be listed on the MSDS.

- The MSDS must list all the chemicals that can cause harm.
- The MSDS must include legal and recommended exposure limits.

Permissible Exposure Limits (PELs)

The Occupational Safety and Health Administration (OSHA) has rules to limit workers' exposure to hundreds of chemicals. These limits are called ***Permissible Exposure Limits***, or ***PELs***. These are the exposure levels that OSHA can legally enforce. In other words, these are the exposure limits that are allowed (permissible).

The amount of a chemical a person is exposed to is usually measured in ***parts per million***, or ***ppm***. For every 1 million parts (molecules) of air, there can be a certain number of parts of a chemical. For example, the permissible exposure limit for carbon monoxide is fifty parts per million. There can be up to fifty molecules of carbon monoxide in every million molecules of air that a worker breathes.

The amount of exposure allowed also depends on how long a worker is exposed to a chemical. PELs are figured on an exposure for a full eight-hour shift. This is called an eight-hour ***Time-Weighted-Average***, or ***TWA***. For example, the permissible exposure level for carbon monoxide is fifty parts per million averaged over an eight-hour day. This means a worker could be exposed to 400 parts per million of carbon monoxide on one shift (50 parts per million x 8 hours).

Short-Term Exposure Limits (STELs) and Ceiling limits

Some chemicals are so dangerous that the exposure limits cannot be averaged over a full shift. Instead, the ***short-term exposure limits*** are based on 15- or 30- minute periods. Other chemicals, like carbon monoxide, have ceiling limits. The ceiling limit for carbon monoxide is 200 parts per million. This means that a worker cannot be exposed to over 200 parts per million at any time.

Recommended Exposure Limits

Other organizations provide guidance on limiting exposure to chemicals. These may differ from OSHA's limits. The National Institute of Occupational Safety and Health (NIOSH) is a government agency that does research on job hazards. NIOSH has a list of ***Recommended Exposure Limits***, or ***RELs***. The American Conference of Governmental Industrial Hygienists' (ACGIH) limits are called ***Threshold Limit Values***, or ***TLVs***.

Section 3. Chemical Characteristics

- The temperature a chemical evaporates into the air ***(boiling point)***.
- Whether a chemical vapor will rise or sink in the air ***(vapor density)***.

Section 4. Fire and Explosion Information

- The way to safely put out a fire if the chemical ignites.
- If there are any unusual fire or explosion hazards.
- The lowest temperature at which a chemical vapor will light on fire ***(flashpoint)***.
- The amount of a chemical vapor that must be in the air to burn (lower explosive or flammable limit and the upper explosive or flammable limit).

Section 5. Reactivity

- Whether a chemical is stable and what conditions to avoid.
- Other chemicals or materials to avoid mixing together.

Section 6. Health Effects

- The way a chemical enters the body (breathing, swallowing, skin absorption).
- The immediate and long-term health effects that might occur.
- Whether the chemical causes cancer or birth defects.

Section 7. Handling and Storing Instructions

- What to do in case of a spill.
- How to safely store and dispose of the chemical.

Section 8. Preventing Exposure (Control Measures)

- The type of ventilation that is needed to work with the chemical.
- The type of respirator, glove, or other equipment to use when working with the chemical.

Training

Workers must receive training on the chemicals they use or work around. The training must include:

- the health hazards caused by the chemical;
- how to handle and store chemicals;
- how employees can protect themselves from exposure to chemicals; and
- employees' rights and employers' duties under the law, such as where to find the MSDS in the workplace and how to read them.

ADMINISTRATIVE CONTROLS

Hazards can be reduced by keeping work areas neat and clean. Part of the workday should be spent keeping the workplace orderly by doing routine tasks, such as:

- putting rags soaked with chemicals in closed containers;
- storing chemicals properly in clearly labeled containers;
- keeping only the amount of chemical on hand that is needed; and
- cleaning chemical spills right away.

Workers need the supplies, time, and procedures to keep chemicals off their clothes and bodies. Employers should provide:

- a separate room away from work areas for eating and storing food;
- soap and water for washing;
- time to wash before breaks, before eating, and at the end of the shift;
- a place to change and store street clothes; and
- a facility to wash work clothes.

Personal Protective Equipment (PPE)

Masks, gloves, and other personal safety gear are the least effective way to protect workers. The protective gear does not remove the chemical hazard from the area. If the gear does not work, then the worker gets no protection. PPE is needed if there is no other solution, or until a better control is installed. Often, PPE is used along with other protections.

OTHER HEALTH RISKS

Heat

- **Health effects include:**
  - Rash, cramps, heat exhaustion, injuries from being less alert;
  - Heat stroke—a deadly condition requiring immediate medical attention.
- **Exposure to heat can be controlled by:**
  - Providing adequate air conditioning and ventilation;
  - Scheduling work early or late in the day, or waiting for cooler weather;
  - Rotating jobs among workers;
  - Taking breaks in an air-conditioned or shady area;
  - Drinking fluids;
  - Getting used to heat gradually;
  - Wearing light colored clothing.

Cold

- **Health effects include:**
  - Frostbite
  - Hypothermia
  - Trench foot
- **Exposure to cold can be controlled by:**
  - Providing frequent rest periods in a warm area;

  - Providing clothing designed to keep cold and wind out and allow heat and perspiration to escape;
  - Building barriers around the worksite, where possible, to block the wind;
  - Supplying workers with warm beverages;
  - Keeping vehicles in good running order to avoid getting stranded in the cold.

Noise

- **Causes temporary or permanent hearing loss.**
- **Noise can be controlled by:**
  - Using equipment that makes less noise.
  - Enclosing or isolating machines to muffle sound.
  - Rotating jobs.
  - Moving workers away from the noise.
  - Keeping equipment lubricated and in good working order.
  - Wearing personal protective equipment, such as ear plugs or ear muffs.
- The OSHA noise standard (29 CFR 1910.95) is 90 decibels (dB) averaged over an eight-hour shift. LEGAL DOES NOT ALWAYS MEAN SAFE! Hearing damage can begin at levels as low as 80 dB over an eight-hour day.

Ionizing Radiation

- **Ionizing radiation includes X-rays, alpha, and beta particles and causes:**
  - Cancer
  - Inability to have children (infertility) and/or birth defects
  - Radiation sickness
- **Preventing exposure to ionizing radiation:**
  - Properly shield and maintain equipment that uses radiation.
  - Isolate radioactive equipment in restricted access areas.
  - Install switches to turn on equipment from outside the room.
  - Move workers as far away from radiation sources as possible.
  - Workers should not hold patients during X-rays.
  - Identify patients with radioactive implants.
  - Wear badges to measure exposure to radiation.
  - Pregnant workers should be temporarily transferred to other jobs without loss of pay, seniority, or other rights or benefits.
  - Use aprons and gloves that shield radiation.

Non-ionizing Radiation

- Non-ionizing radiation includes ultraviolet, infrared, microwave, radio waves, and lasers.
- Health effects include sunburn, skin cancer, cataracts, and other effects.

- Preventing exposure to non-ionizing radiation:
  - Properly shield and maintain equipment that emits non-ionizing radiation.
  - Move workers away from sources of radiation.
  - Wear glasses with UVR filter lenses.
  - Cover skin or use sunscreen when working outdoors.

Electric and Magnetic Fields (EMFs)

- EMFs are produced by power lines, electrical wiring, and electrical equipment, such as computers. The health effects of EMFs are not fully understood. "Prudent avoidance" means avoiding exposure because the risks are unknown.

Electricity

- **Work practices that can prevent shocks and electrocution include:**
  - Have only qualified workers install and maintain electrical systems.
  - Use proper electrical outlets and do not overload extension cords.
  - Keep electrical cords away from water.
  - Ground electrical equipment.
  - Use ground fault circuit interrupters (GFCI).
  - Keep a safe distance from overhead power lines.
  - Locate underground utility wires before digging.
  - Find shelter during rainstorms when there is lightning and thunder.

Physical agents include heat and cold, noise, radiation, electricity, and other dangers. Physical agents can occur naturally or can be produced in the workplace. For example, heat is a physical agent that is the result of the weather and is also produced in kitchens and laundries.

Protecting workers from physical agents requires the same process as other hazards. First, the physical agent and its health effects must be identified. Next, working conditions must be changed to prevent exposure to the hazards.

Mailroom

Workers must constantly be alert for suspicious packages that might contain explosives or bio-chemical weapons. See section on suspicious packages.

HVAC systems

Maintenance of ventilation systems and training of staff are critical for controlling exposure to airborne contaminants, such as CBR agents.

TRAVEL SAFE

General Travel Safety Information

Given the heightened threat of terrorism, security is now tighter in the air, on the railroads and on the highways, and travelers, more than ever, need to keep their eyes open and be aware of what is going on around them.

Planning

Do Your Homework

Before heading on a trip, especially overseas, check for travel warnings, weather forecasts, and advisories. Get a map and make sure you know where you are going. If you are going to be on the road for a few days, make sure you have accommodations booked in advance. You don't want to be driving around strange areas late at night trying to find a room.

If you are traveling overseas, visit the State Department's website. It issues travel advisories and updates these regularly. You can find them on its website www.state.gov, call the State Department's information line at 888-407-4747 for the information or dial 202-647-3000 on your fax machine and follow the prompts to get it faxed to you. The department can also supply you with information sheets on individual countries, which include such helpful information as whether a wide array of pharmaceuticals is generally available in shops. If you are traveling overseas, see our special section at the back of this book.

Medications

Bring Extra Meds

In case your return home is delayed for whatever reason, bring medications for an extra couple of days with you. But handle them with care. Inspectors at border crossings, airports, and the like are checking more bags more carefully, so it's a good idea to keep medications in their original packages.

Airlines have banned passengers from carrying a variety of everyday items, including metal nail files and corkscrews. If you're planning to fly and need to carry needles for medical reasons, you'll need to bring the corresponding medication, with original labeling, with you, says the Federal Aviation Administration (FAA). Since the prescription label is on the outside of boxes of insulin vials, the FAA suggests that passengers with diabetes bring their insulin in this box.

Diabetics can carry on board lancets, as long as they're capped, and can also bring accompanying glucose meters with the manufacturer's name embossed on the meters. For links to more information, see www.diabetes.

org. Since individual airlines may have additional requirements for passengers carrying these medical devices, call yours in advance for details.

A Month Before You Leave

Get Immunized Early

Buy your guidebook and check the Centers for Disease Control and Prevention's destination-specific lists of recommended vaccinations and health precautions. You may need to get some immunizations at least a month before departing.

Get a Checkup

See your primary care physician for vaccinations and a physical to make sure you are OK to travel. If you'll be crossing time zones, ask your doctor whether you should continue taking your meds at your regular, home-zone time, or switch—and how. Ask if any foods you're likely to encounter while away might interact with your meds. Leafy greens, for example, can decrease the effects of some blood-thinning drugs. Get two letters from your doctor if you're crossing international borders. The first should explain any medical problems you have and their treatment. It'll come in handy if you need to consult with health care providers while away. You'll also need one that lists prescription meds (including their generic names) and any alternative remedies you take, as well as why, when, how (whether you use a syringe, for instance), and at what dosage. The letter should note that the quantity you're packing is appropriate for your needs, because every country has its own rules about bringing prescription drugs across its borders. For instance, Valium and similar drugs are prohibited in Mexico. Having all the proper documentation will ease your way through customs and help you get replacement meds should you lose yours en route.

Consider Traveler's Insurance

Call your health insurance company before leaving to find out exactly what's covered when you are traveling abroad. Many policies don't cover evacuation or nonemergency care abroad, so ask.

Packing

Get your numbers straight

Program your doctor's contact information into your cell phone, or jot it in your address book.

Keep Your Meds in Their Original Containers

This goes for all medicines: prescription, over-the-counter, and alternative. If you need a syringe, make sure it's in a clearly labeled container, too. You'll make life easier for security staff at airports and border crossings.

Carry Must-take Meds On Board

If you're flying, put prescription drugs, and any syringes you might need, in your carry-on bag, together with the letter from your doctor, so they're accessible and safe. Storage conditions in cargo holds can affect the stability of certain meds, but you should never put medicines in checked luggage, which may or may not arrive at the same destination.

Remember Your Paperwork

Make a copy of those prescriptions and letters from your doctor; then put one set in your carry-on bag and another in a suitcase. If one gets lost or stolen, you'll have a backup.

Traveling by Car

There are more than 230-million registered motor vehicles in the United States and almost 200-million licensed drivers, and every year they travel 2,890,893,000,000 miles on our roads. It is not surprising, therefore, that every year more than 42,000 people are killed and more than 2.8 million injured in traffic accidents. In addition, there are more than 4.3-million property-damage-only traffic accidents.

Safe Driving Tips

If someone else is in the car, you should refrain from talking with your hands, and realize that you are doing something that could, in one instant, become life threatening. When speaking with others in your car don't feel as if you need to look at them when carrying on a conversation. Keep your eyes on the road. ALWAYS PAY ATTENTION TO THE ROADWAY.

Remember that cell phones and all other distractions increase your chances of having an accident. If you must use your cell phone on a regular basis while driving, consider reducing your risk by purchasing a miniature headset or hands-free device at your local electronics store.

The left lane of any interstate highway is for passing. Staying out of the left lane (and in one of the right lanes) when not passing greatly reduces social stress on the highways, which makes them safer for everyone.

Always remember to buckle your seatbelt—even if the ride is just around the corner.

Keep your children in proper restraint seats, or properly adjusted safety belts if they are older. (Using the phrase "OK everybody—buckle up!" works well before starting the car.) Keep your rear-view, and side mirrors adjusted for maximum visibility.

If you wish to be seen more easily in daily traffic, consider using your headlights during the daytime as well as at night, and always remember to use your headlights when traveling on two-lane highways regardless of the time of day.

Overtaking Tips

Always pass trucks and buses quickly. When you are approaching these vehicles from the rear, always judge your passing speed and don't begin the passing process only to find yourself stuck behind another vehicle—trapped beside the truck or bus. Always wait to let the vehicle in front of you complete their pass before beginning yours.

Be aware that most large commercial vehicles these days have 500 horsepower or more, and are equipped with cruise control. Yet their companies have limited their top speed through their engines' computer system to ensure safe driving and better insurance rates. If you notice one of these vehicles creeping up on you, then make a decision to speed up and pull away, or slow down and let it pass.

Trucks and buses try to maintain a safe distance between themselves and the vehicle in front of them. Yet many automobile drivers commonly invade this "Safety Zone" to increase their position on the road. When passing a truck or bus, always leave at least 50 feet (or five car lengths) of space between you and the other vehicle before merging back into the lane in front. Never zoom around a truck or bus only to pull directly in front of it. Drivers who do this have no control over what might happen in front of them—placing their lives at risk, as well as the lives of others. The larger vehicle will never be able to stop in time should the car driver need to hit the brakes.

Remember that if you can't see a truck's rear-view mirrors, then the driver cannot see you. Let this simple fact be a gauge as to how close behind it you should be. Always stay back far enough so that you can see the mirrors.

Bad Weather Driving Tips

Snow and Ice

Bridges and overpasses freeze first. Slow down and avoid sudden changes in speed or direction. Keep windows clear of snow and ice. Keep your speed steady and slow, but not too slow. In deeper snow, it's often necessary to use the car's momentum to keep moving. Use brakes very cautiously. Abrupt braking can cause brake lock-up, which causes you to lose steering control. Antilock brakes are designed to overcome a loss of steering control on wet or slippery roads. Yet they have little or no effect on ice. To make antilock brakes work correctly, or work at all, you should apply constant, firm pressure to the pedal. During an emergency stop, push the brake pedal all the way to the floor. If you get stuck in snow, straighten the wheels and accelerate slowly. Avoid spinning the tires, because the heat friction caused by spinning tires melts the snow and creates a thin layer of ice. Use sand or cinders under the drive wheels to increase traction if you get a little stuck. Never stand in traffic to push a car that's stuck. Someone else could lose control and seriously injure or even kill you.

High Winds

Use extra care and consider whether a trailer, van or other "high-profile" vehicle should be operated at all.

Rain

The road becomes slippery as water mixes with road oils, grease, and dirt. Also, your car's tires tend to ride on the surface water, reducing traction. . . . Slow down.

Visibility is often impaired. Turn on your headlights at the first sign of rain. Use the defroster or air conditioner to keep windows and mirrors clear.

Fog

Stay to the right of the roadway. Turn on your headlights—day or night—to low beam. If fog thickens, run your hazard flashers to aid others coming up from behind you, increasing the ability for others to see you better. If you're having difficulty seeing the road's edge, pull off at the next exit, well out of the traffic lane, turn on the emergency flashers, and leave your headlights on and your vehicle running.

Severe Weather

Severe hailstorms—find shelter by driving under an overpass or bridge.

Severe thunderstorms—listen to your car radio and stay alert.

If you spot a tornado, don't try to outrun it. Get out of the car; find shelter in a ditch or low-lying area and lie face down to protect yourself from flying debris.

Hurricanes—avoid low areas and move inland while there's still plenty of time.

Cell Phones and Other Distractions

Distracted driving—including the use of cell phones—is a major contributor to automobile crashes. Between 4,000 and 8,000 crashes related to distracted driving occur daily in the United States. In a year, they contribute to as many as one-half of the 7 million US crashes reported annually.

How do hand-held cell phones factor into the equation? Using a cell phone while driving can increase your chances of being involved in a crash. However, cell phones aren't the only problem. Other distractions, such as looking at outside objects and other people in the vehicle, pose a greater risk of contributing to crashes than cell phone use.

Why are hand-held cell phones at the heart of the debate? Hand-held cell phones are readily visible to other drivers. When people chance upon a distracted driver and notice a cell phone, they naturally blame the phone. Most drivers are frustrated when they see inconsiderate, inattentive drivers talking on cell phones. It's often more difficult to determine if a distracted driver is talking to a passenger, tuning the radio, or eating.

What about hands-free phones? Hands-free phones are not risk-free. The hands-free feature is simply a convenience: It does not increase safety. Studies show that hands-free cellular phones distract drivers the same as hand-held phones, because it is the conversation that distracts the driver—not the device.

Will banning hand-held cell phones improve safety? Not according to current research. A study funded by the AAA Foundation for Traffic Safety about the effects of cell phone use on driver attention found that the distraction of using a hands-free cell phone and tuning a radio is similar. Regarding the question of banning specific devices, such as hand-held cell phones, two facts are clear: Banning hand-held phones, but allowing hands-free phones is likely to have little or no effect on safety. No studies show hands-free phones offer safety advantages over hand-held phones. The distracting factor is the conversation—not the device itself. And no one can legislate when and what drivers think.

The bottom line—Drivers need to keep their eyes on the road and their minds on driving. AAA recommends drivers not use their cell phones while driving. However, if using a phone is essential, drivers should follow these safety tips.

TIPS FOR SAFE CELL PHONE USE WHILE DRIVING

- Recognize that driving requires your full attention.
- Use hand-free devices and pull off the road is safe to do so to complete your call.
- Inform the person you're calling that you are speaking from the car.
- Secure your phone in the car, so that it doesn't become a projectile in a crash.

Road Rage

No matter how safe a driver you are, it is the actions of others that you have to be concerned about. One of the most worrying new trends is road rage, which in many cases has led to fatalities.

The aggressive driver has become a menace on our highways to the degree that our lawmakers are involved in trying to find ways to stop the aggressive driver from physically harming others.

It is estimated that about one-third of crashes and two-thirds of the resulting fatalities can be attributed to behavior associated with aggressive driving. For years the highway safety spotlight has been on the impaired driver, the speeding driver, and the unbelted driver. However, today we must add the aggressive driver to the list of major highway-accident producers.

So, who is an aggressive driver? An aggressive driver is anybody whose driving behavior is characterized by impatience and a lack of concern for others. This behavior endangers both people and property. This definition includes a diverse range of driving behaviors, from erratic or abnormal maneuvers to dueling or violence on the road.

Aggressive drivers tend to do the following: Speed, tailgate, fail to yield, weave in, and out of traffic, pass on the right, make quick and unsafe lane changes, run stop signs and lights, make rude hand and facial gestures, scream, honk, and flash their car's headlights. Aggressive drivers behave in this manner because they feel safe and powerful behind the wheel of a big, heavy machine, and they allow their high frustration level to diminish any concern they should have for others.

Driving Tips

- Don't retaliate. Never take the other driver's behavior personally. He/she is only reacting on a road-rage instinct.
- Don't make eye contact with an angry driver.
- Before you react to anything that is done to you, please ask yourself, "Is getting back at that jerk worth my life?"
- Be polite and courteous, even when others are not.
- Always ask yourself: "Could the other driver have possibly made a mistake?"
- If you are harassed by another driver and being followed, do not go home. Go to the nearest police station.
- Slow down and relax!
- Never underestimate other drivers' capacity for mayhem.
- Reduce your driving stress by allowing enough time to get where you are going. Know the roads that are under construction and listen to weather reports that may cause traffic delays. Practice patience and keep your cool.
- Remember that you cannot control the drivers around you, but you can control the way they affect you.

Much of the time, the damaging behavior continues after the aggressive driver has gotten out of his vehicle. Aggressive drivers use a variety of weapons, such as their fists and feet, tire irons and jack handles, baseball bats, knives, razor blades, and defensive sprays, such as mace and guns. If you are involved in an incident with an aggressive driver, stay calm and do not provoke him or her. Wait patiently for the police to arrive.

Subway and Bus Security

As a general precaution, whether you're in the subway, the bus, or even in the street, appear confident. Always look as if you know where you're going, and you're better off not displaying money in public. If you're alert and aware, you can make your subway and bus trips even more secure.

Off-Hours Waiting Areas

Avoid standing at the end of subway platforms or on an empty platform. Instead, wait in the Off-Hours Waiting Area, particularly at night. Many stations have one, generally located near a station booth. Speak to the station agent or other Transit employees if you have a problem. Use a talkback box (mounted on a platform column) to get help in a station where you're not visible to the station agent. When you speak into the box, the agent can speak with you. You can also use a public phone to dial 911 (the police) if you need help. This call is free. Electronic signs in many Off-Hours Waiting Areas indicate when a train is approaching the station. If you wait near the sign, you will have enough time to walk to the platform as the train arrives.

Protect Yourself against Pickpockets

Stay awake. A pickpocket's easiest victim is a sleeping passenger. If you feel drowsy, it's best to get up and stand, or take another seat. When you find yourself alone in an empty subway car, move to a car that has a conductor (usually in the center of the train), a train operator (front car), or other riders.

Being alert and staying in a subway car with other people are always good precautions. Although pickpockets often target people who are alone and asleep, they know how to operate in crowds as well. That's why you should be wary of being pushed or bumped. But even when there aren't many people around you on a bus or in the subway, never keep your wallet or money in a back pocket, and keep all bags, backpacks, and pocketbooks securely closed. Overlooking these things can make you an easy target.

Keep alert if you see or hear a commotion. It could be a pickpocket's trick to divert your attention. And, speaking of remaining alert, be extra cautious if you use a headset. They tend to reduce your awareness. If your pocket is picked while you're on a bus, call out to the bus operator immediately. He or she can request police assistance.

How to Avoid Bag or Chain Snatching

Bag and chain snatchers are more obvious than pickpockets but the result is the same. Following a few precautions can better protect your valuables. Purse snatching is a crime of opportunity. You can eliminate that opportunity. Every female carrying a purse is a potential target. Senior citizens are especially susceptible to these criminals, since they may not be readily able to defend themselves and pursue their assailants. Purse snatching is a crime that can be easily prevented when you take away the opportunity from the thief. Most purse snatchers are juveniles under the age of eighteen, who are waiting for the next opportunity.

WHAT YOU CAN DO

- When you hide the "prize," you protect yourself.
- Shop with a friend, travel together—you're always safer in company.
- If you are in a high-crime area, have only three or four small bills placed inside your purse. Credit cards, currency, driver's license, keys, and jewelry should be carried in a coat or sweater pocket, or concealed on your person to reduce the opportunity of large losses.
- If you carry a purse, don't wrap the strap around your shoulder, neck, or wrist. If your purse is grabbed, a strong strap will not yield easily and you may be injured.
- Never carry anything more valuable than you can afford to lose.
- Always leave all unnecessary credit cards at home.
- When you shop and carry a purse, place it in your shopping bag.
- Never leave your purse on a store counter or in a shopping cart, even for a moment.

DON'T CARRY WEAPONS THAT CAN BE USED AGAINST YOU

You should carry a police whistle and a small flashlight on your key chain where they're readily available. Make it a habit to carry your key chain in a pocket, NOT in your purse. Never put your name and address on your house keys or car keys. This is a simple way of telling the thief who you are and where you live.

CALL THE POLICE IMMEDIATELY

Most victims are attacked from behind. They don't get a good look at their attacker. And when a juvenile gets away with it once, he or she will try it again. If attacked, call the police immediately. Try to remember all details—help the police help you.

Subway

Subway Safety

Most subway accidents result from slips, trips, and falls on stairways when someone is in a rush. The best safety advice we can give: slow down when you are on the stairs, and hold the handrail.

On Platforms

Some trains are shorter (have fewer cars) when it's not rush hour because there are fewer passengers using the subway. Some short trains operate without a conductor. On those trains, the train operator opens and closes the doors, makes station announcements, and assists customers, if needed. Consider waiting at the center of the platform at these times. At most stations, there are signs that read: During Off-Hours, Trains Stop Here. If you stand near the center of the platform near the sign, you won't have to rush when the train arrives. While you're waiting, please stand behind the yellow protective strip, away from the platform edge. No matter what situation arises, keep off the tracks. Tracks contain 600 volts of live electricity. If you drop something on the tracks, go to the station booth and tell the agent.

In Subway Cars

Boarding between subway cars may seem like a time-saver, but it is highly dangerous. It's also dangerous to try to keep subway doors from closing when you are entering or exiting the train. They are not like elevator doors and will not reopen automatically. In addition, make sure that pocketbooks, knapsacks, clothing, packages, umbrellas, and other personal items are clear of the closing doors. When you're inside a moving train, never ride between cars or lean against doors. When you are standing, always hold on.

Using the Emergency Cord

Use the emergency cord only to prevent an accident or injury. For example, if someone gets caught between closing subway-car doors and is being dragged, pull the cord. But if your train is between stations and someone aboard becomes ill, do not pull the emergency cord. The train will stop, preventing medical professionals from reaching the sick passenger. A sick person is better off if the train goes to the nearest station where police and medical services will be waiting, or can be quickly summoned, without interruption.

On Escalators

Never run or walk on escalators; always hold the handrail and face forward. If you're with a child, hold hands. It's not a good idea for small children to hold escalator handrails. Escalator steps are always moving and have spaces that can grab. This means you should avoid resting packages (or yourself) on the stairs. You'll also want to keep clothing and shoes away from the sides. In addition, make sure that laces on footwear are tied. When you leave an escalator, step off, rather than ride off.

On Elevators

Children don't know that elevator doors can hurt them. You need to keep youngsters away from them. So, either hold children's hands or, if you're using a stroller, keep children's hands inside and never use a stroller to block closing doors. Watch clothing, bags, and other personal items—they can get caught in closing doors, too.

With Baby Strollers

Fold strollers so that you can carry infants on stairs or escalators. Strap your child in snugly at all other times. When you're on the platform, keep the stroller away from the edge and apply the stroller brake. That's because platforms tilt toward the tracks to allow for drainage, and the stroller could roll onto the tracks. Never place a stroller between closing subway-car doors. Watch out for the gap between the platform's edge and train when you board. (That's always a good idea, even when you're traveling alone.) And it's better if you board in the center of the train. The conductor is usually there, making it easier to get attention in case of a problem.

Buses

Don't run for the bus—that's when most accidents happen. Slips, trips, and falls are the most common causes of injuries. If you're at the front of the bus, please stay behind the white line. Avoid standing in the stairwell (rear-door step) or leaning against the rear door. And if you're a wheelchair user, please allow the bus operator to secure your chair. While you're riding, keep your head and arms inside bus windows. When you're ready to get off, signal the bus operator two blocks before your stop so that he or she has sufficient time to stop smoothly. We also advise holding the railing when you exit the bus, especially in winter. Bus steps and sidewalks become slippery from snow. As you leave the bus, watch for cars. This is particularly important when the bus operator has not been able to pull completely into the bus stop. Also, avoid crossing in front of the bus after you get off.

Request a Stop

Bus users who travel between 10 p.m. and 5 a.m. can Request-a-Stop. Ask the bus operator to let you off anywhere along the route, even if it isn't a designated stop. The bus operator will comply as long as he or she thinks it's a safe location, and will still make all regularly scheduled stops.

Trains

The US government, private industry, and public agencies all have taken the responsibility of trying to secure our nation's commuter railroads and mass-transit systems very seriously. The Department of Homeland Security has been at the forefront of this initiative by coordinating a comprehensive set of security initiatives that will aid in reducing the risk of a catastrophic attack on both mass transit and/or commuter rail systems. A focused effort on information sharing between all applicable stakeholders and government agencies has allowed for the development of new security measures and plans, increased training/public awareness campaigns and security program funding assistance. The federal government will continue to provide expert guidance and assistance to transit and rail agencies as well as seeking to mitigate any potential terrorist threat, domestic or foreign, intended to cause harm or disruption to this nation's national transportation system.

General Safety Tips

Surrounding Awareness

When traveling, be on the lookout for:

- Unusual behavior and suspicious activity
- Suspicious or unattended packages, devices, baggage, suitcases
- Emergency notifications and procedures given by the station manager or train operator
- Location of emergency exits and intercoms in stations and on trains

Irregular and/or Suspicious Activities and Items

When observing, look for people who:

- Look lost and/or are wandering around
- Appear to be conducting surveillance (using cameras/video)
- Abandon an item and leave the area quickly
- Openly possess a weapon or any prohibited or dangerous item

You should be aware of items or devices that

- are hidden or abandoned
- are connected to wires, timers, tanks, or bottles
- appear to be releasing a mist, gas, vapor, or odor
- appear to be suspicious or dangerous, such as a canister, tank, metal box, bottle, and so on.

What do you do if you notice a threat?

Contact authorities: local law enforcement, security personnel, conductor, or 911. Remain calm and answer questions as best as possible. Avoid using radios and cellular telephones within 50 feet of materials or devices that may be explosives.

Onboard Safety

While onboard any Amtrak train please note the following for a safer and more enjoyable trip:

- Use seat backs and handrails while walking through the train.
- Watch your step when boarding and leaving the train.
- Watch your step moving from car to car while the train is in motion, as the vestibules can be slippery.
- Be sure to step over the gap between the train and platform.
- Leave personal food items and baggage at your seat.
- Wear shoes at all times and use caution when wearing shoes without rubber soles.
- Never attempt to board or exit a moving train.
- Make sure you familiarize yourself with the safety card found in most seat backs.

Safety at the Station

While in any Amtrak station please note the following for a safer, more enjoyable trip:

- Arrive at least thirty minutes before your train is due to depart.
- Check in early to arrange for pre-boarding if you need extra time or assistance.
- Watch your step on station stairs, escalators, and train platforms.
- Please stand back from the edge of the platform.
- Help children when boarding and leaving the train.
- Take care when crossing the gap between the train and platform.
- Use only Red Cap agents to help you with your baggage.
- Please report any suspicious behavior to police, Amtrak Police or station personnel.

Aircraft

Security awareness is everyone's responsibility. To ensure that your travel is safe, secure, and efficient, take time to make security awareness an integral part of your travel experience.

General Guidelines

Be aware of suspicious activity in and around your immediate surroundings. Inform the proper authorities if you are concerned about an unattended item or suspicious activity in and around any facility. Describe the device, do not touch it and do not use a cell phone within 50 feet of a suspicious item. Be extra careful if you work on your laptop. Working takes your attention away from your surroundings. Remember to control all carry-on bags and keep your items and tickets with you at all times. Medicines, cash, jewelry, business papers, and valuables should be packed in carry-on bags. Never leave anything unattended, as it could be subjected to tampering and/or theft. Never carry anything for another person, especially for someone you do not know.

Before the Airport

- Do not pack or bring prohibited items to the airport. Read the Permitted and Prohibited Items list that the airline will provide with your travel documents. Place valuables, such as jewelry, cash, and laptop computers, in **carry-on baggage only**.
- Tape your business card to the bottom of your laptop.

- Avoid wearing clothing, jewelry, and accessories that contain metal. Metal items may set off the alarm on the metal detector.
- Avoid wearing shoes that contain metal or have thick soles or heels. Many types of footwear will require additional screening even if the metal detector does not sound the alarm.
- Put all undeveloped film (yes, some people still use it), thumb drives, SD cards, and the like in your carry-on baggage. Checked baggage screening equipment will damage undeveloped film and may erase electronic files.
- Declare firearms and ammunition (if authorized to carry them) to your airline and place them in your checked baggage.
- If you wish to lock your baggage, use a TSA-recognized lock.
- Do not bring lighters or prohibited matches to the airport.
- Do not pack wrapped gifts and do not bring wrapped gifts to the checkpoint. Wrap on arrival or ship your gifts prior to your departure. TSA may have to unwrap packages for security reasons.

At the Airport

Each adult traveler needs to keep available his/her airline boarding pass and government-issued photo ID until exiting the security checkpoint. Due to different airport configurations, at many airports you will be required to display these documents more than once. Place the following items **IN** your carry-on baggage or in a plastic bag prior to entering the screening checkpoint:

- Mobile phones
- Medications
- Keys
- Loose change
- Money clips
- Large amounts of jewelry
- Metal hair decorations
- Large belt buckles

Take your laptop and video cameras with cassettes **OUT** of their cases and place them in a bin provided at the checkpoint. Take **OFF** all outer coats, suit coats, jackets, and blazers.

Clear Medical and Kids' Companions

After the 9/11 attacks, airlines began allowing only ticketed passengers—with ID—beyond screening checkpoints. If, for medical reasons, you need a companion to accompany you to your seat, or you need an escort to do the same for a child who is traveling alone, airlines will accommodate you. Call the airline in advance to find out what kind of identification that person will need, and what procedures to follow. For more information, go to www.faa.gov.

Allow Extra Time

Bring extra food and drinks, if you have special dietary needs, or just can't stomach air terminal fare. Thanks to tighter airport security and more thorough checks, passengers will need to get to the airport at least two hours before their flights are scheduled to leave.

Report Suspicious Parcels and People

Keep an eye on your bags at all times, and notify security immediately if you spot an unattended bag at the airport, train, or bus station. Do the same if you see a suspicious person. If a fellow passenger is acting strangely in flight, discreetly alert the flight attendant.

Staying Safe during a Hijacking

It's not likely to happen. But should your plane be hijacked, your best bet is to follow the flight crews' instructions. All flight crewmembers have been trained in dealing with these security threats.

Stay Healthy On the Plane

Drinking

Because the air inside planes is drier than outside, your risk of dehydration is higher. Drinking alcohol will dehydrate you further. Drink water or fruit juice whenever you're even slightly thirsty. Always take your own bottle of water and ask for a glass of water each time the flight attendant offers. If she's too busy to give you a refill, you'll have your own supply.

Flex Your Muscles

Sit still for more than two hours, and the blood in your veins can pool and form potentially life-threatening clots. This condition is known as deep-vein thrombosis and was thought to be caused by sitting in cramped conditions in economy class. Now it is known that it is caused by lack of exercise—no matter how much you paid for your seat. Prevent the problem by flexing your muscles, moving your limbs and walking around if the "fasten seat belt" sign is off.

Travel Troubleshooting Guide

Non-emergency Medical Problem

- Traveling in the United States: If you belong to an HMO, call your insurer to find out if there are in-network providers in the area. If not, ask if it will waive out-of-network surcharges.
- Abroad: Contact the US Consulate, which should provide a list of English-speaking doctors. Most insurers won't cover the cost, however.

Medical Emergency

Traveling in the United States: Head straight for the ER. Your insurance should cover the cost for a true emergency. Abroad: Head straight for the nearest ER. Your insurance should cover the tab.

Lost medications

Traveling in the United States: Call your doctor and have him or her phone in a prescription to the nearest pharmacy. Some insurers cover one replacement per lost prescription.

Abroad: Contact the foreign embassy in the United States before leaving to inquire what arrangements would need to be made. Some insurers will cover; others won't.

Safe Cruising

More than 20 million people will go cruising this year and have a fabulous time. Because of heightened security, however, it will take longer to embark and disembark, so plan accordingly. Make sure you have enough time to get to the cruise terminal and complete the embarkation procedure before the ship sails! Expect and plan for longer times to park, check-in and to pass through the security checkpoints. You may experience some inconvenience and delays—please plan to leave for the airport/cruise terminal early. And be patient and understanding. Boarding will be denied without presenting proper documents. Check with your travel agent or the cruise line regarding the documentation you will require.

- US Citizens—A passport or a birth certificate (original or certified copy) plus a picture is needed.
- ID card issued by a federal, state, or local government agency is required. A voter registration card or Social Security card is not considered to be

proof of citizenship. Children under sixteen years of age do not require a picture ID card.

- Non-US Citizens—Valid passports and visas (except for waiver countries) are required, in addition to Alien Registration Cards (ARC or "Green" cards), if an individual is a Resident Alien living in the United States. Certain Canadian and Mexican citizens may travel with alternative documentation depending on their alien status in the United States.
- For customs and immigration purposes, guests are also required to have necessary visa, passport, and other travel authorization documents based upon their nationality and country of residence.

For Cruise Passengers Arriving by Air

- Checking luggage at off-airport sites is prohibited. You may still use these sites to obtain boarding passes and seat assignments, but luggage will need to be checked at the airport. Some cruise lines offer airport check-in at the cruise terminal after disembarkation.
- No form of weapon will be permitted on your person or in carry-on luggage. Regardless of the size, items, such as small penknives, scissors and such, should be left at home or placed in checked baggage.
- Only ticketed passengers will be permitted through the security checkpoints. Make sure that you have acceptable photo identification.
- Travelers with e-tickets must hold one of the following documents indicating a flight departure for a current date: A boarding pass or paper ticket, a receipt for an electronic ticket (e-ticket), or an itinerary generated by an airline or travel agency.
- Check your cruise documentation for up-to-date and appropriate air travel information to comply with security checkpoint measures.
- Check-in at the ticket counter if you are unsure about passing through security.
- Vehicles parked near the airport will be closely monitored and should not be unattended.
- DO NOT leave your luggage unattended at any time.
- You will likely notice additional uniformed, law enforcement officers and FAA canine teams patrolling the airports.

Travel Tips

- Consider purchasing trip/vacation cancellation and interruption insurance from your travel agent or cruise line.
- Review your cruise documents in advance for important information, guidelines, and tips.
- Prior to leaving home, call the airline—or check their website for flight arrival/departure information.

- Be at the airport early—at least two hours unless local circumstances require more time.
- Minimize your carry-on luggage AND be prepared to have it searched. Suspicious items (knives, scissors, clippers, files, etc.) may be confiscated and subject you to a further delay.
- Make sure that your baggage is clearly tagged with your name and contact information and that you have completed and attached the cruise tags provided with your cruise documentation.
- Verify that any required medical items are with you and not in checked luggage.
- Verify that you have the proper identification required for all travelers in your party and that it is on your person—not in checked luggage.
- Expects some delays and be patient.
- Enjoy your cruise and have a great vacation.

Severe Acute Respiratory Syndrome (SARS) and Cruise Travel

Although there have been no reports of SARS on any cruise ship operating in North America to date, the membership of the ICCL has taken increased health and safety measures as a matter of caution.

Monitoring/Prevention of SARS on Cruise Ships

Individual cruise lines are closely monitoring the SARS situation around the clock, and, depending on various factors, such as their employees, passengers and ports of call, have implemented various procedures to ensure the health of everyone onboard. These steps have included:

- Screening of both passengers and crew who are arriving within ten days from CDC identified SARS-travel-alert areas.
- Working to proactively educate both passengers and crew about SARS and its symptoms and will, as appropriate, deny boarding to any passenger that meets certain risk factors for SARS.
- Re-routing arriving passenger and vessel itineraries away from locations for which there are current travel warnings.

Enhanced Cruise Ship Procedures/Protocols

Over the past several years, ICCL member cruise lines have enhanced the safety and security of passengers and crew through the adoption of the most comprehensive and proactive health and sanitation protocols in the travel and tourism industry. Currently, all member cruise lines participate in a voluntary vessel sanitation program administered by the CDC. Cruise-industry health and sanitation protocols, developed in conjunction with the CDC, are effective in reducing the transmission of infectious illnesses aboard ships.

Should a suspected SARS case be identified on a cruise ship, all members of the ICCL have trained medical staff onboard with appropriate isolation and treatment facilities. The disinfectants available for use by ICCL member lines are believed to be the most effective agents available against the virus that is presently thought to cause SARS.

On Vacation

Water Safety

Every year, about 800 children younger than 14 die as a result of unintentional drowning in the United States; another 2,700 are treated in emergency rooms for nearly drowning. And a review by the National Safe Kids Campaign found that in nine out of ten cases, children drowned while under supervision, with 94 percent of parents saying they were watching their kids. The report revealed that parents were also chatting, reading, eating, dialing the phone, having an alcoholic drink, or closing their eyes to relax.

Staying Afloat

Non-swimmers and beginners should wear a personal flotation device (PFD) approved by the US Coast Guard (USCG) when in or near the water. It should be a snug fit, based on the child's weight and age. Use a model with a crotch strap to keep it in place in the water on kids under age 10. It's not a substitute for supervision, but it could help a child keep his or her head up if he or she slips into the pool. Water wings, pool rings, and rafts are OK as toys only, meaning keep your eyes on the child who is playing with them. And don't rely on bathing suits with built-in flotation systems or non-USCG strap-on floats.

Have a Designated Water Watcher

Whenever kids swim, responsible adults or teens should take twenty-minute shifts standing guard. That means no talking, eating, drinking, or playing with the kids while on duty.

Sign Kids up for Swim Class

The Red Cross recommends lessons beginning at age four or five. Classes should include water safety tips. For info on water-watcher tags, swimming classes, and USCG-approved PFDs, visit the American Red Cross website at www.redcross.org.

TRAVELING OVERSEAS

If you are traveling abroad here are the top twelve tips you need to make your trip easier:

1. Make sure you have a signed, valid passport and visas, if required and all necessary vaccinations. Also, before you go, fill in the emergency information page of your passport!
2. If you take prescribed drugs carry a copy of the doctor's prescription.
3. Travel insurance, including repatriation coverage, is recommended.
4. Read the Consular Information Sheets (and Public Announcements or Travel Warnings, if applicable) for the countries you plan to visit.
5. Familiarize yourself with local laws and customs of the countries to which you are traveling. Remember, the US Constitution does not follow you! While in a foreign country, you are subject to its laws.
6. Make two copies of your passport identification page. This will facilitate replacement if your passport is lost or stolen. Leave one copy at home with friends or relatives. Carry the other with you in a separate place from your passport.
7. Leave a copy of your itinerary with family or friends at home so that you can be contacted in case of an emergency.
8. Do not leave your luggage unattended in public areas. Do not accept packages from strangers.
9. Prior to your departure, you should register with the nearest US embassy or consulate through the State Department's travel registration website. Registration will make your presence and whereabouts known in case it is necessary to contact you in an emergency.
10. To avoid being a target of crime, try not to wear conspicuous clothing and expensive jewelry, and do not carry excessive amounts of money or unnecessary credit cards.
11. In order to avoid violating local laws, deal only with authorized agents when you exchange money or purchase art or antiques.
12. If you get into trouble, contact the nearest US embassy.

Resources

REFERENCES AND RESOURCES

American Medical Association

"Physical injuries and fatalities resulting from the Oklahoma City bombing, August 7, 1996"

S. Mallonee, S. Shariat, G. Stennies, R. Waxweiler, D. Hogan, and F. Jordan. *The Journal of the American Medical Association*, Vol. 276, No. 5, pp. 382–387

Abstract at URL: http://jama.ama-assn.org/cgi/content/abstract/276/5/382

Centers for Disease Control and Prevention/National Institute for Occupational Safety and Health

Publication No. 2002-139, "Guidance for Protecting Building Environments from Airborne Chemical, Biological, or Radiological Attacks," May 2002, Cincinnati, OH

http://www.cdc.gov/niosh/bldvent/2002-139.html

Publication No. 2003-136, "Guidance for Filtration and Air Cleaning Systems to Protect Building Environments from Airborne Chemical, Biological, or Radiological Attacks," April 2003, Cincinnati, OH

http://www.cdc.gov/niosh/docs/2003-136/2003-136.html

Central Intelligence Agency

Chemical, Biological, Radiological Incident Handbook, October 1998 http://www.cia.gov/cia/publications/cbr_handbook/cbrbook.htm

Federal Emergency Management Agency

FEMA 386-2, "Understanding Your Risks, Identifying Hazards and Estimating Losses," August 2001 http://www.fema.gov/fima/planning_toc3.shtm

FEMA 426, "Risk Management Series: Reference Manual to Mitigate Potential Terrorist Attacks Against Buildings," December 2003

FEMA 428, "Risk Management Series, Primer to Design Safe School Projects in Case of Terrorist Attacks," December 2003

National Research Council

Protecting Buildings and People from Terrorism: Technology Transfer for Blast-effects Mitigation, 2001, National Academy Press, Washington, D.C., ISBN 0-309-08286-2 http://books.nap.edu/books/0309082862/html/index.html

Useful Organizations

American Lifelines Alliance: http://www.americanlifelinesalliance.org

The American Institute of Architects (AIA), Security Resource Center: http://www.aia.org/security

Applied Technology Council: http://www.atcouncil.org

Architectural Engineering Institute (AEI) of ASCE: http://www.asce.org/instfound/aei.cfm

Battelle Memorial Institute, National Security Program: http://www.battelle.org/natsecurity/default.stm

Center for Strategic and International Studies (CSIS): http://www.csis.org

Centers for Disease Control and Prevention (CDC)/National Institute for Occupational Safety and Health (NIOSH): http://www.cdc.gov/niosh

Central Intelligence Agency (CIA): http://www.cia.gov

Federal Aviation Administration (FAA): http://www.faa.gov

Federal Bureau of Investigation: http://www.fbi.gov/publications/terror/terroris.htm

Federal Emergency Management Agency (FEMA): http://www.fema.gov

Federal Facilities Council – See National Academy of Sciences

The Infrastructure Security Partnership (TISP): http://www.tisp.org

International CPTED [Crime Prevention through Environmental Design] Association (ICA): http://new.cpted.net/home.amt

National Academy of Sciences: http://www4.nationalacademies.org/nas/nashome.nsf

- Federal Facilities Council (FFC) Standing Committee on Physical Security and Hazard Mitigation: http://www7.nationalacademies.org/ffc/Physical_Security_Hazard_Mitigation.html
- National Research Council: http://www.nationalacademies.org/nrc

National Defense Industrial Association (NDIA): http://www.ndia.org

National Institute of Justice (NIJ): http://www.ojp.usdoj.gov/nij

Office of Domestic Preparedness (ODP): http://www.ojp.usdoj.gov/odp

Public Entity Risk Institute: http://www.riskinstitute.org

Security Industry Association (SIA): http://www.siaonline.org/

U.S. Army Corps of Engineers: http://www.usace.army.mil

U.S. Department of Justice: http://www.usdoj.gov

U.S. Marshals Service (USMS): http://www.usdoj.gov/marshals

Index

About the Author

Don Philpott has been writing, reporting, and broadcasting about security, civil unrest, and terrorism for almost fifty years. For twenty years he was a senior correspondent with Reuters-Press Association, the wire service, traveling the world reporting on major events and trouble spots including Northern Ireland, Lebanon, Israel, South Africa, and the Far East.

He was editor of Homeland Defense Journal, the leading homeland security publication in the United States, and he writes for magazines and newspapers in the United States and Europe.

He is the author of more than two hundred books and is a regularly contributor to radio and television programs on security issues. His latest books to be published in the spring of 2019 are The Workplace Violence Prevention Handbook and Public School Emergency Preparedness, both published by Bernan Press.

www.ingramcontent.com/pod-product-compliance
Lightning Source LLC
Chambersburg PA
CBHW061607180926
51615CB00006B/11